Date Due			

Discrete Computational Structures

This is a volume in
COMPUTER SCIENCE AND APPLIED MATHEMATICS

A Series of Monographs and Textbooks

Editor: WERNER RHEINBOLDT

A complete list of titles in this series appears at the end of the volume.

DISCRETE
COMPUTATIONAL
STRUCTURES

Robert R. Korfhage
SOUTHERN METHODIST UNIVERSITY

ACADEMIC PRESS New York and London
A Subsidiary of Harcourt Brace Jovanovich, Publishers

ACADEMIC PRESS, INC.
111 Fifth Avenue, New York, New York 10003

United Kingdom Edition published by
ACADEMIC PRESS, INC. (LONDON) LTD.
24/28 Oval Road, London NW1

Library of Congress Cataloging in Publication Data

Korfhage, Robert R
        Discrete computational structures.

        (Computer science and applied mathematics)
        1.   Electronic digital computers–Programming.
2.   Electronic data processing–Mathematics.   I.   Title.
QA76.6.K68   001.6'4'044   73-9432
ISBN 0–12–420850–9

AMS (MOS) 1970 Subject Classifications: 02B01, 05C01, 06A01, 68A01
ACM Subject Classifications: 1.1, 5.30

PRINTED IN THE UNITED STATES OF AMERICA

Contents

Preface   ix
Acknowledgements   xiii

Chapter 1   **Basic Forms and Operations**

1. Introduction   1
2. Elements and Sets   2
3. Subsets   3
4. Venn Diagrams and Set Complements   7
5. Computer Representation of Sets   8
6. Set Operations   11
7. Set Algebra   12
8. Computer Operations on Sets   14
9. Product Sets   17
10. Relations, Mappings, Functions   18
11. Equivalence and Order Relations   26
12. The Lattice of Subsets   31
13. Vectors and Matrices   32

Chapter 2   **Undirected Graphs**

1. Graph Theory   37
2. Basic Definitions   38
3. Special Classes of Graphs   42
4. Matrix Representation of Graphs   49
5. Relations among Graph Matrices   53

 6. Invariants and Graph Isomorphism   58
 7. Cycle Basis   67
 8. Maximal Complete Subgraphs   72
 9. Storage Minimization for Matrices   78
10. Bandwidth of Cubic Graphs   83
11. Bandwidth of Bipartite Graphs   92
12. Planar Graphs and the Four Color Conjecture   97
     References   100

Chapter 3   **Gorn Trees**

 1. Introduction   101
 2. Tree Domains   104
 3. Trees   109
 4. Prefix Representation and Tree Forms   114
 5. Explicit Definitions   119
 6. Searching, Subroutines, and Theorem Proving   126
     References   129

Chapter 4   **Directed Graphs**

 1. Introduction   130
 2. Basic Definitions   130
 3. Special Classes of Graphs   134
 4. Matrix Representation of Directed Graphs   135
 5. Flowcharts   136
 6. Networks   139
 7. Minimal Cost Flows   147
 8. Pruning Branches to Find the Shortest Path   149
 9. Critical Paths   154
10. Graphs of Multiprocessing Systems   157
11. Information Networks   158
     Reference   165

Chapter 5   **Formal and Natural Languages**

 1. Introduction   166
 2. Semigroups   167
 3. Formal Languages   168
 4. Backus Naur Form and Algol-Like Languages   176
 5. Semantics of Formal Languages   180
 6. Natural Languages   181
     References   182

Chapter 6   **Finite Groups and Computing**

 1. Definitions of Groups and Subgroups   183
 2. Groups of Graphs   186

3. Graphs of Groups   189
4. Generators and Relations   190
5. Permutations and Permutation Groups   197
6. Permutation Generators   203

Chapter 7 **Partial Orders and Lattices**

1. Introduction   208
2. Partial Orders   208
3. Lattices   211
4. Specialized Lattices   217
5. Atomic Lattices   223
   Reference   225

Chapter 8 **Boolean Algebras**

1. Introduction   226
2. Properties of Boolean Algebras   227
3. Boolean Algebras and Set Algebras   229
4. Boolean Functions   230
5. Switching Circuits   236
6. Boolean Function Minimization   239
7. Computer Arithmetic   249
   Reference   251

Chapter 9 **The Propositional Calculus**

1. Introduction   252
2. Fundamental Definitions   253
3. Truth Tables   255
4. Well-Formed Formulas   261
5. Minimal Sets of Operators   263
6. Polish Notation   266
7. Proofs in Logic   272
8. Sets and Wordsets   280
   Reference   280

Chapter 10 **Combinatorics**

1. Introduction   281
2. Permutations of Objects   281
3. Combinations of Objects   287
4. Enumerators for Combinations   291
5. Enumerators for Permutations   298
6. Stirling Numbers   300
7. Cycle Classes of Permutations   303

8. Partitions and Compositions   306
    References   311

Chapter 11 **Systems of Distinct Representatives**

1. Introduction and History   312
2. The Third Question, General Case   318
3. The Third Question, Partition Case   322
4. Summary   325
    References   326

Chapter 12 **Discrete Probability**

1. Probabilities on a Discrete Set   327
2. Conditional Probability and Independence   333
3. Computation of Binomial Coefficients   337
4. Distributions   341
5. Random Numbers   351
    Reference   355

Answers and Hints for Selected Exercises   356

Index   375

Preface

The digital computer is a machine which is inherently finite in all of its aspects which are of importance to a programmer or user. Whether the problem posed to the computer relates to numerical computation, symbol manipulation, information retrieval, or picture generation, the procedures and results are restricted by the finiteness of word length and memory size, and by the discrete time steps in which a computer operates. Hence for a thorough understanding of the capabilities and limitations of computers, it is important to be aware of the impact of these restrictions. Thus the purpose of this text is to bring together many of the concepts from discrete mathematics which are important to computing. Throughout I have tried to keep two questions in mind: (1) How does this topic influence the theory and practice of computing? (2) How is the computer used to solve problems in this topic? These questions cannot always be given equally good answers, and there are topics, particularly early in the book, where any relation between the topic and computing is largely suppressed. Hopefully, these topics are few and brief, and in the nature of a mathematical background necessary to the remainder of the book.

The text is designed as a sophomore- or junior-level book, corresponding to the course B3, Introduction to Discrete Structures, in the ACM Curriculum 68. Thus I assume that the student knows and can use at least one high level programming language, and I use this assumption in both the text and the exercises. I view B3 as the more mathematical half of a one-year course. Thus the student who has covered the material presented here should be ready to use these mathematical concepts in courses which deal more specifically and directly with computers. He should have the basic mathematics to enable him to feel comfortable in undergraduate courses in data structures (the natural follow-up to B3), switching circuits, and automata theory. With this in mind, topics such as lists, circuit design, and finite state machines have been largely omitted, to be covered in succeeding courses.

This book is not intended as a formal, rigorous introduction to the theory of discrete structures, although I believe that the mathematical content is accurate and not misleading. It has been my experience that most computer science students at this level do not appreciate the subtleties of the theory. Hence I have tried to motivate the theoretical constructs as thoroughly as possible from computing, and to present any arguments, however formal, in an informal setting. Nevertheless, I firmly believe that the instructor should be willing and able to put the challenge of formal work to those students who can absorb more theory.

The text is quite loosely structured, with ample material so that the instructor can pick and choose. Very little mathematical background is required of the student, although students who have had a course in the foundations of modern mathematics will find much of Chapters 1 and 9 to be review material. Basically there are five sections to the text, and almost any permutation of these sections makes sense. However, my experience with the material has led me to the present order, both among and within sections.

Chapter 1 constitutes the first section. This covers the basic mathematical background needed, and should be studied first. It also includes some discussion of the computer representation of sets.

The section on graph theory consists of Chapters 2–4. The organization progresses from the simplest concepts toward more complex ones. The sections on storage minimization and bandwidth present an example of the close interaction which is possible between mathematics and computing, to the benefit of both. Chapter 3 develops a conceptual framework which has been little publicized ouside of research papers, as does the last section of Chapter 4.

The algebra section, Chapters 5–7, concentrates on semigroups, groups, and lattices. This does not imply that rings, fields, and universal algebras are unimportant, but rather that at this level of text, and for the purposes of this course, these latter structures can well be omitted.

The concepts of Boolean algebra and propositional calculus are covered, in rather traditional fashion, in Chapters 8 and 9. A new tabular method of Boolean function minimization is described. Discussion of the Polish notation in Chapter 9 provides a nice tie back to the concepts of Chapter 3. The predicate calculus is omitted, for much the same reason that certain algebraic structures are omitted: little is lost by postponing discussion of it until a later course.

The final section of the book, centered around combinatorics and probability, is perhaps the most novel. Chapter 10 introduces the student to the

general process of enumerating objects, through a number of examples showing different techniques. Polya's theory, while extremely important, is sufficiently complex that it cannot properly be covered in a course at this level.
Hence I have reluctantly omitted any reference to it. Chapter 11 is totally in
the nature of a case study. The problem chosen is a complex one which has
numerous ties into applications in operations research, and which illustrates
both the use of a computer, and the importance of knowing something
about the combinatorial nature of a problem before doing the programming.
Finally, Chapter 12 introduces the student to some of the probabilistic concepts which he will need repeatedly in his career.

A few exercises, mostly programs, have been marked (*) or (**). The
single asterisk identifies a difficult problem, while the double asterisk marks
a problem which the students will probably find extremely difficult or impossible to do. These problems are included both to challenge the brighter
students, and, more importantly, to provide material for class discussion.
That is, I suggest assigning a (**) problem for the students to spend a week
on (without really expecting them to solve it), and then spending time in
class discussing the difficulties which they have encountered, and how to
work around them.

Acknowledgments

The writing of any book is an iterative process, involving, in the case of textbooks, discussions with colleagues, trying the material out on several classes, and so forth. Much of this background is inherently vague and unrecognizable. However, I would like to thank the several classes of students at Purdue University and Southern Methodist University who have suffered through the preliminary versions of this text, and also to especially thank A. T. Berztiss, S. T. Hedetnemi, R. E. Nance, and M. B. Wells, among my colleagues, for their comments and criticisms. Any errors which remain after their careful scrutiny are, of course, my responsibility.

The production of this book has been greatly facilitated by my able typists, Talitha Williams and Mary Kateley, and by the staff of Academic Press.

CHAPTER 1

Basic Forms
and Operations

1. INTRODUCTION

This is a book about structures. Between this page and the final one we shall examine a wide variety of mathematical frameworks. We shall compare them—finding at times that we have two or more quite distinct ways of describing a framework. We shall explore and map them, so that we can more easily find our position in a given framework, and can reference and discuss various locations within the framework. We shall clothe these frameworks with data and with programs, and examine what can be said about the resulting structures. Always, since our interest is in digital computation, our structures will be discrete. That is, they will have no more parts than there are integers.

A computer is basically a data transformer. Given certain input data, a computer will, by means of its programs and its internal circuitry, transform the data into other data which constitute the output of the machine. As users, we may interpret the data to represent numbers, words, music, pictures, and so on. The data become information. We propose to examine the concept of data as it relates to computers, to determine the basic forms

1

that data can take regardless of the interpretation, and to study these forms for properties which affect computation. We shall begin with the simplest form, the set, and progress into data forms with more complex structures.

2. ELEMENTS AND SETS

The simplest structure that we encounter is the set. Without defining *set* precisely, a set is a mere collection of "things": books, computers, concepts, colors, words, numbers, cars, and so forth. These "things" in a set are called the *elements* or *members* of the set. A set, as a set, is almost devoid of properties: there is no order to the set, no relationship among the elements other than that they all belong to the set. In fact, the only basic property which we shall assume is that, given a "thing" x and a set A, we can determine whether x is an element of A. We shall denote set membership by the symbol $\in$, writing $x \in A$ if x is an element of A, and $x \notin A$ if x is not an element of A.

A set may be described either by listing all of its elements within braces, or by stating a property that determines whether a "thing" is an element of the set.

Example 2.1

Describe the sets consisting of the elements (a) 1, 2, 3; (b) a, e, i, o, and u; (c) red, yellow, and blue.

ANSWER.   (a) $\{1, 2, 3\}$, (b) $\{a, e, i, o, u\}$, (c) {red, yellow, blue}.

Whenever the meaning is clear, we may use ellipses (...) to abbreviate the list. Thus the set of all integers between 1 and 1,000,000 may be written $\{1, 2, 3, \ldots, 1,000,000\}$, and the set of all negative integers may be written $\{-1, -2, -3, \ldots\}$.

It should be noted that a change in the order in which the elements of a set are listed does not change the set. Nor does multiple listing of elements change the set, any more than multiple listing of houses for sale changes the set of houses available. Thus the listings $\{1, 2, 3\}$, $\{2, 3, 1\}$, and $\{1, 1, 3, 2\}$ all describe the same set.

The form used to describe a set by a property is $\{x \mid P(x)\}$, read "the set of all x such that $P(x)$ holds," where P is the defining property.

Example 2.2

Describe the sets of Example 2.1 by properties.

> ANSWER. (a) $\{x \mid x$ is a positive integer less than $\pi\}$, (b) $\{x \mid x$ is a vowel$\}$, (c) $\{x \mid x$ is a primary color$\}$.

Note. Other defining properties could have been chosen. For example, the first set could also be described as $\{x \mid x = 1$ or x is a prime number less than 4$\}$. The important thing is that all the elements of the set, and only these, have the defining property.

3. SUBSETS

We are interested in the relationships between sets. The most obvious relationship occurs when every element of a set A is also an element of a set B. For example, every man is a member of the set of human beings, as well as of the set of men. If every element of A is also in B, we say that *A is a subset of B* and write $A \subseteq B$. Clearly every element of A is an element of A. Thus A is a subset of itself, $A \subseteq A$.

In general, A is not the only subset of A. For example, the sets $\{1\}$, $\{1, 3\}$, and $\{3, 2, 4\}$ are all subsets of the set $\{1, 2, 3, 4\}$. We call such subsets *proper*. More specifically, *A is a proper subset of B*, $A \subset B$, if A is a subset of B, and B is not a subset of A. That is, $A \subset B$ if every element of A is an element of B, but not every element of B is an element of A.

One very special set is the *empty set* or *null set* $\emptyset$ which contains no elements. This set can be defined as $\emptyset = \{x \mid x \neq x\}$, or $\emptyset = \{x \mid x$ is an even prime greater than 2$\}$, for example. Note that since $\emptyset$ has no elements, the condition for $\emptyset$ to be a subset of any set A is automatically satisfied.

> For any set A, either $\emptyset = A$ or $\emptyset \subset A$.

If $A \subseteq B$ and $B \subseteq A$, we say that *A and B are equal sets* and write $A = B$. This is of importance when A and B are in fact the same set, described differently. A classic example is the set of *Fibonacci numbers*, which may be defined by two quite different properties.

Example 3.1

Let

$$A = \left\{ F_n \mid F_n = \frac{1}{\sqrt{5}} \left[\left(\frac{1+\sqrt{5}}{2} \right)^n - \left(\frac{1-\sqrt{5}}{2} \right)^n \right], \; n \geqslant 0 \right\}$$

and let

$$B = \{ f_n \mid f_0 = 0; \; f_1 = 1; \text{ and for } n > 1,$$
$$f_n = f_{n-1} + f_{n-2} \}.$$

Show that $A = B$.

Solution. Set equality is most easily shown by double inclusion: $A \subseteq B$ and $B \subseteq A$. To show $A \subseteq B$, observe that $F_0 = 0$ and $F_1 = 1$. Hence these correspond to f_0 and f_1, respectively. Next compute $F_{n-1} + F_{n-2}$.

$$
\begin{aligned}
F_{n-1} + F_{n-2} &= \frac{1}{\sqrt{5}} \left[\left(\frac{1+\sqrt{5}}{2} \right)^{n-1} - \left(\frac{1-\sqrt{5}}{2} \right)^{n-1} + \left(\frac{1+\sqrt{5}}{2} \right)^{n-2} \right. \\
&\qquad \left. - \left(\frac{1-\sqrt{5}}{2} \right)^{n-2} \right] \\
&= \frac{1}{\sqrt{5}} \left[\left(\frac{1+\sqrt{5}}{2} \right)^{n-2} \left(\frac{1+\sqrt{5}}{2} + 1 \right) \right. \\
&\qquad \left. - \left(\frac{1-\sqrt{5}}{2} \right)^{n-2} \left(\frac{1-\sqrt{5}}{2} + 1 \right) \right] \\
&= \frac{1}{\sqrt{5}} \left[\left(\frac{1+\sqrt{5}}{2} \right)^{n-2} \left(\frac{3+\sqrt{5}}{2} \right) - \left(\frac{1-\sqrt{5}}{2} \right)^{n-2} \left(\frac{3-\sqrt{5}}{2} \right) \right] \\
&= \frac{1}{\sqrt{5}} \left[\left(\frac{1+\sqrt{5}}{2} \right)^{n-2} \left(\frac{1+\sqrt{5}}{2} \right)^2 - \left(\frac{1-\sqrt{5}}{2} \right)^{n-2} \left(\frac{1-\sqrt{5}}{2} \right)^2 \right] \\
&= \frac{1}{\sqrt{5}} \left[\left(\frac{1+\sqrt{5}}{2} \right)^n - \left(\frac{1-\sqrt{5}}{2} \right)^n \right] \\
&= F_n.
\end{aligned}
$$

Thus the elements of A satisfy the conditions defining elements of B, and hence $A \subseteq B$.

To show $B \subseteq A$, begin by examining the numbers $f_0, f_1, f_2, \ldots$, and the differences $f_i - f_{i-1}$ for $i = 1, 2, \ldots$:

i	f_i	$f_i - f_{i-1}$
0	0	
		1
1	1	
		0
2	1	
		1
3	2	
		1
4	3	
		2
5	5	
		3
6	8	
		5
7	13	
		8
8	21	
		13
9	34	

Observe that the differences reproduce the original values. Now e^x has the property that its derivative is itself. This suggests that some exponential should satisfy the recurrence relation $f_n = f_{n-1} + f_{n-2}$. Obviously e^n does not $(e^n \neq e^{n-1} + e^{n-2})$. Try $f_n = a^n$: then $a^n - a^{n-1} - a^{n-2} = 0$, or $a^2 - a - 1 = 0$. This equation has two roots, $a_1 = (1 + \sqrt{5})/2$, and $a_2 = (1 - \sqrt{5})/2$. Thus either a_1^n or a_2^n or some linear combination $c_1 a_1^n + c_2 a_2^n$ could be f_n. Any such combination satisfies the recursion. Now use the initial conditions $f_0 = 0, f_1 = 1$:

$$(f_0) \qquad c_1 a_1^0 + c_2 a_2^0 = 0,$$

$$c_1 + c_2 = 0. \qquad (1)$$

$$(f_1) \qquad c_1 a_1^1 + c_2 a_2^1 = 1,$$

$$c_1 \left(\frac{1 + \sqrt{5}}{2} \right) + c_2 \left(\frac{1 - \sqrt{5}}{2} \right) = 1,$$

$$\frac{1}{2}(c_1 + c_2) + \frac{\sqrt{5}}{2}(c_1 - c_2) = 1.$$

Hence

$$\frac{\sqrt{5}}{2}(c_1 - c_2) = 1,$$

or

$$c_1 - c_2 = \frac{2}{\sqrt{5}}. \tag{2}$$

From (1) and (2), $c_1 = -c_2 = 1/\sqrt{5}$. Thus

$$f_n = \frac{1}{\sqrt{5}}\left[\left(\frac{1+\sqrt{5}}{2}\right)^n - \left(\frac{1-\sqrt{5}}{2}\right)^n\right] = F_n.$$

Hence the numbers that satisfy both the initial conditions and the recurrence relation are all of the form F_n. Thus $B \subseteq A$.

ANSWER.   The sets are equal.

The Fibonacci numbers defined in Example 3.1 have a number of interesting and easily discovered properties. (See Exercises 1 and 7, and Exercise 3, Section 10, Chapter 2.)

Whenever we talk of sets, there is at least an implied limitation of the universe of discourse. In discussing the numerical solution of equations we assume without so stating that we are talking about a set of numbers. In text editing there is a tacit assumption that the editing program will be handling a set of words in English or some other language, rather than paintings, musical tones, or automobiles. This implicit limitation leads us to define a *universal set* $\mathscr{U}$, which contains all "things" that we wish to consider. Note that $\mathscr{U}$ does not contain everything—only all things pertinent to the discussion.

Example 3.2

Determine the set $A = \{x \mid x^2 + 1 = 0\}$, (a) if $\mathscr{U}$ is the set of real numbers, and (b) if $\mathscr{U}$ is the set of complex numbers.

Solution.   Since there is no real number that satisfies the given equation if $\mathscr{U}$ is the set of real numbers, then $A = \varnothing$. However, there are complex numbers that satisfy the equation, namely i and $-i$. Thus if $\mathscr{U}$ is the set of complex numbers, A is not empty.

ANSWER.   (a) $A = \varnothing$, (b) $A = \{i, -i\}$.

EXERCISES

1. Let $f_0, f_1, f_2, \ldots$ be the Fibonacci numbers (Example 3.1). Show that any nonnegative integer n has a unique representation of the form

$$n = \sum_{i=1}^{\infty} a_i f_i,$$

   where the a_i are all 0 or 1, and $a_i \cdot a_{i+1} = 0$, $i = 1, 2, \ldots$.

Determine A if the universal set is (a) the positive integers, (b) the integers, (c) the real numbers, (d) the complex numbers:

2. $A = \{x \mid x \text{ is of the form } 3y+1 \text{ for } y \in \mathcal{U}\}$.

3. $A = \{x \mid x \text{ is of the form } y-4 \text{ for } y \in \mathcal{U}\}$.

4. $A = \{x \mid x^2 + x - 2 = 0\}$.

5. $A = \{x \mid x^2 - x - 1 \leqslant 0\}$.

6. $A = \{x \mid x^3 - 2x^2 - 2x - 3 = 0\}$.

7. Expand $(1 - t - t^2)^{-1}$ in a series. What are the coefficients? What about the coefficients of $(1 + t - t^2)^{-1}$? (See Example 3.1.)

4. VENN DIAGRAMS AND SET COMPLEMENTS

Sets may be conveniently represented by Venn diagrams. For such a diagram, a rectangle represents the universal set $\mathcal{U}$. Any set A is represented by a closed region, usually circular, within the rectangle (Fig. 4.1). The

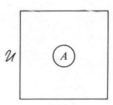

Figure 4.1

points inside and on the circle represent elements of A, while those outside represent objects not in A. Thus the subset relation $A \subseteq B$ is easily represented by region A lying entirely within region B, as in Fig. 4.2. In each case, B is the entire large circular region.

The set of all elements of $\mathcal{U}$ that are not in A is called the *complement of A*, written $\overline{A}$. In a Venn diagram, $\overline{A}$ is represented by the points outside the region. Similarly, the *relative complement of A with respect to B*, written

$B - A$, consists of those elements of B that are not elements of A. Note that $B - A$ is defined even if A is not a subset of B.

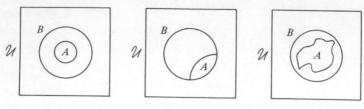

Figure 4.2

Example 4.1

Let $\mathcal{U}$ be the set of natural numbers, $\mathcal{U} = \{1, 2, 3, 4, \ldots\}$, let $A = \{x \,|\, x \text{ is a prime number}\}$, and let $B = \{x \,|\, x = 3k + 1$ for some natural number $k\}$. Define $\bar{A}$, $A - B$, and $B - A$, and diagram the latter two sets.

Solution.   Since A is the set of prime numbers, $\bar{A}$ is the set consisting of 1 and the composite numbers, $\bar{A} = \{1, 4, 6, 8, 9, 10, 12, 14, 15, \ldots\}$. Similarly, the set $A - B$ consists of prime numbers that do not satisfy the equation given for B, while the set $B - A$ consists of all solutions to the equation that are not prime numbers. The Venn diagram representing these sets is given in Fig. 4.3.

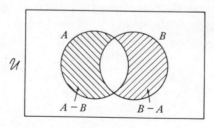

Figure 4.3

ANSWER. $\bar{A} = \{1, 4, 6, 8, 9, \ldots\}$,
$$A - B = \{2, 3, 5, 11, 17, 23, \ldots\},$$
$$B - A = \{4, 10, 16, 21, 24, \ldots\}.$$

5. COMPUTER REPRESENTATION OF SETS

Although a set has no inherent order among its elements, an order is imposed on any set as soon as we place it in a computer memory. This is true whether the memory used is core, magnetic tape, disk, drum, or any other device currently in use. By the sequence of storage cells, and by the linear order of a tape, there will automatically be a first, second, ... element in a given set as it is represented in the computer. While the set operations that we are discussing are independent of the order in which elements are listed, by utilizing the imposed order properly we can often simplify the implementation of various set operations.

The use of this order is apparent in the concept of a vector. (See Section 13.) We do not usually name the elements of a set A in a haphazard fashion. Rather, we think of these elements as forming a vector, that is, an ordered list, and refer to them as $a_1, a_2, ..., a_n$. This enables us to write simple programs to perform the set operations, as the following examples indicate.

Example 5.1

Write a program fragment to test whether x is an element of A, a set with n elements.

Solution. Assume that the language does not have set operations, for if it does there is no problem. A typical program fragment might look like this:[†]

```
⋮
for i = 1 to n:
    if x = aᵢ:
        (program to be executed for elements of the set)
(continuation of main program)
⋮
```

Example 5.2

Assume $\mathscr{U} = \{1, 2, ..., 10\}$, and $A \subseteq \mathscr{U}$. Write a program fragment to compute the complement of A.

Solution. Denote the complement of A by $\bar{A}$, and assume that the elements of A are given in increasing order, for example, $A = \{1, 4, 6, 7\}$, along with the number n of elements in A.

[†] In the language used here, a simplified form of Madcap, loop control is by indentation. That is, the "for" statement in this example defines a loop containing all subsequent indented lines (the lines "if $x = a_i$:", and "(program ... set)"). Similarly, the "if" statement defines a loop containing all subsequent lines which are further indented (the line "(program ... set)").

$$\vdots$$
Last $= 1$
$j = 1$
for $k = 1$ to n
    if $A_k > $ Last
        for $m = $ Last to $A_k - 1$: $\overline{A}_j = m; j = j + 1$
    Last $= A_k + 1$
\#1   for $k = j$ to $10 - n$, as $i = $ Last, Last $+ 1, \ldots$: $\overline{A}_k = i$
$$\vdots$$

In this example, statement \#1 illustrates the use of coincident indices. The value of i changes with the value of k. For example, if $j = n = 4$ and Last $= 6$ upon encountering this statement, the resulting computation is

for $k = 4, i = 6$: $\overline{A}_4 = 6$
for $k = 5, i = 7$: $\overline{A}_5 = 7$
for $k = 6, i = 8$: $\overline{A}_6 = 8$

If the universal set has few enough elements, then great economies can be achieved both in storage and in execution time through representing each set as a *wordset*. In this representation, the universal set is represented by a single computer word, with each bit in the word representing a particular element. For example, in a computer with a 12-bit word, any set of at most 12 elements can be represented by a single word. It is customary in this representation to sequence the bits from right to left (Fig. 5.1).

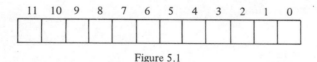

Figure 5.1

Example 5.3

Let $\mathscr{U} = \{a, b, \ldots, k, l\}$. Represent the sets $\{a, b, c, d\}$, $\{a, e, i\}$, $\{b, d, f, h, j, l\}$, and $\{b, d, e, k\}$ as wordsets in a 12-bit computer.

Solution. $\mathscr{U}$ is represented in right-to-left order, so that bit 0 represents a, bit 1 represents b, and so forth.

ANSWER. $\{a, b, c, d\}$:   000000001111,
$\{a, e, i\}$:   000100010001,
$\{b, d, f, h, j, l\}$:   101010101010,
$\{b, d, e, k\}$:   010000011010.

The representation of sets as wordsets permits use of the logical operations of the computer for set manipulations. We shall say more on this in Section 8 and in Chapter 9.

We can of course represent larger sets in this way if the computer has longer words. In addition, the wordset type of representation can be used with more than one word representing the larger sets, although some of the economy is lost.

EXERCISES

1. Write a Fortran program fragment to test whether x is an element of A, a set with n elements.

2. Write a Snobol program fragment to test whether x is an element of A, a set with n elements.

3. Assume $\mathscr{U} = \{a, b, ..., k\}$, and $A \subseteq \mathscr{U}$. Write a Fortran program fragment to compute $\bar{A}$.

4. Assume $\mathscr{U} = \{a, b, ..., k\}$, and $A \subseteq \mathscr{U}$. Write a Snobol program fragment to compute $\bar{A}$.

5. Assume $\mathscr{U} = \{1, 2, ..., 100\}$, and A and B are subsets of $\mathscr{U}$. Write a program fragment (any language) to test whether $A \subseteq B$.

6. Assume $\mathscr{U} = \{1, 2, ..., 100\}$, and $A \subseteq \mathscr{U}$. Write a program fragment (any language) to find all elements x such that $\{x, x+1\} \subseteq A$.

6. SET OPERATIONS

We have already considered the complement of a set, and the relative complement of two sets. We now wish to examine the other set operations that are at our disposal.

We begin by noting that the general Venn diagram for two sets divides the rectangle into four regions (Fig. 6.1).

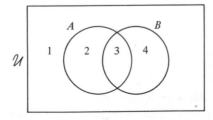

Figure 6.1

We can readily identify certain combinations of the regions:

Set	Regions	Set	Regions
$\mathcal{U}$	1, 2, 3, 4	$A - B$	2
A	2, 3	$\overline{A - B}$	1, 3, 4
$\bar{A}$	1, 4	$B - A$	4
B	3, 4	$\overline{B - A}$	1, 2, 3
$\bar{B}$	1, 2	$\varnothing$	No region included

In fact, all 16 combinations of regions can be identified with expressions involving only A, B, their complements, and various relative complements. However, it is easier to identify the remaining six region combinations by means of the three set operations, union ($\cup$), intersection ($\cap$), and symmetric difference ($\triangle$).

The *union* or *join* of two sets, $A \cup B$, is the set whose elements are either in A, or in B, or in both A and B. Looking at Fig. 6.1, we see that $A \cup B$ corresponds to regions 2, 3, and 4; and that its complement $\overline{A \cup B}$ corresponds to the region 1.

The *intersection* or *meet* of two sets, $A \cap B$, is the set whose elements are in both A and B. Thus $A \cap B$ corresponds to region 3 of Fig. 6.1, while its complement $\overline{A \cap B}$ corresponds to regions 1, 2, and 4.

Finally, the *symmetric difference* of two sets, $A \triangle B$, is the set whose elements are either in A or in B, but not in both. Thus $A \triangle B$ corresponds to regions 2 and 4 of the figure, while $\overline{A \triangle B}$ corresponds to regions 1 and 3.

EXERCISE

1. Consider Fig. 6.1. Find expressions for each of the following region combinations in terms of A, B, their complements, and various relative complements.

    (a) region 1   (b) region 3
    (c) regions 1, 3   (d) regions 2, 4
    (e) regions 1, 2, 4   (f) regions 2, 3, 4.

7. SET ALGEBRA

The operations that we have defined on sets have a number of interesting properties. These may be verified by Venn diagrams, or proven in a more rigorous fashion. We list some of the more important ones:

$$\overline{(\overline{A})} = \overline{\overline{A}} = A$$

$$\left.\begin{array}{l} A \cup A = A \\ A \cap A = A \end{array}\right\} \quad \text{idempotent laws}$$

$$A \bigtriangleup A = \varnothing$$

$$A \cup \mathcal{U} = \mathcal{U}, \qquad A \cap \mathcal{U} = A, \qquad A \bigtriangleup \mathcal{U} = \overline{A}$$

$$A \cup \varnothing = A, \qquad A \cap \varnothing = \varnothing, \qquad A \bigtriangleup \varnothing = A$$

$$\left.\begin{array}{l} \overline{A \cup B} = \overline{A} \cap \overline{B} \\ \overline{A \cap B} = \overline{A} \cup \overline{B} \end{array}\right\} \quad \text{DeMorgan's laws}$$

$$\left.\begin{array}{l} A \cup (A \cap B) = A \\ A \cap (A \cup B) = A \end{array}\right\} \quad \text{absorption laws}$$

$$\left.\begin{array}{l} A \cup B = B \cup A \\ A \cap B = B \cap A \\ A \bigtriangleup B = B \bigtriangleup A \end{array}\right\} \quad \text{commutative laws}$$

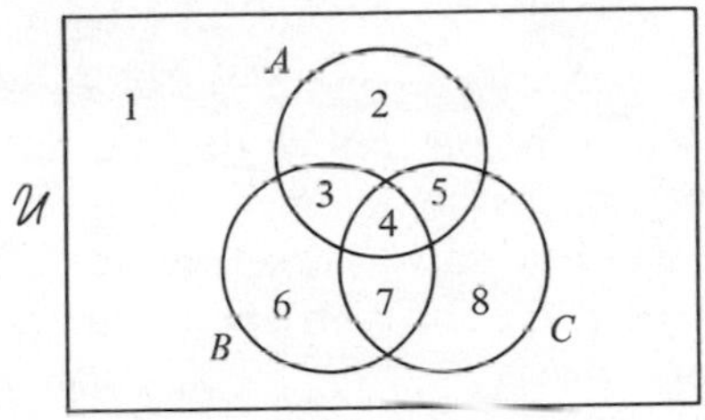

Figure 7.1

As we add a third set C the number of regions in the Venn diagram doubles to eight (Fig. 7.1). A little thought shows that any combination of these eight regions corresponds to some expression involving the operations which we have already defined. For example, regions 2, 3, and 8 together correspond to the expression $(A-C) \cup (C-(A \cup B))$. Thus we do not need to define new set operations in order to handle three or more sets. We can always manipulate the sets two at a time, falling back on the operations that we have already defined.

With the addition of the third set, we can verify that the following properties hold for any sets A, B, and C, whether or not they are distinct.

$$\left.\begin{array}{l} A \cup (B \cup C) = (A \cup B) \cup C \\ A \cap (B \cap C) = (A \cap B) \cap C \\ A \bigtriangleup (B \bigtriangleup C) = (A \bigtriangleup B) \bigtriangleup C \end{array}\right\} \quad \text{associative laws}$$

$$\left.\begin{array}{l} A \cup (B \cap C) = (A \cup B) \cap (A \cup C) \\ A \cap (B \cup C) = (A \cap B) \cup (A \cap C) \\ A \cap (B \bigtriangleup C) = (A \cap B) \bigtriangleup (A \cap C) \end{array}\right\} \quad \text{distributive laws}$$

By defining the operations of union, intersection, symmetric difference, and complementation, we have imposed a structure on the collection of sets, an algebraic structure known as a *set algebra*. We can use the defined operations on sets in a manner analogous to the use of the algebraic operations of addition, subtraction, multiplication, and division on numbers. We shall return to these set algebras in Chapters 8 and 9. For the moment, it suffices to understand the basic operations and to be convinced of the properties that we have ascribed to them.

EXERCISES

1. By Venn diagrams, verify the properties stated in this section.

2. Show $A \cup (B \triangle C) \neq (A \cup B) \triangle (A \cup C)$.

8. COMPUTER OPERATIONS ON SETS

Let us notice that each of the operations that we have defined uses two operands. Even forming $\bar{A}$, the complement of a set A, is forming the complement of A relative to a universal set $\mathcal{U}$, and hence uses two sets. Furthermore, the basic decisions for each operation are made on the basis of matching elements in two sets.

Let us assume a fixed universal set, represented by a vector $V_1, V_2, \ldots$ of elements in the computer. We say that V_i *precedes* V_j if $i < j$, and write $V_i < V_j$ to denote this. Thus in numerical order 3 precedes 17, or $3 < 17$; in alphabetical order f precedes m, or $f < m$. Whenever we compare two sets, we assume that their elements are given in the same order—that order which is determined by the universal set.

With these assumptions we can draw a general purpose flowchart for the set operations that we have defined on sets A and B. See Fig. 8.1. The ⒶA is taken whenever one or the other of the lists is exhausted. The action taken at the four decision points determines which set operation is being performed (Table 8.1). For example, at decision point 1, the element of

TABLE   8.1

	$A \cup B$	$A \cap B$	$A \triangle B$	$B - A$. or $\bar{A}$
Decision point 1	A in	A out	A in	A out
Decision point 2	A, B in	A, B in	A, B out	A, B out
Decision point 3	B in	B out	B in	B in
Decision point 4	remainder in	remainder out	remainder in	remainder of B in, A out

set A being tested is included in the output set if we are computing $A \cup B$ or $A \triangle B$, but not otherwise. At decision point 4, if one of the sets is not exhausted, the remaining elements are included or excluded as indicated in Table 8.1.

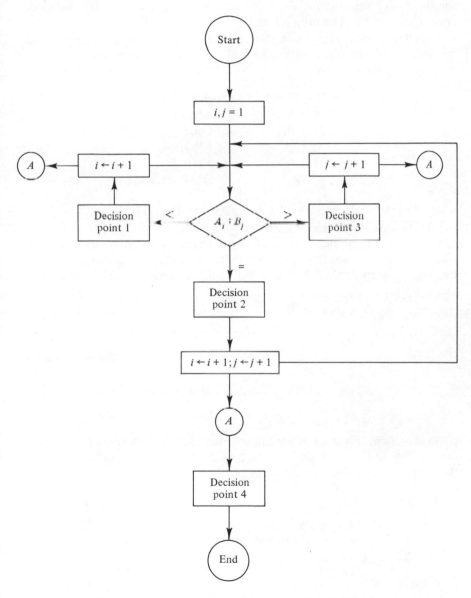

Figure 8.1

Notice that for $A \cup B$ everything in both vectors A and B is included in the resultant vector. For $A \cap B$, only those elements that occur in both vectors A and B are included in the resultant vector. For $A \triangle B$, elements that are in either vector A or vector B, but not in both are included. And for $B - A$, the elements of vector B that are not in vector A are included. (For $\bar{A}$, vector B is the universal set.)

For sets represented by wordsets, these same operations are accomplished by means of the computer's logical operations, with the following correspondence:

Set operation	Logical operation
$A \cap B$	A AND B
$A \cup B$	A OR B
$A \triangle B$	A EXOR B
$\bar{A}$	COMP A
$B - A$	(COMP A) AND B,

where COMP A denotes the 1's complement of A, that is, the computer word obtained by interchanging 0's and 1's in A.

While the logical significance of these operations will be discussed in Chapter 9, here we are interested in the fact that they are performed bit by bit, according to Table 8.2. Recall that a bit (1) denotes the presence of a

TABLE   8.2

$A \backslash B$	0	1	0	1	0	1	
0	0	0	0	1	0	1	1
1	0	1	1	1	1	0	0
	AND		OR		EXOR		COMP

particular element in a set, while the absence of a bit (0) denotes the absence of that element from the set. Thus each of these operations performs one of the set operations simultaneously on the entire set.

Example 8.1

Let $A = \{b, d, f, h, j, l\}$ and $B = \{b, d, e, k\}$ be represented by wordsets, as in Example 5.3. Compute $A \cap B$, $A \cup B$, $A \triangle B$, $\bar{B}$, and $A - B$.

Solution.   The set A is represented by 101010101010, and the set B is represented by 010000011010.

$A \cap B$: Use 101010101010 AND 010000011010. The result is 000000001010.

$A \cup B$: Use 101010101010 OR 010000011010. The result is 111010111010.

$A \triangle B$: Use 101010101010 EXOR 010000011010. The result is 111010110000.

$\bar{B}$: Use COMP 010000011010. The result is 101111100101.

$A - B$: Use (COMP 010000011010) AND 101010101010. The result is 101111100101 AND 101010101010 = 101010100000.

ANSWER. $A \cap B = \{b, d\}$,
$A \cup B = \{b, d, e, f, h, j, k, l\}$,
$A \triangle B = \{e, f, h, j, k, l\}$,
$\bar{B} = \{a, c, f, g, h, i, j, l\}$,
$A - B = \{f, h, j, l\}$.

EXERCISES

1. In developing the general purpose flowchart, Fig. 8.1, we assumed that the elements of sets A and B are given in the same order. Draw the corresponding general purpose flowchart if this assumption is not made.

2. (cont.) Determine the effect of the ordering assumptions on the number of comparisons $A_i : B_j$ needed to compute $A \cup B$.

9. PRODUCT SETS

Many of the data processed by a computer consist of associated items of information: a man's name and his social security number, a part number and its name, a document title and its list of descriptors. Formally, such associations are handled by product sets. Given two sets A and B, the *product set* $A \otimes B$ (read "A cross B") is the set of all ordered pairs $\langle a, b \rangle$ where $a \in A$ and $b \in B$. Note that the first component of each pair is an element of A and the second is an element of B. Thus $B \otimes A = \{\langle b, a \rangle \mid b \in B, a \in A\}$, for which the first component of each pair is an element of B, not A, is quite distinct from $A \otimes B$.

Example 9.1

Let $A = \{1, 2, 3\}$ and $B = \{a, b\}$. Form the sets $A \otimes B$ and $B \otimes A$.

ANSWER. $A \otimes B = \{\langle 1, a \rangle, \langle 1, b \rangle, \langle 2, a \rangle, \langle 2, b \rangle, \langle 3, a \rangle, \langle 3, b \rangle\}$;
$B \otimes A = \{\langle a, 1 \rangle, \langle b, 1 \rangle, \langle a, 2 \rangle, \langle b, 2 \rangle, \langle a, 3 \rangle, \langle b, 3 \rangle\}$.

If $|A|$ is the *cardinality* or number of elements in the set A, and $|B|$ is the cardinality of B, then it is easy to show that $A \otimes B$ and $B \otimes A$ each contain $|A| \cdot |B|$ elements.

Similarly, if $A_1, A_2, \ldots, A_n$ are sets, the product set $A_1 \otimes A_2 \otimes \cdots \otimes A_n$ is the set of all ordered n-tuples $\langle a_1, a_2, \ldots, a_n \rangle$ such that $a_i \in A_i$ for $i = 1, 2, \ldots, n$. Denoting the cardinality of A_i by $|A_i|$, it is easy to show that for any finite n,

$$|A_1 \otimes A_2 \otimes \cdots \otimes A_n| = |A_1| \cdot |A_2| \cdot \cdots \cdot |A_n|.$$

Very often the entire product set $A \otimes B$ is not only not wanted, but indeed causes difficulties. For example, suppose A consists of 10^6 books in a library, and B consists of 10^4 terms that the library uses to describe the books. If we wish to develop a computer information retrieval system for the library, then the product set calls for massive amounts of computer memory. What we really want to store are only those ordered pairs $\langle a, b \rangle$ for which descriptor b properly applies to book a. If there is an average of 15 descriptors per book, then we really need only store 15×10^6 ordered pairs—only 0.15 percent of the pairs in the product set $A \otimes B$.

In this example, and indeed in most cases, we are interested in a subset of the product set rather than the entire product set. Such subsets are called *relations*, and are discussed in the next section.

(We shall have occasion also to refer to sets whose elements are *unordered* pairs of elements of a given set A; that is, for which $\langle a, b \rangle$ and $\langle b, a \rangle$ represent the same element. We shall denote such a set by $A \oplus B$, rather than $A \otimes B$.)

EXERCISES

1.   Show that $|A \otimes B| = |B \otimes A| = |A| \cdot |B|$.

2.   Show that for any finite n

$$|A_1 \otimes A_2 \otimes \cdots \otimes A_n| = |A_1| \cdot |A_2| \cdot \cdots \cdot |A_n|.$$

3.   Since $A \oplus B$ consists of unordered pairs of elements, (a, b) and (b, a) are two representations of the same element of $A \oplus B$, where $a \in A, b \in B$. Does this imply $|A \oplus B| < |A \otimes B|$? If possible, give examples for which $|A \oplus B| < |A \otimes B|$, and for which $|A \oplus B| = |A \otimes B|$.

10. RELATIONS, MAPPINGS, FUNCTIONS

A *relation* is an arbitrary subset of a product set. Most important are the *binary relations*, subsets of $A \otimes B$, and the *ternary relations*, subsets of $A \otimes B \otimes C$. In general, a subset of $A_1 \otimes A_2 \otimes \cdots \otimes A_n$ is called an *n-ary relation*.

Among the common binary relations, those dealing with pairs of elements, are the numerical relationships of equality and inequality, the familial relationships ("son of," "parent of," and so on), and the relationships which label or identify an object ("social security number of," "title of," and the like).

Example 10.1

Give six examples of binary relations.

Solution. Let $R_1, R_2, \ldots, R_6$ denote the relations. Among the many binary relations that exist choose, for example, these six. In all examples $a \in A$ and $b \in B$.

Numerical binary relations. Let A and B be sets of real numbers.
R_1, equals: $\langle a, b \rangle \in R_1$ if and only if $a = b$.
R_2, less than: $\langle a, b \rangle \in R_2$ if and only if $a < b$.
R_3, square of: $\langle a, b \rangle \in R_3$ if and only if $a = b^2$.
Nonnumerical binary relations.
R_4, son of: Let A and B be sets of people. Then $\langle a, b \rangle \in R_4$ if and only if a is a son of b.
R_5, alumnus of: Let A be a set of people and B be a set of colleges. Then $\langle a, b \rangle \in R_5$ if and only if a is an alumnus of b.
R_6, driver's license number of: Let A be a set of integers and B be a set of people. Then $\langle a, b \rangle \in R_6$ if and only if a is the driver's license number of b.

One other type of binary relation is especially important in computing. This is the relation that identifies memory locations and the contents of these locations, either in terms of the address–content pair, or in terms of the position–value pair for a list or other data structure.

The most common ternary relations are those associated with the ordinary arithmetic operations. For example, if a, b, and c are numbers, then one such relation is S, the sum, with $\langle a, b, c \rangle \in S$ if and only if $c = a + b$. Such relations appear explicitly in three-address programming schemes, where every operation requires three operands. The computational operators then define a relation between the contents of the three specified memory locations.

Just as sets have no particular property other than that of set membership, so also relations have no general property, other than that certain *n*-tuples are included in a relation R, and others are not. Three classes of relations, however, are sufficiently important to have been singled out for mathematical study. These are the equivalence and order relations, which we discuss in Section 11, and functions.

An n-ary relation R on $A_1 \otimes A_2 \otimes \cdots \otimes A_n$ is a *function* or *mapping* if whenever $\langle a_1, a_2, \ldots, a_{n-1}, x \rangle \in R$ and $\langle a_1, a_2, \ldots, a_{n-1}, y \rangle \in R$, then $x = y$.

That is, R is a function or mapping precisely if the first $n-1$ components of the n-tuple determine the last one.

Thus the relation R_1, equals, of Example 10.1 is a function, while R_2, less than, is not. (One number is less than a great many other numbers.) Similarly, the ternary relations associated with arithmetic operations are functions. The relation R_3, square of, may or may not be a function, depending on the set B. If B consists of nonnegative numbers, then R_3 is a function. However, if B is the set of all integers or all real numbers, then R_3 is not a function since it contains all pairs $\langle x^2, x \rangle$ and $\langle x^2, -x \rangle$: the second component is not uniquely determined by the first.

The term *mapping* relates to the concept of assigning to an element in one set a unique element in another set (Fig. 10.1). It is useful to think

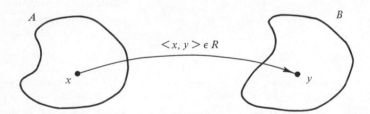

$$\langle x, y \rangle \in R$$

Figure 10.1

in these terms whenever one of the sets consists of labels, identification numbers, or descriptors to be assigned to the elements of the other set.

Consider now the situations depicted in Fig. 10.2 as being typical for the relations F_1, F_2, and F_3. Think of A as being a set of identification numbers, and B a set of individual objects, say, books. Then while F_1 is a perfectly good relation on $A \otimes B$, it has the property of assigning the same identification number x to two different books y and z. This may or may not be a desirable property. The relations F_2 and F_3 do not have this property. Rather, they associate with each identification number a unique book. Thus F_2 and F_3 are functions, while F_1 is not.

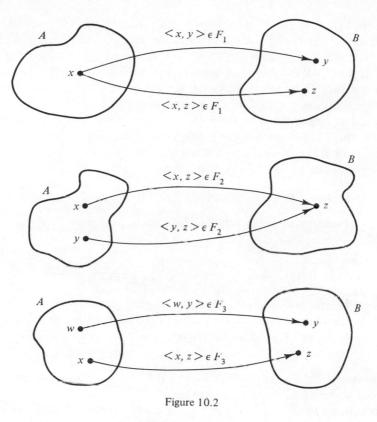

Figure 10.2

Now reverse the process. For each relation R on $A \otimes B$ there is an *inverse relation* R^{-1} on $B \otimes A$, defined by $\langle b, a \rangle \in R^{-1}$ if and only if $\langle a, b \rangle \in R$. Thus each of the relations F_1, F_2, and F_3 has an inverse relation, F_1^{-1}, F_2^{-1}, and F_3^{-1} respectively. These are also represented by Fig. 10.2, but with the arrows reversed.

Notice which relations are functions. While F_1 is not a function, F_1^{-1} is: given y (or z), x is uniquely determined. However, F_2^{-1} is not a function even though F_2 is since $\langle z, x \rangle \in F_2^{-1}$ and $\langle z, y \rangle \in F_2^{-1}$. Only in the third case, F_3 and F_3^{-1}, are both a relation and its inverse functions.

A relation R is *one-to-one* $(1:1)$ if both R and R^{-1} are functions.

If the relation f on $A \otimes B$ is a function, we often write $f: A \rightarrow B$ to indicate the concept of f *mapping* A *into* B. The subset of A for which f

is defined (that is, the set of all $a \in A$ such that $\langle a, y \rangle \in f$ for some $y \in B$) is the *domain* of f, and the subset of B for which f is defined (that is, the set of all $b \in B$ such that $\langle x, b \rangle \in f$ for some $x \in A$) is called the *range* of f. If $f: A \rightarrow B$ is a function whose range is B, we say that f is mapping *onto* B, rather than into. If $\langle a, b \rangle \in f$, we also write $f(a) = b$, and $f: a \rightarrow b$.

A 1:1 function f with domain A and range B is called a *set isomorphism* or a *set bijection* from A to B. Set isomorphisms are important since they define a complete identification of the elements of A with the elements of B: given any $a \in A$, a set isomorphism f uniquely determines a corresponding element $b \in B$, and conversely.

Example 10.2

Determine which of the following mappings are set isomorphisms:

(a) $f: x \rightarrow 2x$;   (b) $g: x \rightarrow x + 3$;
(c) $h: x \rightarrow x^2$;   (d) $R: x \rightarrow |x|$.

For (a) and (b) both the domain and the range consist of the real numbers; for (c) and (d) the domain is the set of real numbers, while the range is the set of nonnegative real numbers.

Solution.   (a) Given a value of x, the value of $2x$ is uniquely determined. Thus f is a function. Now given a value of $2x$, the value of x is also uniquely determined, namely one-half of the given value. Thus f is a set isomorphism since $f^{-1}: x \rightarrow \frac{1}{2}x$ is also a function. (b) The inverse to g is the mapping $g^{-1}: y \rightarrow y - 3$. Since this is a function, g is a set isomorphism. (c) In this case the inverse relation is not a function. Observe that under h both x and $-x$ map into x^2. Hence h^{-1} will relate a number to both of its square roots. Thus although h is a function it is not a set isomorphism. (d) Here the problem is similar to that in (c). Under R two distinct real numbers map onto the same number, and under R^{-1} each number other than 0 will have two images. Thus R is a function but not a set isomorphism.

ANSWER.   (a) and (b) are set isomorphisms; (c) and (d) are not.

Observe also that the inverse mapping f^{-1} to a set isomorphism f is a set isomorphism: for any $b \in B$ the value of $f^{-1}(b)$ is a unique $a \in A$. Thus given either the identification number or the book, we can determine the other one.

While this discussion is phrased in terms of binary relations, it applies equally well to n-ary relations, provided that we think of these as binary relations on more complex sets. For example, a relation on $A_1 \otimes A_2 \otimes A_3 \otimes A_4$ may be considered to be any of the three binary relations

$$A_1 \otimes (A_2 \otimes A_3 \otimes A_4), \qquad \text{or} \qquad (A_1 \otimes A_2) \otimes (A_3 \otimes A_4), \qquad \text{or}$$

$$(A_1 \otimes A_2 \otimes A_3) \otimes A_4 .$$

Of all such decompositions, the extreme ones

$$A_1 \otimes (A_2 \otimes \cdots \otimes A_{n-1} \otimes A_n) \qquad \text{and} \qquad (A_1 \otimes A_2 \otimes \cdots \otimes A_{n-1}) \otimes A_n$$

are the most common, corresponding to the relations between an individual (book, person, part, ...) and a list of its descriptors.

We frequently have occasion to compose, or chain together, mappings or other relations. For example, we may use one relation to associate with a book its list of descriptors, followed by a second relation which associates with the descriptor list other books having the same or similar descriptor lists.

If R is a relation on $A \otimes B$ and S is a relation on $C \otimes D$, then the *composition* $R \circ S$ is defined by

$$R \circ S = \{\langle x, z \rangle | \langle x, y \rangle \in R \quad \text{and} \quad \langle y, z \rangle \in S \quad \text{for some} \quad y \in B \cap C\}.$$

See Fig. 10.3. Note in particular that $R \circ S = \emptyset$ if $B \cap C = \emptyset$. For functions or mappings, $f: A \to B$ and $g: C \to D$, composition results in a function. (See Exercise 1.) The only general statement that can be made about the domain and range of $f \circ g$ is that the domain is A and the range is D if f and g are onto and $B = C$. Notationally, observe that $(f \circ g)(x) = g(f(x))$.

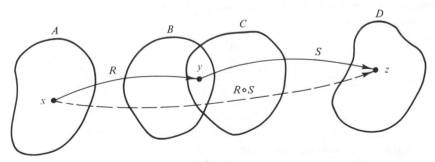

Figure 10.3

Example 10.3

Suppose the domain and range of the relations R, S, f, and g are subsets of the integers. Determine the composite relations $R \circ S$ and $f \circ g$ if

$$R = \{\langle x, y \rangle \mid 1 + x \leqslant y\},$$

$$S = \{\langle x, y \rangle \mid 1 - x \leqslant y \leqslant 3 - x\},$$

$$f = \{\langle x, y \rangle \mid y = 3x + 1\},$$

$$g = \{\langle x, y \rangle \mid y = x^2 - 2x + 7\}.$$

Solution.   Consider first $R \circ S$, and change the variables in the description of S:

$$S = \{\langle y, z \rangle \mid 1 - y \leqslant z \leqslant 3 - y\}.$$

Given a value x_0, R determines a set of y values, $y_{01} = x_0 + 1$, $y_{02} = x_0 + 2$, $y_{03} = x_0 + 3$, ... Each value $y_{0i} = x_0 + i$ determines a set of z values, $1 - y_{0i}$, $2 - y_{0i}$, $3 - y_{0i}$, or $1 - x_0 - i$, $2 - x_0 - i$, $3 - x_0 - i$. Thus from x_0, the entire set of z values determined by the composite relation is

$$
\begin{array}{ccc}
-x_0, & 1 - x_0, & 2 - x_0, \\
-1 - x_0, & -x_0, & 1 - x_0, \\
-2 - x_0, & -1 - x_0, & -x_0, \\
\end{array}
$$

$$\ldots,$$

that is, all $z \leqslant 2 - x_0$. Thus, $R \circ S = \{\langle x, z \rangle \mid z \leqslant 2 - x\}$.

Similarly, change variables in the description of g:

$$g = \{\langle y, z \rangle \mid z = y^2 - 2y + 7\}.$$

Since f is defined by the equation $y = 3x + 1$, it follows that $f \circ g$ is defined by the equation

$$z = (3x + 1)^2 - 2(3x + 1) + 7$$

$$= 9x^2 + 6.$$

ANSWER. $R \circ S = \{\langle x, z \rangle \mid z \leqslant 2 - x\}$,
$f \circ g = \{\langle x, z \rangle \mid z = 9x^2 + 6\}.$

For any (finite) set of mappings, there is a natural diagrammatic representation. For example, Fig. 10.4 represents the mappings $f: A \to B$, $g: B \to C$, and $h: A \to C$. Similarly, Fig. 10.5 is a diagram of four mappings

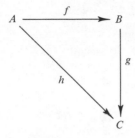

Figure 10.4

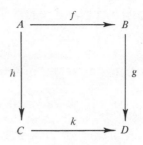

Figure 10.5

(plus two formed by composition). While in general there is no implied relationship between the arcs (arrows) of such diagrams, it is entirely possible that as functions $h = f \circ g$ (in the first case) or $h \circ k = f \circ g$ (in the second). Such diagrams are said to be commutative: a diagram is *commutative* if any two paths from one given set to another in the diagram represent the same (composite) mapping.

A few special mappings are used sufficiently to warrant special names.

For any set A, the *identity map* is $I_A = \{\langle x, x \rangle \mid x \in A\}$.

For $P = A_1 \otimes A_2 \otimes \cdots \otimes A_n$ and each $i = 1, \ldots, n$, the mapping θ_i: $P \to A_i$ is the *projection mapping* of P on A_i if for each $x = \langle x_1, x_2, \ldots, x_n \rangle$, $\theta_i(x) = x_i$.

Let $f: A \to B$, and $A' \subset A$. The *restriction of f to A'*, $f \mid A'$, is a mapping $g: A' \to B$ such that for $a \in A'$, $g(a) = f(a)$. That is, $f \mid A'$ is simply f with its domain changed from A to the smaller set A'. Note that the range may also change as a result of this restriction. If $f: A \to B$, $f': A' \to B$, $A' \subset A$, and $f' = f \mid A'$, then f is an *extension* of f'. A function has many extensions.

EXERCISES

Both the domain and the range are subsets the integers. Determine the composite relation $R \circ S$:

1. $R = \{\langle x, y \rangle \mid y = 2x - 1\}$
   $S = \{\langle x, y \rangle \mid y = x + 3\}$

2. $R = \{\langle x, y \rangle \mid y = x - 4\}$
   $S = \{\langle x, y \rangle \mid y = x^2 + 3x + 1\}$

3. $R = \{\langle x, y \rangle \mid y = 2^x\}$
   $S = \{\langle x, y \rangle \mid x = 2^y\}$

4. $R = \{\langle x, y \rangle \mid y = 2^x\}$
   $S = \{\langle x, y \rangle \mid y = x^2\}$

5. $R = \{\langle x, y \rangle \mid y \leqslant x^2 - 7\}$
   $S = \{\langle x, y \rangle \mid y = x^2 + x + 1\}$

6. $R = \{\langle x, y \rangle \mid xy \leqslant 100\}$
   $S = \{\langle x, y \rangle \mid xy \leqslant 5\}$

7. Let $f: A \to B$ be a mapping. Show that
   (a) f is onto if and only if $f^{-1} \circ f = I_B$;
   (b) f is 1:1 if and only if $f \circ f^{-1} = I_A$.

11. EQUIVALENCE AND ORDER RELATIONS

Three sets of properties are often used to classify relations. Two combinations of these properties define the equivalence and order relations.

Let A be an arbitrary set, and ρ a binary relation on A. That is, $\rho \subseteq A \otimes A$. We shall adopt the notation $a\rho b$ for $\langle a,b \rangle \in \rho$. The relation ρ is *reflexive* if $a\rho a$ for all $a \in A$. It is *irreflexive* if $a\rho a$ for no $a \in A$. The relation ρ is *symmetric* if whenever $a\rho b$, then $b\rho a$. It is *antisymmetric* if whenever both $a\rho b$ and $b\rho a$, then $a = b$. The relation ρ is *transitive* if whenever $a\rho b$ and $b\rho c$, then $a\rho c$.

The numerical relations of equality and inequality provide standard examples for these properties. The relation "equals" ($=$) is reflexive, symmetric, and transitive. The relation "less than or equal to" ($\leqslant$) is reflexive and transitive, but antisymmetric. Clearly if $a \leqslant b$ and $b \leqslant a$, then $a = b$. However, although $3 \leqslant 5$, it is not true that $5 \leqslant 3$. The relation "less than" ($<$) is irreflexive and transitive, but is not symmetric. Note that $<$ is vacuously antisymmetric. That is, it is antisymmetric since there are no numbers a and b for which $a < b$ *and* $b < a$. Thus the conditions for $<$ to be antisymmetric automatically hold.

A relation ρ is an *equivalence relation* if it is reflexive, symmetric, and transitive. It is a *partial order* if it is reflexive, antisymmetric, and transitive. And it is a *simple* or *linear order* if it is a partial order such that for any a and b in A, $a\rho b$ or $b\rho a$.

Of the three numerical relations that we have discussed, $=$ is an equivalence relation, and $\leqslant$ and $<$ are both linear orders. The "subset" relation $\subseteq$ provides an example of a partial order that is not linear. For example, $\{1,2\} \subseteq \{1,2,3\}$, and $\{2,3\} \subseteq \{1,2,3\}$, but neither $\{1,2\} \subseteq \{2,3\}$ nor $\{2,3\} \subseteq \{1,2\}$ holds.

Less trivial examples of an equivalence relation are provided by numerical congruences. For any integers a, b, and n we say that a *is congruent to* b *modulo* n, written $a \equiv b \pmod{n}$, if $a-b$ is exactly divisible by n. For example, $17 \equiv 442 \pmod{5}$ since $17-442 = -425$ is divisible by 5. Such a congruence is an equivalence relation. Clearly for any a and b, $a \equiv a \pmod{n}$, and if $a \equiv b \pmod{n}$ then $b \equiv a \pmod{n}$. Now suppose that $a \equiv b \pmod{n}$ and $b \equiv c \pmod{n}$. Then $a-b = k_1 n$ and $b-c = k_2 n$ for some integers k_1 and k_2. Thus

$$a - c = (a-b) + (b-c)$$

$$= k_1 n + k_2 n$$

$$= (k_1 + k_2)n.$$

Hence $a-c$ is divisible by n, or $a \equiv c$ (mod n). Thus we have proven an important theorem.

> For any integer n, congruence modulo n is an equivalence relation on the set of integers.

A *partition* of a set S is a collection of subsets $A_1, A_2, ..., A_n, ...$ which *cover* S, that is, $A_1 \cup A_2 \cup \cdots \cup A_n \cup \cdots = S$, and such that if $i \neq j$, then $A_i \cap A_j = \varnothing$. An equivalence relation ρ on a set S defines a set of *equivalence classes* which partition S. For $a \in S$, let $R(a) = \{x \in S \mid a\rho x\}$. The sets $R(a)$ partition S. Observe that for any $a \in s$, $a \in R(a)$. Hence the sets cover S. Next suppose that $a \neq b$, and $R(a) \cap R(b) \neq \varnothing$. Take $x \in R(a) \cap R(b)$, and y any element of $R(a)$. Then $a\rho y$, hence $y\rho a$. But also $a\rho x$, so that by transitivity, $y\rho x$. But also $b\rho x$, or $x\rho b$, so that $y\rho b$, and hence $b\rho y$. Thus if $y \in R(a)$ then $y \in R(b)$, or $R(a) \subseteq R(b)$. Similarly $R(b) \subseteq R(a)$. That is, if $R(a) \cap R(b) \neq \varnothing$, then $R(a) = R(b)$. Hence the equivalence classes partition S. We leave it as an exercise to show that any partition of a set defines an equivalence relation.

Example 11.1

Determine the equivalence classes of integers for the relation "congruence modulo 5."

Solution. By the theorem just proven, congruence modulo 5 is an equivalence relation. Determine the equivalence classes $R(i)$ by finding numbers equivalent to certain chosen numbers. For example, $0 \in R(0)$. Also $5 \in R(0)$ since $0 \equiv 5$ (mod 5). Similarly, $-5, 10, 15, ...$ are elements of $R(0)$. In fact $R(0)$ contains all numbers exactly divisible by 5. However, $1 \notin R(0)$ since $0 \not\equiv 1$ (mod 5). But $1 \in R(1)$. Hence $R(1) \neq R(0)$: it is a different equivalence class. The class $R(1)$ contains $6, 11, ...$, in fact, all integers leaving a remainder of 1 upon division by 5. In like manner, determine that $R(2)$, $R(3)$, and $R(4)$ are distinct equivalence classes. Finally, since any integer leaves a remainder of $0, 1, 2, 3$, or 4 upon division by 5, the set of integers has been covered.

$$\text{ANSWER.} \quad \begin{aligned} R(0) &= \{..., -10, -5, 0, 5, 10, ...\}, \\ R(1) &= \{..., -9, -4, 1, 6, 11, ...\}, \\ R(2) &= \{..., -8, -3, 2, 7, 12, ...\}, \\ R(3) &= \{..., -7, -2, 3, 8, 13, ...\}, \\ R(4) &= \{..., -6, -1, 4, 9, 14, ...\}. \end{aligned}$$

The three properties (reflexive, symmetric, transitive) of an equivalence relation ρ on a set A allow us to establish a diagram of such a relation.

Suppose the set A is ordered in some arbitrary way, for example, numerical or alphabetical order, and take $x, y, z \in A$. The diagram is begun by plotting the set A on both the x and y axes.

Reflexive: Since $x\rho x$, all points $\langle x, x \rangle$ must be in the diagram of ρ. That is, all points $\langle x, x \rangle$ on the diagonal $x = y$ for which $x \in A$ are part of the diagram. See Fig. 11.1.

Symmetric: If $x\rho y$, then $y\rho x$; thus the diagram of ρ is symmetric about the line $x = y$. See Fig. 11.2.

Transitive: If $x\rho y$ and $y\rho z$, then $x\rho z$. See Fig. 11.3.

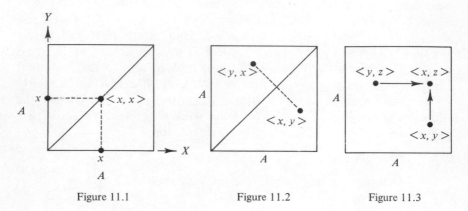

Figure 11.1   Figure 11.2   Figure 11.3

Note that this also implies the points $\langle z, x \rangle$, $\langle y, x \rangle$, and $\langle z, y \rangle$ are in the diagram, as are $\langle x, x \rangle, \langle y, y \rangle, \langle z, z \rangle$ which are on the diagonal (Fig. 11.4). In summary, the diagram of ρ consists of points, lines, and rectangular blocks, symmetrically located with respect to $x = y$, and points on the line $x = y$. See Fig. 11.5.

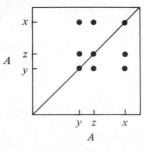

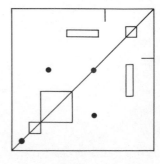

Figure 11.4   Figure 11.5

Example 11.2

Suppose $\{1,4,5\}$, $\{2\}$, $\{3,6\}$, and $\{7,8\}$ are equivalence classes for a relation R on $1, \ldots, 8$. Diagram the relation.

Solution. From $\{1,4,5\}$, the points $\langle 1,4 \rangle$, $\langle 1,5 \rangle$, and $\langle 4,5 \rangle$ together with their diagonal reflections $\langle 4,1 \rangle$, $\langle 5,1 \rangle$, and $\langle 5,4 \rangle$ must be in the diagram, along with $\langle 1,1 \rangle$, $\langle 4,4 \rangle$, and $\langle 5,5 \rangle$. The class $\{2\}$ yields only the point $\langle 2,2 \rangle$, while the class $\{3,6\}$ yields points $\langle 3,3 \rangle$, $\langle 3,6 \rangle$, $\langle 6,3 \rangle$, and $\langle 6,6 \rangle$. Similarly, from $\{7,8\}$ the points $\langle 7,7 \rangle$, $\langle 7,8 \rangle$, $\langle 8,7 \rangle$, and $\langle 8,8 \rangle$ are in the diagram.

ANSWER.   See Fig. 11.6.

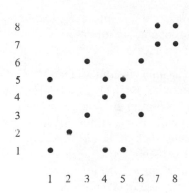

Figure 11.6

Example 11.3

One problem in computer-assisted instruction is that of handling misspelled words or variant spellings. This is generally done by a phonetic reduction routine, the first step of which is a mapping that defines an equivalence relation. For example, the routine in Planit uses the following mapping:

$$f \downarrow \quad \begin{array}{l} ABCDEFGH\,I\,JKLMN\,OPQRS\,TUVWXYZ \\ ABCDABCH\,ACC\,LMM\,ABCRCD\,ABH\,CAC \end{array}$$

That is, $f(A) = A$, $f(E) = A$, $f(F) = B$, and so on. Determine the equivalence classes defined by ρ: $a\rho b$ if and only if $f(a) = f(b)$.

Solution. Since $f(A) = A$, determine first all letters x such that $f(x) = A$. These form an equivalence class, $R(A) = \{A, E, I, O, U, Y\}$. (Note that $R(A) = R(E) = \cdots = R(Y)$.) Similarly, $R(B) = \{B, F, P, V\}$, and so on. Since the classes partition the alphabet, ρ is an equivalence relation.

$$
\begin{aligned}
\text{ANSWER.} \quad R(A) &= \{A, E, I, O, U, Y\}, \\
R(B) &= \{B, F, P, V\}, \\
R(C) &= \{C, G, J, K, Q, S, X, Z\}, \\
R(D) &= \{D, T\}, \quad R(H) = \{H, W\}, \\
R(L) &= \{L\}, \quad R(M) = \{M, N\}, \\
R(R) &= \{R\}.
\end{aligned}
$$

The Planit routine has three more steps:

(2) for any consonant α, map αH into α (for example, THE $\rightarrow$ DHA $\rightarrow$ DA)
(3) for any consonant α, map $\alpha\alpha$ into α
(4) eliminate all occurrences of A.

For example, *digital computer* becomes DCDL CMBDR; *discrete computational structures* becomes DCRD CMBDDML CDRCDRC. By this reduction process many common spelling errors are wiped out before any word comparisons are made. For example, *their* and *there* both reduce to DR. Double letter variants such as *labeled* and *labelled* reduce to the same string, in this case LBLD. Thus any search for equivalent words, such as matching a document descriptor with a request term, is greatly simplified. However, this is at the expense of equating some words with very different meanings. For example, *Thomas, tonic,* and *think* all reduce to DMC. In deciding upon a reduction routine, one must balance the cost of such mismatches against the economy of the simplified matching that results.

EXERCISES

1. Let π be a partition of a set S into classes $\pi_1, \pi_2, \pi_3, \ldots$. Show that the relation ρ, defined by $a\rho b$ if and only if a and b are in the same partition class π_i, is an equivalence relation. *Hint:* You do not need anything exotic. Use the relation ρ as defined.

2. Prove or disprove: all points within any one rectangle of the diagram of an equivalence relation are equivalent. Is the converse true?

3. Write a program to realize the four-step Planit routine of Example 11.3.

4. (cont.) It is of interest to examine the ambiguity introduced by use of the Planit reduction technique. Extend your routine to accept an arbitrarily long text segment, and to print out a list of the reduced strings produced, together with the words producing each. For example, the list generated by the first sentence of this exercise is

B	OF, BY
BLMD	PLANIT
C	IS, USE
CMM	EXAMINE
D	IT, TO, THE
DCMC	TECHNIQUE
MBCD	AMBIGUITY
MDRCD	INTEREST
MDRDCD	INTRODUCED
RDCDM	REDUCTION

12. THE LATTICE OF SUBSETS

We have earlier introduced the scheme of representing a subset of a given (universal) set as a wordset: a binary numeral with ones denoting elements in the subset, and zeroes denoting elements in its complement. Since each position in the word representing the universal set may have one of two possible values, the number of subsets for a given set is easy to determine:

> A set of n elements has exactly 2^n subsets.

Example 12.1

Determine the subsets of the set $S = \{a, b, c\}$.

Solution. Represent S as a wordset, in the order c, b, a. Then the subsets of S are given by their representations:

111:	$\{a,b,c\}$	011:	$\{a,b\}$
110:	$\{b,c\}$	010:	$\{b\}$
101:	$\{a,c\}$	001:	$\{c\}$
100:	$\{c\}$	000:	$\varnothing$.

Note that this three-element set has $2^3 = 8$ subsets.

One relation R on subsets of a given set is defined by $\langle A, B \rangle \in R$ if and only if B is constructed by adding one element of S to A. In the wordset representation, this amounts to changing a 0 in the representation of A to a 1 to form the representation of B.

It is customary to represent the relation R just defined as a diagram in which $\langle A, B \rangle \in R$ is depicted by B lying above A and connected to A by a line. These diagrams for sets of one, two and three elements are given in

Fig. 12.1. The structure represented by such a diagram is called a *lattice of subsets*. It is an example of a general class of structures, lattices, to be studied in Chapter 7.

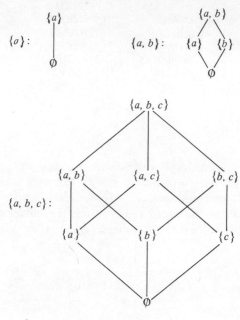

Figure 12.1

EXERCISES

Draw the lattice of subsets for the set:

1. $\{1\}$.   2. $\{1, 3, 7\}$.

3. $\{a, b, c, d\}$.   4. $\{1, 3, 7, a, e\}$.

5. $\{2, 5, a, \rho, 13\}$.

13. VECTORS AND MATRICES

Vectors and matrices are mathematical objects that may be defined in a number of ways. For our purpose, the representation of data structures, rather simple definitions are sufficient.

A *vector* of *n components* is an ordered *n*-tuple, or a list of *n* quantities. These quantities may be all numerical, or a mixture of numerical and nonnumerical data, as one might find in a descriptor list. If V is a vector, then the ith component of V is customarily represented by V_i or $V(i)$.

A *matrix* is a two-dimensional rectangular data array whose components are doubly indexed, with the first index indicating a row number and the second a column number (Fig. 13.1). A component of an array is often called an *entry*.

$$
\begin{array}{c}
\begin{array}{cccc}
\text{col 1} & \text{col 2} & & \text{col } n
\end{array} \\
\begin{array}{c}
\text{row 1} \\
\text{row 2} \\
\\
\\
\\
\text{row } m
\end{array}
\left[
\begin{array}{cccc}
a_{1,1} & a_{1,2} & \cdots & a_{1,n} \\
a_{2,1} & a_{2,2} & \cdots & a_{2,n} \\
\\
\\
\\
a_{m,1} & a_{m,2} & & a_{m,n}
\end{array}
\right]
\end{array}
$$

Figure 13.1

A matrix with m rows and n columns is called an $m \times n$ *matrix*. For many purposes an n-component vector can be identified with either a $1 \times n$ or an $n \times 1$ matrix.

We shall reserve the term *matrix* for an array whose components have an associated algebra, usually that of the real numbers. With this agreement, we can define algebraic operations on matrices.

Addition. Two matrices $A = (a_{ij})$ and $B = (b_{ij})$ may be added if and only if they are of the same size, $m \times n$, and their components have the same associated algebra. The sum is a new matrix $A + B = C = (c_{ij})$ defined by

$$c_{ij} = a_{ij} + b_{ij}, \qquad i = 1, \dots, m; \quad j = 1, \dots, n.$$

Multiplication. Two matrices A and B may be multiplied in order $A \times B$ if and only if the number of columns in A equals the number of rows in B, and their components have the same associated algebra. If A is an $m \times k$ matrix and B is $k \times n$, the product is a new matrix $A \times B = D = (d_{ij})$ of size $m \times n$ defined by

$$
d_{ij} = \sum_{t=1}^{k} a_{it} b_{tj}
$$
$$
= a_{i1} b_{1j} + a_{i2} b_{2j} + \cdots + a_{ik} b_{kj}.
$$

Note that even if $A \times B$ is defined, $B \times A$ may not be. Furthermore, $A \times A$ is not defined unless A is square $(n \times n)$.

Example 13.1

Compute $A + B$, $A \times C$, and $B \times C$, where

$$A = \begin{bmatrix} 1 & 0 & 2 & 5 \\ -1 & 3 & 4 & 2 \\ 2 & 4 & 1 & 3 \end{bmatrix},$$

$$B = \begin{bmatrix} 0 & 0 & -1 & -5 \\ 1 & 2 & -3 & -7 \\ -4 & 0 & 2 & -1 \end{bmatrix},$$

$$C = \begin{bmatrix} 1 & 2 \\ -3 & 4 \\ 0 & -1 \\ 2 & 5 \end{bmatrix}.$$

Solution. A and B are the same size, 3×4. Hence the sum is determined by adding corresponding components:

$$A + B = \begin{bmatrix} 1+0 & 0+0 & 2-1 & 5-5 \\ -1+1 & 3+2 & 4-3 & 2-7 \\ 2-4 & 4+0 & 1+2 & 3-1 \end{bmatrix}$$

$$= \begin{bmatrix} 1 & 0 & 1 & 0 \\ 0 & 5 & 1 & -5 \\ -2 & 4 & 3 & 2 \end{bmatrix}.$$

Since the number of columns in A and B is the same as the number of rows in C, the products $A \times C$ and $B \times C$ are defined. (Note that $C \times A$ and $C \times B$ are not defined.)

$$A \times C = \begin{bmatrix} 1 \cdot 1 + 0 \cdot - 3 + 2 \cdot 0 + 5 \cdot 2 & 1 \cdot 2 + 0 \cdot 4 + 2 \cdot - 1 + 5 \cdot 5 \\ -1 \cdot 1 + 3 \cdot - 3 + 4 \cdot 0 + 2 \cdot 2 & -1 \cdot 2 + 3 \cdot 4 + 4 \cdot - 1 + 2 \cdot 5 \\ 2 \cdot 1 + 4 \cdot - 3 + 1 \cdot 0 + 3 \cdot 2 & 2 \cdot 2 + 4 \cdot 4 + 1 \cdot - 1 + 3 \cdot 5 \end{bmatrix}$$

$$= \begin{bmatrix} 11 & 25 \\ -6 & 16 \\ -4 & 34 \end{bmatrix}.$$

Similarly,

$$B \times C = \begin{bmatrix} -10 & -24 \\ -19 & -22 \\ -6 & -15 \end{bmatrix}.$$

The *identity matrix* I_n of size n is a square matrix such that $a_{ii} = 1$, $i = 1, \ldots, n$, and $a_{ij} = 0$ for all i and j, $j \neq i$. See Fig. 13.2. Identity matrices

$$\begin{bmatrix} 1 & 0 & 0 & \ldots & 0 \\ 0 & 1 & 0 & \ldots & 0 \\ 0 & 0 & 1 & \ldots & 0 \\ \cdot & & & & \cdot \\ \cdot & & & \ddots & \cdot \\ \cdot & & & & \cdot \\ 0 & 0 & 0 & \ldots & 1 \end{bmatrix}$$

Figure 13.2

have the property that $I_m \times A = A$ and $A \times I_n = A$ whenever the products are defined (that is, whenever A is an $m \times n$ matrix).

If A is a square matrix, $n \times n$, then there may or may not exist a matrix B such that $B \times A = A \times B = I_n$. If such a B exists, then A is called *nonsingular* and B is its *inverse* A^{-1}. Note also that $A = B^{-1}$. Any nonsquare matrix is singular. See Exercise 4.

Not every square matrix is nonsingular. In fact, a square matrix is nonsingular if and only if its *determinant* is nonzero, where the determinant is defined as follows. Let $A = (a_{ij})$ be an $n \times n$ matrix, and let $\pi(1)$, $\pi(2), \ldots, \pi(n)$ be a permutation, or a particular ordering, of the integers, $1, 2, \ldots, n$. (See Section 5, Chapter 6.) Then the determinant of A is

$$\det(A) = \sum \operatorname{sgn}(\pi) a_{1, \pi(1)} a_{2, \pi(2)} \cdots a_{n, \pi(n)},$$

where $\operatorname{sgn}(\pi)$ is $+1$ if $\pi(1), \ldots, \pi(n)$ is an even permutation and -1 if $\pi(1), \ldots, \pi(n)$ is an odd permutation, and where the sum is taken over all possible permutations of $1, \ldots, n$. (The terms *even* and *odd* refer to the number of interchanges necessary to put the sequence back into numerical order.)

Example 13.2

Compute the formulas for the determinant of a 2×2 matrix, A, and a 3×3 matrix, B.

Solution.   If $A = (a_{ij})$ is a 2×2 matrix, use the permutations of $1, 2$. There are two of these: $1, 2$, which is even, and $2, 1$, which is odd. Thus

$$\det(A) = a_{1,1} a_{2,2} - a_{1,2} a_{2,1}.$$

For the 3×3 matrix $B = (b_{ij})$ use permutations of $1, 2, 3$. There are three even permutations, 123, 231, and 312, and three odd permutations, 132, 213, and 321. Thus

$$
\begin{aligned}
\det(B) = \; & b_{1,1} b_{2,2} b_{3,3} - b_{1,1} b_{2,3} b_{3,2} \\
& + b_{1,2} b_{2,3} b_{3,1} - b_{1,2} b_{2,1} b_{3,3} \\
& + b_{1,3} b_{2,1} b_{3,2} - b_{1,3} b_{2,2} b_{3,1}.
\end{aligned}
$$

ANSWER. $\det(A) = a_{11} a_{22} - a_{12} a_{21},$

$$
\begin{aligned}
\det(B) = \; & b_{11} b_{22} b_{33} - b_{11} b_{23} b_{32} \\
& + b_{12} b_{23} b_{31} - b_{12} b_{21} b_{33} \\
& + b_{13} b_{21} b_{32} - b_{13} b_{22} b_{31}.
\end{aligned}
$$

Note that this is computationally a very poor definition. The determinant of an $n \times n$ matrix, as defined, involves a sum of $n!$ terms, each of which is a product of n factors. Thus a total of $(n-1)n! + n! - 1 = n \cdot n! - 1$ operations are required. For $n = 10$, this amounts to $36, 287, 999$ operations. The cumulative error in this computation is very high. Better methods for computing determinants exist, but are not of immediate interest to us.

The *transpose* of an $m \times n$ matrix $A = (a_{ij})$ is the $n \times m$ matrix $B = (b_{ij})$ defined by $b_{ij} = a_{ji}$ for all i and j. The transpose of A is denoted by A^{T}.

The matrix $A = (a_{ij})$ is *symmetric* if $a_{ij} = a_{ji}$ for all i and j. It is *zero-symmetric* if whenever $a_{ij} = 0$, then also $a_{ji} = 0$, for all i and j. That is, in a zero-symmetric matrix the zero entries are symmetrically located, but the nonzero entries are not necessarily symmetric.

EXERCISES

1. Compute $A + B$, $A \cdot B^{\mathrm{T}}$, $A^{\mathrm{T}} \cdot B$, $B \cdot A^{\mathrm{T}}$, $B^{\mathrm{T}} \cdot A$.

$$
A = \begin{bmatrix} 1 & 2 \\ 3 & 7 \\ 0 & 4 \end{bmatrix}, \qquad
B = \begin{bmatrix} -2 & 1 \\ 4 & 0 \\ 1 & 3 \end{bmatrix}.
$$

2. Compute A^{-1}.

$$
A = \begin{bmatrix} 1 & 2 \\ 3 & 4 \end{bmatrix}.
$$

3. Write a program to compute A^{-1} if A is a 3×3 matrix. Test your result with the given matrix by computing $A A^{-1}$ and $A^{-1} A$:

$$
A = \begin{bmatrix} 1 & 2 & 3 \\ 2 & 3 & 1 \\ 3 & 1 & 2 \end{bmatrix}.
$$

4. It is conceivable that if A is a nonsquare matrix of size $m \times n$, a matrix B of size $n \times m$ exists such that $A \times B = I_m$ and $B \times A = I_n$. Show that this cannot happen. *Hint:* Show that on the one hand $\sum_i \sum_k a_{ik} b_{ki} = m$, while on the other hand $\sum_k \sum_i b_{ki} a_{ik} = n$.

CHAPTER 2

Undirected Graphs

1. GRAPH THEORY

The theory of graphs is concerned with the analysis of certain structures called *linear graphs*. These are not the graphs of functions, but rather are structures such as the edges and vertices of a polyhedron, or the diagram of a lattice of subsets—that is, structures made up of points joined by line segments.

The study of graphs may be divided in a number of ways, of which the following are basic:

Directed versus Undirected Graphs. In a directed graph, each line segment has a direction assigned to it. Such graphs are useful for the analysis of structures involving flow, such as flowcharts and PERT diagrams. Undirected graphs have no such assignment on their edges. They relate to more static situations, such as the study of data base structures. See Fig. 1.1.

Finite versus Infinite Graphs. It is entirely possible for a graph to have infinitely many vertices or edges, or both. However, most of the theory deals with graphs in which the set of edges and the set of vertices are both finite, called *finite graphs*. We shall consider only finite graphs.

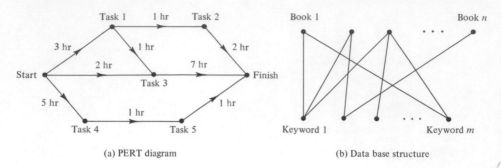

(a) PERT diagram   (b) Data base structure

Figure 1.1

Labeled versus Unlabeled Graphs. The basic theory of graphs does not assume that either the vertices or the edges are labeled in any way (except possibly for identification). Yet in many applications values of some sort are assigned to the edges or vertices. For example, the branches of a flowchart may be labeled with the conditions under which each branch is taken. The edges of a PERT chart may be labeled with the time or cost of traversing each edge. And the vertices in the diagram of a subset lattice may be labeled with the sets they denote. Thus, the theory must deal with both unlabeled and labeled graphs.

We begin with the study of finite, undirected, and unlabeled graphs.

2. BASIC DEFINITIONS

Let $G = \langle V, E, f \rangle$, where $V = \{v_i\}$ is the set of *vertices* of G, $E = \{e_j\}$ is the set of *edges* of G, and $f : E \to V \oplus V$ is a function mapping edges into *unordered* pairs of vertices. Then G is called an *undirected graph*. For $e \in E$, if $f(e) = (v, w)$, we say that e is *incident* to v and to w. For the edge $f(e) = (v, w)$, we also use the notation $e = vw$. Note that to each e_k is assigned a pair (v_i, v_j) of vertices, but that generally there are vertex pairs which are assigned to no edge.

Example 2.1

Draw the graph with vertices $V = \{v_1, \ldots, v_7\}$, edges $E = \{e_1, \ldots, e_7\}$, and f defined by Table 2.1.

Solution. In drawing graphs, the incidence of edges to vertices is the important consideration. An edge need not be straight, nor of any particular length or direction. One drawing of this graph is given in Fig. 2.1, where the edges and vertices are labeled for identification only.

TABLE 2.1

e	$f(e)$
e_1	(v_1, v_3) (or (v_3, v_1))
e_2	(v_1, v_2)
e_3	(v_2, v_4)
e_4	(v_4, v_4)
e_5	(v_4, v_5)
e_6	(v_4, v_5)
e_7	(v_6, v_7)

The graph of Fig. 2.1 has some special features. First there is an edge $e_4 = v_4 v_4$. Such an edge, beginning and ending at the same vertex, is called a *loop* or a *self-loop*. Second, there are *multiple edges* between v_4 and v_5. Third, the graph consists of two disjoint pieces. Finally, this is a finite graph since V and E are both finite sets.

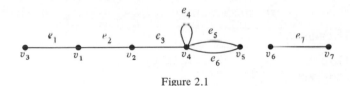

Figure 2.1

Assumption 1. The undirected graphs that we study are finite, without loops or multiple edges.

Under the assumption of no multiple edges, the edge $v_i v_j$ is well defined (if it exists). That is, the phrase does not refer to two or more distinct edges.

If G and H are graphs, H is a *subgraph* of G if all vertices and edges of H are vertices and edges of G. Formally, a subgraph $H = \langle V', E', f' \rangle$ of a graph $G = \langle V, E, f \rangle$ is any graph such that $V' \subseteq V$, $E' \subseteq E$, and $f' = f \,|\, E'$, where $f \,|\, E'$ denotes the function f with its domain restricted to E'. H is a *spanning subgraph* of G if it is a subgraph of G containing all the vertices of G. Note that if v and w are vertices of H, it is not required that edge vw be in H even though it is in G. If V' is the vertex set of H, a subgraph of G, then H is an *induced subgraph* of G if it contains all the edges of G which join vertices in V'. Thus in Fig. 2.2 H_1 is a subgraph of G on five vertices, but not induced, while H_2 is the induced subgraph of G on the same five vertices. (Note that while two edges of G cross in the drawing, there is no vertex where they cross.)

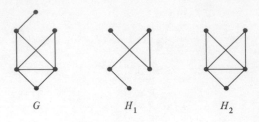

Figure 2.2

Now suppose G is a finite graph. A *walk* in G is an alternating sequence of vertices and edges $v_1 e_1 v_2 e_2 \cdots v_{n-1} e_{n-1} v_n$ such that edge e_i is incident to vertices v_i and $v_{i+1}, i = 1, 2, \ldots, n-1$. The walk may also be represented either by the vertex sequence $v_1 v_2 \cdots v_n$ or the edge sequence $e_1 e_2 \cdots e_{n-1}$. A *trail* is a walk with no repeated edges. A *cycle* is a trail (walk) such that $v_1 = v_n$, and there are no other repeated vertices. A *path* is a trail (walk) with no repeated vertices. The *length* of a walk, trail, cycle, or path is the number of edges it contains, repetitions counted. A *k-cycle* is a cycle of length k.

Example 2.2

Give examples of walks, trails, cycles, and paths for G, shown in Fig. 2.3.

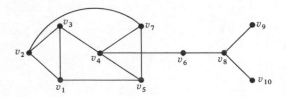

Figure 2.3

Solution. Note that the drawing has one "false" crossing which is not a vertex of the graph. The walks, trails, cycles, and paths of the graph include

$v_1 v_3 v_4 v_6 v_8$	(path)	$v_1 v_3 v_2 v_7 v_4 v_6$	(path)
$v_6 v_8 v_9 v_8$	(walk)	$v_5 v_1 v_2 v_7 v_5$	(cycle)
$v_2 v_1 v_5 v_7 v_2 v_3 v_4$	(trail)	$v_4 v_5 v_1 v_2 v_3 v_4$	(cycle)

and many others.

An undirected graph G is said to be *connected* if for any two distinct vertices v_1 and v_2 of G there is a path from v_1 to v_2. Each maximal connected piece of a graph is called a *component*. That is, a component of a graph G is a connected subgraph H with the property that no vertex not in H is connected to any vertex in H. Thus the graph of Fig. 2.3 is connected, while that of Fig. 2.1 is not, but has two components.

Assumption 2. The undirected graphs that we study are connected.

Since virtually all graph manipulation is componentwise, this assumption has no deleterious effects.

The connectivity of a graph can be defined in terms of either vertices or edges. The *vertex connectivity* $v(G)$ of G is the minimum number of vertices that must be deleted to either disconnect G or reduce it to a single vertex. *The edge connectivity* $\varepsilon(G)$ is the minimum number of edges that must be deleted to either disconnect G or to reduce it to a single vertex. It can be shown that $v(G) \leqslant \varepsilon(G)$. We say that G is *k-connected* if $v(G) = k$. A 1-connected graph is called *separable*.

Example 2.3

Compute $v(G)$ and $\varepsilon(G)$ for G of Fig. 2.4.

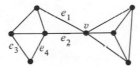

Figure 2.4

Solution. From the figure, removal of vertex v (and the associated edges) disconnects G. Hence $v(G) = 1$. However, removal of any single edge does not disconnect G: two edges must be removed. Either e_1 and e_2, or e_3 and e_4 will do. Hence $\varepsilon(G) = 2$.

ANSWER. $v(G) = 1$, $\varepsilon(G) = 2$.

Let v be a vertex of an undirected graph G. The *degree of v*, deg(v), is the number of edges incident to v. The *degree of G* is $\max_{v \in G} \deg(v)$. A vertex of degree 0 is called *isolated*.

Example 2.4

Compute the degree of the graph of Fig. 2.3.

Solution. Vertices $v_1, v_2, \ldots, v_{10}$ have degrees 3, 3, 3, 4, 3, 2, 3, 3, 1, 1 respectively. Hence deg(G) = 4.

ANSWER.   deg(G) = 4.

Since the degree of a vertex is a strictly local property, any loop is counted twice. Thus the vertex shown in Fig. 2.5 has degree 4.

Figure 2.5

We could also define the *degree of an edge* as the number of vertices incident to it, but since this is always two, the concept is rather meaningless. However, it will be of use in discussing graph matrices.

EXERCISES

1. Show $v(G) \leqslant \varepsilon(G) \leqslant \deg(G)$.

2. If $G_1, G_2, \ldots, G_k$ are the connected components of G, determine $\deg(G)$ in terms of $\deg(G_i)$.

3. Show that if G is k-connected then for any two vertices v and w of G there exist at least k paths from v to w, no two of which have any points in common other than the end points.

Assume that G is an undirected graph of $n \leqslant 100$ edges and no isolated vertices, and that the input consists of n and an $n \times 2$ array $E = (e_{ij})$, with the pairs (e_{i1}, e_{i2}) denoting edges. Write a program:

4. To compute $\deg(G)$.   5. To determine if G is connected.

*6. To compute $v(G)$.   *7. To compute $\varepsilon(G)$.

3. SPECIAL CLASSES OF GRAPHS

Since undirected graphs come in all sizes, shapes, and colors, much of the work in graph theory has been related to a few easily defined classes of graphs. Those classes that have been most thoroughly studied include regular, complete, and bipartite graphs, and trees.

A graph G is *regular of degree k* or *k-regular* if each of its vertices has degree k. That is, a regular graph is one with the same number of edges incident to each vertex.

Example 3.1

Draw connected regular graphs of degrees 0, 1, 2, and 3.

Solution. For degree 0, there can be no edges. Hence the only connected graph of this degree consists of a single vertex. It is regular.

For degree 1, begin with a vertex v_1. Incident to this there is a single edge, which ends in another vertex v_2. There can be no other edges attached to these vertices because of the degree. Hence the only 1-regular connected graph consists of a single edge and its end points.

For degree 2, begin also with a single vertex v_1. Two edges are incident to it, each ending in another vertex, say v_2 and v_3. Since the graph is to be 2-regular, v_2 and v_3 are each incident to another edge. It could be the edge $v_2 v_3$, completing a 3-cycle, or there could be two different edges. Continuing in this way, all cycles and only these are connected and 2-regular (Fig. 3.1).

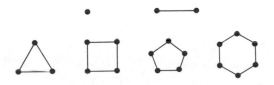

Figure 3.1

The 3-regular connected graphs, called the *cubic graphs*, constitute the first nontrivial set of regular graphs. We leave it as an exercise to show that any cubic graph has an even number of vertices. With four vertices, because of the degree each vertex is connected to every other vertex by an edge (Fig. 3.2). With six vertices, three of them may form a triangle. If so, the other three do also (Fig. 3.3). However, it is also possible that no three vertices (with their edges) form a triangle (Fig. 3.4).

Figure 3.2   Figure 3.3   Figure 3.4

Some cubic graphs with eight or more vertices are shown in Fig. 3.5.

As is evident from Example 3.1, the regular graphs of degrees 0, 1, and 2 are quite simple and uninteresting. However, the cubic graphs already constitute a complex class of graphs. We shall later explore some of the properties of cubic graphs.

The complexity of the class of cubic graphs has tended to discourage investigation of regular graphs of higher degree. Much more work has been done on other, simpler classes of graphs.

8 vertices

10 vertices

+ 16 others

12 vertices

+ 85 others

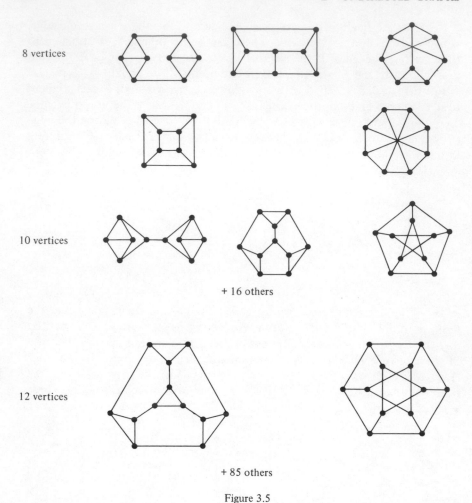

Figure 3.5

A graph G on n vertices is *complete* if each possible pair of vertices (v_i, v_j) $(i \neq j)$ is an edge. The complete graph on n vertices is denoted by K_n. Note that K_n is connected and regular of degree $n - 1$.

Example 3.2

Draw the complete graphs K_n, $n = 1, 2, 3, 4, 5, 6$.

ANSWER.   See Fig. 3.6.

If H is a subgraph of a connected graph G, the *complement of H with respect to G, G − H*, consists of all edges of G that are not in H, and the vertices

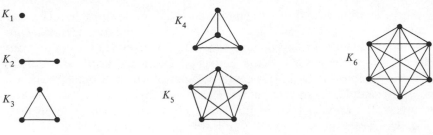

Figure 3.6

incident to these edges. In particular, the *complement of H*, $\overline{H}$, is the complement of H with respect to K_n, where H has n vertices.

A graph G is *bipartite* if its vertices fall into two disjoint sets V and W such that no vertex in V (in W) is joined to any other in V (in W) by an edge; G is a *complete bipartite graph* if in addition each vertex in V is joined to every vertex in W by an edge. If V consists of m vertices and W consists of n vertices, the complete bipartite graph on vertices $V \cup W$ is denoted by $K_{m,n}$. To standardize, we assume $m \leqslant n$. Note that $K_{m,n}$ is connected, and $K_{n,n}$ is regular of degree n.

Bipartite graphs relate to matching problems, and play a role in the manipulation of sparse matrices. (See Section 11.)

Example 3.3

Draw $K_{m,n}$ for $m = 1, 2, 3$ and $n = m, m+1, m+2$.

ANSWER.   See Fig. 3.7, where V is on the left and W on the right.

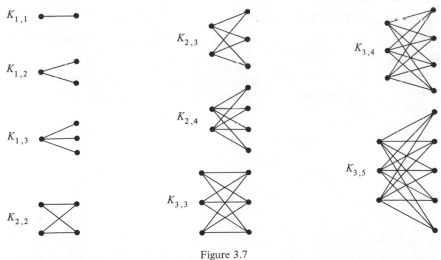

Figure 3.7

A *tree* is a finite, undirected, connected graph with no cycles. The concept of a tree is an extremely important one in computational work. Data sets are frequently arranged, and searched, as trees. For example, any classification scheme such as the Universal Decimal Classification of the libraries organizes data (in this case books) into a tree. A binary search is another example of a tree. In such a search each question has two possible answers, for example, yes and no, and the data are divided accordingly (Fig. 3.8). Trees are sufficiently significant that we shall devote an entire chapter (Chapter 3) to their study.

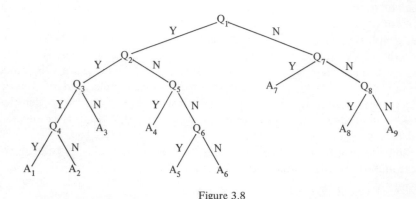

Figure 3.8

A tree may be characterized in many ways. We shall show two of these and leave others as exercises.

Example 3.4

Show that a tree is a connected graph with n vertices and exactly $n-1$ edges.

Solution. By definition, a tree is a connected graph with no cycles. Observe that the removal of a single edge (leaving the vertices) breaks a tree into exactly two components.

Suppose G is a tree with n vertices. If $n = 1$ or $n = 2$, clearly G has $n-1$ edges. Now suppose that if $n \leqslant k$, G has exactly $n-1$ edges, and let G' be a tree with $k+1$ vertices. Remove an edge. This results in graphs G_1 and G_2 with $n_1 \leqslant k$ and $n_2 \leqslant k$ vertices, respectively, where $n_1 + n_2 = k+1$. Since G_1 and G_2 are trees (why?), they have $n_1 - 1$ and $n_2 - 1$ edges, respectively. Hence G' has $(n_1 - 1) + (n_2 - 1) + 1 = n_1 + n_2 - 1 = k$ edges.

Now suppose G_0 is a connected graph with n vertices and m edges. If G_0 has a cycle, locate one, and remove an edge, forming a new graph G_1. Then G_1 is connected, and has $m-1$ edges. Similarly, if G_1 has a cycle, locate one and remove an edge, forming G_2, connected, with $m-2$ edges. Continue thus, breaking cycles, until a connected graph G_k is formed having $m-k$ edges and no cycles. By definition, G_k is a tree with n vertices and $m-k$ edges. But by the first part of this example, $m-k = n-1$. Hence if the original graph G_0 has exactly $n-1$ edges, $m = n-1$ and $k = 0$. That is, since k is the number of cycles, G_0 has no cycles. Hence a connected graph with n vertices and $n-1$ edges is a tree.

Example 3.5

Show that G is a tree if and only if it has no cycles, but the addition of any one edge (not a loop) forms a cycle.

Solution. Suppose G is a tree with n vertices. By Example 3.4 it has exactly $n-1$ edges. Add an edge. Then by Example 3.4 the resulting graph G' is not a tree. Since G is connected, the addition of an edge must have formed a cycle.

Now suppose G is a graph with no cycles, but that the addition of any edge to G forms a cycle. Then G is connected, since otherwise an edge joining two components of G could be added without forming a cycle. Thus G is a connected, cycle-free graph, that is, a tree.

A finite graph is a tree if and only if either
(1) it is connected and acyclic; or
(2) it is connected, with n vertices and $n-1$ edges; or
(3) it is acyclic, but the addition of any one edge forms a cycle.

Within any finite connected graph it is possible to find a tree containing all vertices of the graph. Such a tree is called a *spanning tree* for the graph.

Example 3.6

Give two examples of spanning trees for the graph of Fig. 3.9.

ANSWER. See Fig. 3.10.

As we shall see later, spanning trees are important in studying the structure of a graph.

The complement of a spanning tree of G is a *cotree* of G. It is interpreted to include all vertices of G. If G has n vertices and m edges, since any spanning tree of G contains $n-1$ edges, any cotree contains $m-n+1$ edges. This value, $m-n+1$, is called the *cyclomatic number* of G.

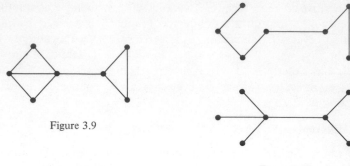

Figure 3.9

Figure 3.10

A *rooted tree* is a tree in which one vertex, called the *root*, has been especially marked.

One further class of graphs is of use in certain network applications. (See Section 11, Chapter 4.) An *Euler graph* is a graph that has an Euler trail, that is, a trail whose end point is its starting point, and which contains all edges of the graph. It is clear that every Euler graph is connected and has no vertices of odd degree. (As one traverses the Euler trail, each edge entering a vertex is paired with another edge leaving it.) The converse is also true. If G is a connected graph with only vertices of even degree, let C_1 be a cycle in G. (See Exercise 18.) If $C_1 = G$, then G is an Euler graph. Otherwise $G_1 = G - C_1$ has only vertices of even degree. It must contain a cycle C_2 having a vertex in common with C_1. Then clearly the graph consisting of C_1 and C_2 is an Euler graph. Continue in this way, gradually adding cycles of G until G is entirely used up. In the process you have constructed an Euler trail through G.

> A graph is an Euler graph if and only if all of its vertices are of even degree.

EXERCISES

1. Show that the sum of the degrees of the vertices of a graph is twice the number of edges of the graph.

2. Show that each cubic graph has an even number of vertices. *Hint:* Count the edges.

3. Show that there are exactly five cubic graphs on eight vertices.

4. Find the remaining cubic graphs on 10 vertices.

5. Determine the number of edges in K_n.

6. Determine the degrees of the vertices of $K_{m,n}$. Hence determine the number of edges.

7. Show that a graph G on n vertices is a tree if and only if any one of these properties holds:
   (1) G is connected, but the deletion of any one edge disconnects G;
   (2) for any two vertices v_1 and v_2 of G, there is exactly one path from v_1 to v_2;
   (3) G has exactly $n-1$ edges and no cycles.

8. Find the minimum, maximum possible degrees for a tree on n vertices.

9. Find all spanning trees for the graph in Fig. 3.9.

10. Prove that G is connected if and only if it has a spanning tree.

11. Does a connected graph G necessarily have a spanning tree that is a path?

12. How many spanning trees are there

   (a) for a tree?   (b) for a connected regular graph of degree 2?
   (c) for K_n?   (d) for $K_{m,n}$?

13. Show that a connected k-regular graph on n vertices has at most

$$\binom{\frac{nk}{2}}{n-1} = \frac{\frac{nk}{2} \cdot \left(\frac{nk}{2} - 1\right) \cdots \left(\frac{nk}{2} - n + 2\right)}{1 \cdot 2 \cdots (n-1)}$$

   spanning trees. Find a k-regular graph having fewer then this number. *Hint:* Look at 3-regular graphs on eight vertices.

14. Show that a graph is bipartite if and only if it has no cycles of odd length.

15. Find all cubic bipartite graphs on six, eight, and ten vertices.

16. Find all regular graphs of degree 4 on five, six, and seven vertices.

17. Write a program to generate a spanning tree for a graph of at most 20 vertices and 100 edges. Assume the input is the list of vertex pairs for the edges, together with the number of vertices and edges.

18. Show that if a finite graph has no vertices of odd degree and no isolated vertices, then it must contain a cycle.

4. MATRIX REPRESENTATION OF GRAPHS

There are many ways to represent graphs in a form that is suitable for computational work. Matrix representations are the most popular, and of these, two are particularly important.

Let $G = \langle V, E, f \rangle$ be a finite undirected graph (not necessarily connected), without loops or multiple edges. Let $V = \{v_1, v_2, \ldots, v_m\}$ and $E = \{e_1, e_2, \ldots, e_n\}$. An *incidence matrix* for G is an $m \times n$ matrix $M = (m_{ij})$ such that

$$m_{ij} = \begin{cases} 1 & \text{if } v_i \text{ is incident to } e_j, \\ 0 & \text{otherwise.} \end{cases}$$

A (*vertex*) *adjacency matrix* for G is an $m \times m$ matrix $A = (a_{ij})$ such that

$$a_{ij} = \begin{cases} 1 & \text{if } (v_i, v_j) \text{ is an edge of } G, \\ 0 & \text{otherwise.} \end{cases}$$

Example 4.1

Find incidence and adjacency matrices for the graph shown in Fig. 4.1.

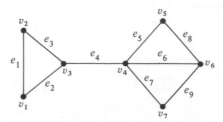

Figure 4.1

ANSWER.

$$M = \begin{bmatrix} 1 & 1 & 0 & 0 & 0 & 0 & 0 & 0 & 0 \\ 1 & 0 & 1 & 0 & 0 & 0 & 0 & 0 & 0 \\ 0 & 1 & 1 & 1 & 0 & 0 & 0 & 0 & 0 \\ 0 & 0 & 0 & 1 & 1 & 1 & 1 & 0 & 0 \\ 0 & 0 & 0 & 0 & 1 & 0 & 0 & 1 & 0 \\ 0 & 0 & 0 & 0 & 0 & 1 & 0 & 1 & 1 \\ 0 & 0 & 0 & 0 & 0 & 0 & 1 & 0 & 1 \end{bmatrix},$$

$$A = \begin{bmatrix} 0 & 1 & 1 & 0 & 0 & 0 & 0 \\ 1 & 0 & 1 & 0 & 0 & 0 & 0 \\ 1 & 1 & 0 & 1 & 0 & 0 & 0 \\ 0 & 0 & 1 & 0 & 1 & 1 & 1 \\ 0 & 0 & 0 & 1 & 0 & 1 & 0 \\ 0 & 0 & 0 & 1 & 1 & 0 & 1 \\ 0 & 0 & 0 & 1 & 0 & 1 & 0 \end{bmatrix}.$$

Note that for any graph with m vertices and n edges there are $m! \, n!$ incidence matrices (not necessarily distinct) and $m!$ adjacency matrices (not necessarily distinct) since the numbering of the vertices and edges (rows and columns) is immaterial. Note also that the column sums of an incidence matrix are all 2, while the row sums are the degrees of the vertices. In contrast, the adjacency matrix is a symmetric matrix (that is, $a_{ij} = a_{ji}$) whose row and column sums are both the degrees of the vertices.

The incidence and adjacency matrices each completely specify a graph since they display all of the edges and vertices and the relationships between them. One can, however, define matrices that relate to other features of a graph. While most of these do not completely specify a graph, they are useful in various ways. We describe a few of the more interesting.

Edge adjacency matrix, $E = (e_{ij})$: rows denote edges, columns denote edges,

$$e_{ij} = \begin{cases} 1 & \text{if edges } e_i \text{ and } e_j \text{ are incident to a common vertex,} \\ 0 & \text{otherwise.} \end{cases}$$

Cycle matrix, $C = (c_{ij})$: rows denote cycles, columns denote edges,

$$c_{ij} = \begin{cases} 1 & \text{if cycle } C_i \text{ contains edge } e_j, \\ 0 & \text{otherwise.} \end{cases}$$

Path matrix, $P_{vw} = (p_{ij})$: rows denote paths from vertex v to vertex w, columns denote edges,

$$p_{ij} = \begin{cases} 1 & \text{if path } P_i \text{ contains edge } e_j, \\ 0 & \text{otherwise.} \end{cases}$$

Spanning tree matrix, $T = (t_{ij})$: rows denote spanning trees, columns denote edges,

$$t_{ij} = \begin{cases} 1 & \text{if spanning tree } t_i \text{ contains edge } e_j, \\ 0 & \text{otherwise.} \end{cases}$$

Maximal complete subgraph matrix, $S = (s_{ij})$: rows denote vertices, columns denote maximal complete subgraphs, that is, subgraphs which are complete and are contained in no other complete subgraph,

$$s_{ij} = \begin{cases} 1 & \text{if vertex } v_i \text{ is contained in the maximal complete} \\ & \text{subgraph } M_j, \\ 0 & \text{otherwise.} \end{cases}$$

Bipartite matrix, $B = (b_{ij})$: if G is a bipartite graph on vertex sets V and W, rows denote vertices in V, columns denote vertices in W,

$$b_{ij} = \begin{cases} 1 & \text{if } (v_i, w_j) \text{ is an edge of } G, \\ 0 & \text{otherwise.} \end{cases}$$

EXERCISES

1. What is the adjacency matrix of K_n, of $K_{m,n}$?

2. Find incidence matrices for the graphs of Figs. 4.2 and 4.3.

3. Find adjacency matrices for the graphs of Figs. 4.2 and 4.3.

4. Find edge adjacency matrices for the graphs of Figs. 4.2 and 4.3.

5. Find cycle matrices for the graphs of Figs. 4.2 and 4.3.

6. Find path matrices for paths from v_1 to v_6 in the graphs of Figs. 4.2 and 4.3.

7. Find spanning tree matrices for the graphs of Figs. 4.2 and 4.3.

8. Find maximal complete subgraph matrices for the graphs of Figs. 4.2 and 4.3.

9. Find bipartite matrices for the graphs of Figs. 4.3 and 4.4.

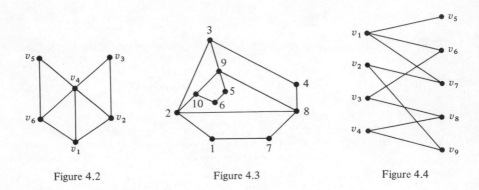

Figure 4.2   Figure 4.3   Figure 4.4

Let G be a finite undirected graph of at most 20 vertices and 100 edges. Write the required program, assuming the input to be either an incidence or an adjacency matrix for G, a code to indicate which matrix is supplied, and the size of the matrix:

10. To compute $\deg(G)$.

11. To determine if G is connected.

*12. To compute $v(G)$.

*13. To compute $\varepsilon(G)$.

14. To generate a spanning tree for G.

15. To compute E, an edge adjacency matrix for G.

16. To compute C, a cycle matrix for G.

17. To determine if G is bipartite, and if it is, to compute B, a bipartite matrix for G.

5. RELATIONS AMONG GRAPH MATRICES

The various matrices that we have defined are not independent, but are related in rather simple ways. Each matrix defines between the rows and columns a relation that holds for a particular graph, a relation which we may picture as going *from* rows *to* columns. For example, the incidence matrix M relates each vertex to all of the edges with which it is incident. The transpose of the cycle matrix C^T relates each edge to all of the cycles that contain it. Thus $M \times C^T$ may be pictured as relating each vertex, through its incident edges, to the cycles containing it (Fig. 5.1).

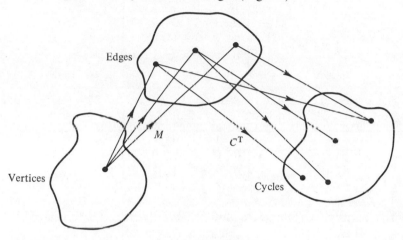

Figure 5.1

In fact, the matrix $M \times C^T$ does just this. Let us examine the graph interpretation of this and other matrix products by means of an example.

Example 5.1

For the graph G in Fig. 5.2, compute and interpret the six matrix products $M \times C^T$, $M \times P_{23}^T$, $M \times T^T$, $S^T \times M$, $M \times M^T$, and $M^T \times M$.

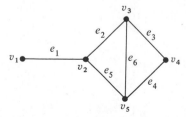

Figure 5.2

Solution. First determine the matrices M, C, P_{23}, T, and S. Also determine the matrices A and E since these will be used later:

$$A = \begin{bmatrix} 0 & 1 & 0 & 0 & 0 \\ 1 & 0 & 1 & 0 & 1 \\ 0 & 1 & 0 & 1 & 1 \\ 0 & 0 & 1 & 0 & 1 \\ 0 & 1 & 1 & 1 & 0 \end{bmatrix}, \qquad E = \begin{bmatrix} 0 & 1 & 0 & 0 & 1 & 0 \\ 1 & 0 & 1 & 0 & 1 & 1 \\ 0 & 1 & 0 & 1 & 0 & 1 \\ 0 & 0 & 1 & 0 & 1 & 1 \\ 1 & 1 & 0 & 1 & 0 & 1 \\ 0 & 1 & 1 & 1 & 1 & 0 \end{bmatrix},$$

$$M = \begin{bmatrix} 1 & 0 & 0 & 0 & 0 & 0 \\ 1 & 1 & 0 & 0 & 1 & 0 \\ 0 & 1 & 1 & 0 & 0 & 1 \\ 0 & 0 & 1 & 1 & 0 & 0 \\ 0 & 0 & 0 & 1 & 1 & 1 \end{bmatrix}.$$

The cycles of G are $C_1 = e_2 e_5 e_6$, $C_2 = e_3 e_4 e_6$, and $C_3 = e_2 e_3 e_4 e_5$. Thus

$$C = \begin{bmatrix} 0 & 1 & 0 & 0 & 1 & 1 \\ 0 & 0 & 1 & 1 & 0 & 1 \\ 0 & 1 & 1 & 1 & 1 & 0 \end{bmatrix}.$$

The paths of G joining v_2 and v_3 are $P_1 = e_2$, $P_2 = e_5 e_6$, and $P_3 = e_5 e_4 e_3$. Thus

$$P_{23} = \begin{bmatrix} 0 & 1 & 0 & 0 & 0 & 0 \\ 0 & 0 & 0 & 0 & 1 & 1 \\ 0 & 0 & 1 & 1 & 1 & 0 \end{bmatrix}.$$

The spanning trees of G are $t_1 = e_1 e_2 e_3 e_4$, $t_2 = e_1 e_2 e_3 e_5$, $t_3 = e_1 e_2 e_3 e_6$, $t_4 = e_1 e_2 e_6 e_4$, $t_5 = e_1 e_5 e_4 e_3$, $t_6 = e_1 e_5 e_6 e_3$, $t_7 = e_1 e_5 e_4 e_6$, and $t_8 = e_1 e_2 e_5 e_4$. Thus

$$T = \begin{bmatrix} 1 & 1 & 1 & 1 & 0 & 0 \\ 1 & 1 & 1 & 0 & 1 & 0 \\ 1 & 1 & 1 & 0 & 0 & 1 \\ 1 & 1 & 0 & 1 & 0 & 1 \\ 1 & 0 & 1 & 1 & 1 & 0 \\ 1 & 0 & 1 & 0 & 1 & 1 \\ 1 & 0 & 0 & 1 & 1 & 1 \\ 1 & 1 & 0 & 1 & 1 & 0 \end{bmatrix}.$$

The maximal complete subgraphs of G are $S_1 = v_1 v_2$, $S_2 = v_2 v_3 v_5$, and $S_3 = v_3 v_4 v_5$. Thus

$$S = \begin{bmatrix} 1 & 0 & 0 \\ 1 & 1 & 0 \\ 0 & 1 & 1 \\ 0 & 0 & 1 \\ 0 & 1 & 1 \end{bmatrix}.$$

Now compute the products. In each case refer to the product matrix as (a_{ij}).

(a)

$$M \times C^{\mathrm{T}} = \begin{bmatrix} 1 & 0 & 0 & 0 & 0 & 0 \\ 1 & 1 & 0 & 0 & 1 & 0 \\ 0 & 1 & 1 & 0 & 0 & 1 \\ 0 & 0 & 1 & 1 & 0 & 0 \\ 0 & 0 & 0 & 1 & 1 & 1 \end{bmatrix} \begin{bmatrix} 0 & 0 & 0 \\ 1 & 0 & 1 \\ 0 & 1 & 1 \\ 0 & 1 & 1 \\ 1 & 0 & 1 \\ 1 & 1 & 0 \end{bmatrix} = \begin{bmatrix} 0 & 0 & 0 \\ 2 & 0 & 2 \\ 2 & 2 & 2 \\ 0 & 2 & 2 \\ 2 & 2 & 2 \end{bmatrix}$$

Interpretation. This matrix relates vertices to cycles. A 2 in the i,j position indicates that vertex v_i is counted twice in cycle c_j—as an end point of the two edges incident to v_i. That is, a_{ij} is the degree of v_i in cycle c_j.

(b)

$$M \times P_{23}^{\mathrm{T}} = \begin{bmatrix} 0 & 0 & 0 \\ 1 & 1 & 1 \\ 1 & 1 & 1 \\ 0 & 0 & 2 \\ 0 & 2 & 2 \end{bmatrix}.$$

Interpretation.
$$a_{ij} = \begin{cases} 0: & v_i \text{ does not occur on path } p_j, \\ 1: & v_i \text{ is an endpoint of path } p_j, \\ 2: & v_i \text{ occurs on, but is not an end point of, path } p_j. \end{cases}$$

That is, a_{ij} is the degree of v_i in path p_j.

(c)

$$M \times T^{\mathrm{T}} = \begin{bmatrix} 1 & 1 & 1 & 1 & 1 & 1 & 1 & 1 \\ 2 & 3 & 2 & 2 & 2 & 2 & 2 & 3 \\ 2 & 2 & 3 & 2 & 1 & 2 & 1 & 1 \\ 2 & 1 & 1 & 1 & 2 & 1 & 1 & 1 \\ 1 & 1 & 1 & 2 & 2 & 2 & 3 & 2 \end{bmatrix}.$$

Interpretation.   Again, a_{ij} is the degree of v_i in spanning tree t_j.

(d)

$$
S^T \times M = \begin{bmatrix} 1 & 1 & 0 & 0 & 0 \\ 0 & 1 & 1 & 0 & 1 \\ 0 & 0 & 1 & 1 & 1 \end{bmatrix} \begin{bmatrix} 1 & 0 & 0 & 0 & 0 & 0 \\ 1 & 1 & 0 & 0 & 1 & 0 \\ 0 & 1 & 1 & 0 & 0 & 1 \\ 0 & 0 & 1 & 1 & 0 & 0 \\ 0 & 0 & 0 & 1 & 1 & 1 \end{bmatrix}
$$

$$
= \begin{bmatrix} 2 & 1 & 0 & 0 & 1 & 0 \\ 1 & 2 & 1 & 1 & 2 & 2 \\ 0 & 1 & 2 & 2 & 1 & 2 \end{bmatrix}.
$$

Interpretation.   Once again, the matrix may be interpreted in terms of degrees, with a_{ij} being the degree of edge e_j in the maximal complete subgraph s_i. Thus, e_j is an edge of s_i if and only if $a_{ij} = 2$ (both end points of e_j are in s_i).

(e)

$$
M \times M^T = \begin{bmatrix} 1 & 1 & 0 & 0 & 0 \\ 1 & 3 & 1 & 0 & 1 \\ 0 & 1 & 3 & 1 & 1 \\ 0 & 0 & 1 & 2 & 1 \\ 0 & 1 & 1 & 1 & 3 \end{bmatrix} ; \quad M^T \times M = \begin{bmatrix} 2 & 1 & 0 & 0 & 1 & 0 \\ 1 & 2 & 1 & 0 & 1 & 1 \\ 0 & 1 & 2 & 1 & 0 & 1 \\ 0 & 0 & 1 & 2 & 1 & 1 \\ 1 & 1 & 0 & 1 & 2 & 1 \\ 0 & 1 & 1 & 1 & 1 & 2 \end{bmatrix}.
$$

Interpretation.   Note that except for the main diagonal, $M \times M^T = A$ and $M^T \times M = E$. The diagonal elements are the degrees of the vertices in $M \times M^T$, and the degrees of the edges (always 2) in $M^T \times M$.

From the example we observe that the incidence matrix, since it relates vertices to edges, enables us to pass back and forth between vertex versus structure matrices and edge versus structure matrices. However, the results must be interpreted with care. Although the significance of a_{ij} can be stated in terms of degree of the vertex or edge relative to a structure, extended products generally correspond to rather complex relations.

Example 5.2

Using the graph of Example 5.1, compute and interpret $M^T \times M \times P_{23}^T$.

Solution.   Since $M^T \times M$ is essentially the edge adjacency matrix, this product should relate edges to paths, probably in a more complicated way than P_{23}^T does. This is indeed the case. The value is

$$
M^T \times M \times P_{23}^T = \begin{bmatrix} 1 & 1 & 1 \\ 2 & 2 & 2 \\ 1 & 1 & 3 \\ 0 & 2 & 4 \\ 1 & 3 & 3 \\ 1 & 3 & 3 \end{bmatrix}.
$$

For example, in this matrix $a_{52} = 3$ is interpreted as: edge e_5 has three end points in path p_2. Recall that $p_2 = e_5 e_6$. The end points of e_5 are counted: v_2—once in p_2 as an end point of e_5; v_5 —twice in p_2 as an end point of both e_5 and e_6. Hence the count of 3.

Because of this complexity of interpretation, extended matrix products are little used. In one case, however, the extended matrix product has a simple interpretation.

> If A is the (vertex) adjacency matrix of graph G, and $A^n = (a_{ij}^{(n)})$, then $a_{ij}^{(n)}$ is the number of walks of length n between v_i and v_j.

The reason for this is that each term in the product represents a walk of length n. This is certainly true for $A^1 = A$. For A^2, each matrix entry is a sum of terms of the form $a_{ik}^{(1)} a_{kj}^{(1)}$. But this term is equal to one if and only if $a_{ik}^{(1)} = a_{kj}^{(1)} = 1$, that is, if and only if there are walks of length one from v_i to v_k and from v_k to v_j. These combine to form a walk $v_i v_k v_j$ of length two. Similarly, the terms of A^n, $n > 2$, represent walks of length n.

EXERCISES

For the graph of Fig. 5.3, compute and interpret the matrix product:

1. $M \times C^T$.
2. $M \times P_{14}^T$.
3. $M \times T^T$.
4. $S^T \times M$.
5. $M \times M^T$.
6. $M^T \times M$.
7. $S \times S^T$.
8. $S^T \times S$.
9. $C \times C^T$.
10. $C^T \times C$.
11. $P_{14}^T \times P_{14}$.
12. $T^T \times T$.
13. $P_{14} \times T^T$.
14. A^5.

15. E^2.    16. E^3.

17. $A \times M \times E$.   18. $A^2 \times M \times E^2$.

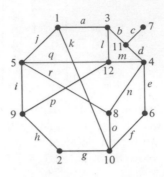

Figure 5.3

6. INVARIANTS AND GRAPH ISOMORPHISM

We say that two graphs $G = \langle V, E, f \rangle$ and $G' = \langle V', E', f' \rangle$ are *iso-morphic* if there exists a $1:1$ mapping ϕ of V onto V' and E onto E' such that $\phi \circ f = f' \circ \phi$. That is, the mapping preserves incidence: if $e_1 = v_1 v_2$, then $\phi e_1 = (\phi v_1)(\phi v_2)$. The mapping diagram in Fig. 6.1 commutes.

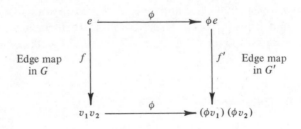

Figure 6.1

Testing for graph isomorphism is in practice often quite simple, although theoretical difficulties abound. One cannot start by checking all mappings $\phi: V \xrightarrow{1:1 \text{ onto}} V'$. There are simply too many mappings. (For 20 vertices there are $20! = 2432\,90200\,81766\,40000$ such mappings.) In practice, one begins by computing *invariants* for the graph, quantities that are fixed, independent of representation of the graph. We shall define some of these invariants in this section and discuss their computation.

Any representation of a graph can be used to compute the invariants. Unfortunately, many of the most easily computed quantities change with the representation and hence are not invariants. For example, the matrices associated with a graph change as we renumber vertices, edges, cycles, and so on. Even the drawings of a graph can change drastically. The two graphs in Fig. 6.2 are isomorphic, although the drawings appear to be quite different.

Figure 6.2

The most obvious invariants, and the simplest to determine, are the numbers of vertices and edges. Since isomorphism involves a 1:1 mapping of vertices onto vertices and edges onto edges, if G and G' are isomorphic, they must have the same number of vertices and the same number of edges.

But this numerical equality is not enough. The vertices and edges must match up in corresponding ways. Thus we are led to the next invariant, the *degree spectrum* of a graph. This is a vector $\langle d_0, d_1, ..., d_{n-1} \rangle$ (for a graph of n vertices) with d_i equal to the number of vertices of degree i. For example, the degree spectrum of the graph in Fig. 6.3 is $\langle 1, 3, 2, 2, 1, 1, 0, 0, 0, 0 \rangle$.

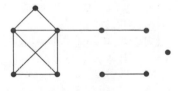

Figure 6.3

Any graph that is isomorphic to this graph must have the same degree spectrum.

Even this is not enough. The two graphs of Fig. 6.4 have the same degree spectrum $\langle 0, 3, 2, 1, 0, 0, 0 \rangle$, but are not isomorphic. As a matter of fact, no *good* complete set of invariants exists for determining graph isomorphism. Each invariant that we find aids in determining whether graphs are isomorphic, but does not completely solve the problem. Through the use of simply computed invariants, one can detect most nonisomorphic graph pairs.

The degree spectrum of a graph is easily computed from either the incidence or the adjacency matrix since the row sums in each matrix are the degrees of the vertices. It is often useful to arrange these matrices in order of nonincreasing row sums. This facilitates comparisons since all vertices of the same degree occur together and in the same location in the matrices of G and G'.

Example 6.1

For the graphs of Fig. 6.4, use the row sums of the matrices to test isomorphism.

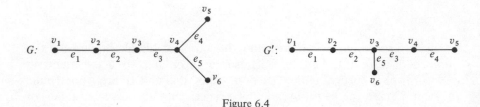

Figure 6.4

Solution. The incidence matrices for G and G', with the given vertex and edge numbering are

$$
M(G) = \begin{bmatrix} 1 & 0 & 0 & 0 & 0 \\ 1 & 1 & 0 & 0 & 0 \\ 0 & 1 & 1 & 0 & 0 \\ 0 & 0 & 1 & 1 & 1 \\ 0 & 0 & 0 & 1 & 0 \\ 0 & 0 & 0 & 0 & 1 \end{bmatrix} \quad \begin{matrix} \text{Row sums} \\ 1 \\ 2 \\ 2 \\ 3 \\ 1 \\ 1 \end{matrix}
$$

$$
M(G') = \begin{bmatrix} 1 & 0 & 0 & 0 & 0 \\ 1 & 1 & 0 & 0 & 0 \\ 0 & 1 & 1 & 0 & 1 \\ 0 & 0 & 1 & 1 & 0 \\ 0 & 0 & 0 & 1 & 0 \\ 0 & 0 & 0 & 0 & 1 \end{bmatrix} \quad \begin{matrix} 1 \\ 2 \\ 3 \\ 2 \\ 1 \\ 1 \end{matrix}
$$

To place the row sums in nonincreasing order, interchange rows 1 and 4 in $M(G)$, and permute rows 1, 3, and 4 in $M(G')$ into rows 4, 1, and 3 respectively. Then the transformed matrices $M'(G)$ and $M'(G')$ are

$$M'(G) = \begin{bmatrix} 0 & 0 & 1 & 1 & 1 \\ 1 & 1 & 0 & 0 & 0 \\ 0 & 1 & 1 & 0 & 0 \\ 1 & 0 & 0 & 0 & 0 \\ 0 & 0 & 0 & 1 & 0 \\ 0 & 0 & 0 & 0 & 1 \end{bmatrix} \quad ; \quad M'(G') = \begin{bmatrix} 0 & 1 & 1 & 0 & 1 \\ 1 & 1 & 0 & 0 & 0 \\ 0 & 0 & 1 & 1 & 0 \\ 1 & 0 & 0 & 0 & 0 \\ 0 & 0 & 0 & 1 & 0 \\ 0 & 0 & 0 & 0 & 1 \end{bmatrix} .$$

These matrices represent the same graphs as do the original matrices, with the vertices renumbered (Fig. 6.5). Similarly, a column interchange or permutation corresponds to an edge renumbering. For these particular graphs,

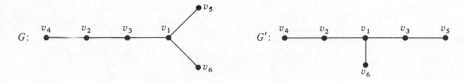

Figure 6.5

the non-isomorphism is now apparent from the second and third rows—the vertices of degree 2. In G, the edge e_2 joins the two vertices of degree 2; in G' no edge joins the two vertices of degree 2.

ANSWER. G and G' are not isomorphic.

The degree spectrum is of the most help when the vertices have a wide variety of degrees. It is useless for a regular graph, when the vertices all have the same degree. To measure the utility, we first observe that if G and G' are isomorphic, then the isomorphism is degree-preserving: vertices of degree k map into vertices of degree k. Now if G and G' both have n vertices, there are $n! = n(n-1)(n-2) \cdots 3 \cdot 2 \cdot 1$ mappings of the vertices of G into the vertices of G'. For the first vertex of G, we may choose any of the n vertices of G'; for the second we choose any of the remaining $n-1$, and so forth. But if there are p_i vertices of degree $i, i = 0, 1, 2, \ldots$, then we know that a vertex of G of degree i maps into one of these p_i vertices, *not* into any other. Hence the number of mappings that are important for isomorphism is $p_0! p_1! \cdots p_{n-1}!$, not $n!$ This is generally much smaller. For example, the graphs of Fig. 6.4 have six vertices. Hence there are $6! = 720$ mappings of the vertices of G onto the vertices of G'. But since the degree spectrum is $\langle 0, 3, 2, 1, 0, 0, 0 \rangle$. only $0! \cdot 3! \cdot 2! \cdot 1! \cdot 0! \cdot 0! \cdot 0! = 12$ of these mappings are important for isomorphism testing.

Another vector invariant is the *cycle vector* of a graph. This is a vector $\langle c_1, c_2, \ldots, c_n \rangle$ (for a graph of n vertices) with c_i equal to the number of cycles of length i. The first two components c_1 and c_2 of the vector denote the number of loops and pairs of multiple edges respectively. When we are working with graphs having no loops or multiple edges, these components have the value zero and are customarily omitted. We will discuss the computation of the cycle vector in Section 7. For now it suffices to note that under isomorphism cycles of length k map into cycles of length k, so that the cycle vector is of the use in solving graph isomorphism problems.

As with the other invariants, graphs with equal cycle vectors need not be isomorphic. The graphs in Fig. 6.6 provide a simple example of this. Each

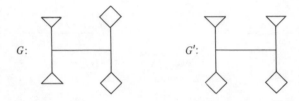

Figure 6.6

has two 3-cycles and two 4-cycles, and no other, but they are not isomorphic. We call non-isomorphic graphs having the same cycle vector *metamorphic*. Metamorphic graphs have not been much studied, although the known families of such graphs appear to arise from rather simple relationships, basically rearrangements or reorientations of portions of the graphs.

The remaining invariants that we shall discuss are, like the numbers of vertices and edges, single numbers which apply to a graph. Some are simple to compute, while for others no efficient computational method is known. The importance of the latter invariants lies in the research which arises from their study, more than in their utility.

From the cycle vector one can identify two graph parameters. The first of these, the *girth*, is the length of the shortest cycle in a graph. Intuitively, the girth measures (crudely) how closely connected the graph is. This interpretation has some computational validity since it is known that the girth of a graph is related to minimum storage required for the graph. (See Section 10.)

At the other end of the cycle vector we find the longest cycle. If the longest cycle contains all of the vertices, the graph is called *Hamiltonian*. Although Hamiltonian graphs have been extensively studied, there is no known simple method to determine directly whether an arbitrary graph is Hamiltonian.

The *number of components* (maximal connected pieces) of a graph is another invariant that is relatively simple to compute. If the adjacency matrix of a graph is properly arranged, then the components are manifested as distinct blocks along the diagonal (Fig. 6.7).

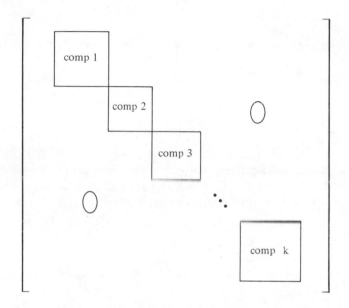

Figure 6.7

For a connected graph G, let us define the *distance* $d(u, v)$ between vertices u and v to be the length of the shortest path from u to v. The *eccentricity* $e(v)$ of a vertex v is the maximum distance from v to any other point of the graph. Then the *radius*, which is the minimum eccentricity for G, and the *diameter*, which is the maximum eccentricity for G, are two further invariants.

It should be noted that eccentricities, and hence the radius and diameter, are easily computed from the adjacency matrix A. The eccentricity of vertex v_i is the least value of n such that the ith row of $A + A^2 + A^3 + \cdots + A^n$ contains no zeros.

We call a graph *planar* if and only if it can be drawn so that there are no false crossings, that is, lines crossing where there are no vertices. The minimal number of false crossings is the *crossing number* of a graph. For example, while the graph in Fig. 6.8a has a false crossing, it can be redrawn as Fig. 6.8b, with no false crossings, and hence is planar. The graphs in Figs. 6.8c and 6.8d are known to be nonplanar, and each has crossing number 1. Figure 6.8e is an example of a graph with crossing number 2.

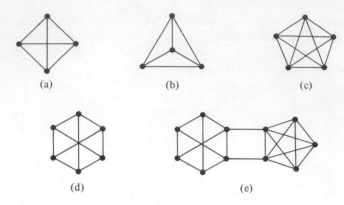

Figure 6.8

In 1930 Kuratowski proved that each nonplanar graph contains a sub-graph that differs from either $K_{3,3}$ or K_5 by at most some vertices of degree 2. See Fig. 6.9. However, this characterization of nonplanar graphs is rather

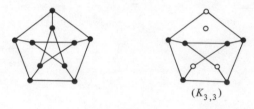

$(K_{3,3})$

Figure 6.9

difficult to use as a practical test of planarity. One algorithm that uses it generally detects nonplanar graphs very rapidly, but must examine virtually all cycles of a graph before establishing conclusively that it is planar [1]. More efficient planarity algorithms, such as that of Tarjan [2], are based on quite different principles.

Two other parameters associated with planarity are the thickness and coarseness of a graph. The *thickness* of a graph G is the minimum number of planar subgraphs whose union is G, while the *coarseness* of G is the maximum number of edge-disjoint nonplanar subgraphs contained in G. These planarity parameters are of importance in the design of computer circuits. If a circuit has a planar graph, it can be printed in a single layer on a circuit board. If it is not planar, then it must be printed in more than one layer, the minimum required number being the thickness of the graph.

Unfortunately the planarity parameters are generally rather difficult to compute. Algorithms for determining whether a graph is planar, which can often be modified to compute the various parameters, usually involve working with most of the cycles of a graph.

Another classical graph parameter, also difficult to compute in general, is the *chromatic number* of a graph. Let the vertices of a graph G be colored in such a way that no two adjacent vertices have the same color. The chromatic number of $G, \chi(G)$, is the minimum number of colors for which this is possible.

The famous four color conjecture states that the regions of any plane map can be colored with at most four colors, so that no two regions with an edge in common have the same color. In Section 12 we show that this is equivalent to stating that for any planar graph G, $\chi(G) \leqslant 4$. While much is known about coloring maps on nonplanar surfaces, the proof of the four color conjecture has eluded mathematicians for over 100 years. All that is known for a planar graph G is that $\chi(G) \leqslant 5$, and if G has at most 40 vertices, then $\chi(G) \leqslant 4$.

Graph coloring has an application to scheduling problems. For example, suppose we are scheduling classes for a school. Let the vertices of a graph denote the classes, and join two vertices A and B by an edge if and only if there is at least one student who must take both classes A and B simultaneously. Saying that A and B must be scheduled at different times then corresponds to saying that A and B must have different colors.

One good, but not perfect, coloring algorithm is that of Welch and Powell [3]. Number the vertices of G in order of decreasing degree. Assign to vertex 1 the first color. Proceeding sequentially, assign this color to any vertex that is not already adjacent to a vertex of the same color. Proceed similarly for the remaining colors until all vertices are colored.

Example 6.2

Apply the Welch–Powell algorithm to color the graph of Fig. 6.10.

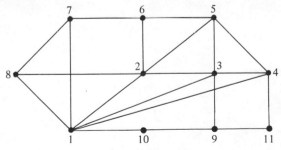

Figure 6.10

Solution. The vertex numbering in the figure is in order of descending degree. Let vertex 1 have the first color, say R. Then vertices 2, 3, and 4 must have a different color. Assign color R to vertex 5, and (continuing) to vertex 9. Choose a new color, say G, and assign this in order to vertices 2, 4, 7, and 10. Finally assign color B to vertices 3, 6, 8, and 11.

ANSWER. R: 1, 5, 9; G: 2, 4, 7, 10;
B: 3, 6, 8, 11.

EXERCISES

For the graph, compute the girth, radius, diameter, crossing number, and chromatic number:

1. Fig. 6.11.   2. Fig. 6.12.

3. Fig. 6.13.   4. Fig. 6.14.

5. Fig. 6.15.   6. Fig. 6.16.

7. Fig. 6.17.   8. Fig. 6.18.

9. Fig. 6.19.   10. Fig. 6.20.

Figure 6.11

Figure 6.12

Figure 6.13

Figure 6.14

Figure 6.15

Figure 6.16

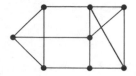

Figure 6.17

Figure 6.18

Figure 6.19

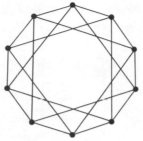

Figure 6.20

Assume the same input as specified in Section 4, and write a program:

   11. To compute the number of components of G.

   12. To compute the eccentricity of each vertex, and the radius and diameter of G.

  *13. To determine the planarity of G.

 **14. To compute the crossing number of G.

 **15. To compute the thickness of G.

   16. To implement the Welch–Powell algorithm.

17. Find a graph for which the Welch–Powell algorithm does not produce the minimum number of colors.

7. CYCLE BASIS

   Since the cycle vector counts the cycles of various sizes, if we are to use this invariant we must be able to compute all of the cycles of a graph G. The first step in this computation is to find a spanning tree T for G. Then the addition of any single edge to T results in the formation of a cycle, which can be found by pruning away all vertices of degree one, until none are left. The set of all cycles found in this way is a *cycle basis* for G.

   Example 7.1

   Compute a cycle basis for the graph of Fig. 7.1.

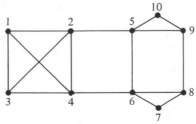

Figure 7.1

Solution. Let the spanning tree T consist of edges $(1,2), (2,3), (2,4), (2,5),$ $(5,6), (6,7), (6,8), (8,9), (9,10)$. See Fig. 7.2. There are in G seven edges that

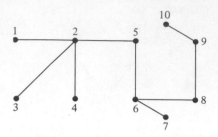

Figure 7.2

are not in T, namely $(1,3), (1,4), (3,4), (4,6), (7,8), (5,9),$ and $(5,10)$. The addition of each of these forms a cycle, listed in Table 7.1. These seven cycles form a cycle basis for G.

TABLE 7.1

Edge	Cycle formed						
$(1,3)$	1	2	3	1			C_1
$(1,4)$	1	2	4	1			C_2
$(3,4)$	2	3	4	2			C_3
$(4,6)$	2	4	6	5	2		C_4
$(7,8)$	6	7	8	6			C_5
$(5,9)$	5	6	8	9	5		C_6
$(5,10)$	5	6	8	9	10	5	C_7

ANSWER. The cycles $C_1, ..., C_7$ of Table 7.1.

If G has p vertices and q edges, then any spanning tree of G has $p-1$ edges. Hence there are $q-p+1$ cycles in any cycle basis for G. Note that this is the cyclomatic number of G. (See Section 3.) The *cycle space* of G consists of all cycles of G. It is formed by adding the cycles of the cycle basis together as edge sets, modulo 2. For example, cycle C_1 of Table 7.1 consists of edges $e_1 = (1,2), e_2 = (2,3), e_3 = (1,3)$, while cycle C_2 of the table consists of edges $e_1, e_4 = (2,4),$ and $e_5 = (1,4)$. We may write $C_1 = e_1 + e_2 + e_3$ and $C_2 = e_1 + e_4 + e_5$. Then

$$C_1 + C_2 = (e_1 + e_2 + e_3 + e_1 + e_4 + e_5) \bmod 2$$
$$= (2e_1 + e_2 + e_3 + e_4 + e_5) \bmod 2$$
$$= e_2 + e_3 + e_4 + e_5.$$

That is, the sum $C_1 + C_2$ is the cycle 2 3 1 4 2. Intuitively, if one traverses C_1 in a particular direction, say $1 \to 2 \to 3 \to 1$, and C_2 in the direction which is opposite along the common edge, $1 \to 4 \to 2 \to 1$, then the edges of the sum cycle are consistently traversed, $2 \to 3 \to 1 \to 4 \to 2$. Since each of the basic cycles may be included or excluded in forming a sum, if the cycle basis contains k cycles, there are exactly $2^k - 1$ possible nonzero sums. Unfortunately, structures that are not cycles are also obtained by this process, and must be eliminated. Referring again to Table 7.1, for example, neither $C_1 + C_7$ nor $C_4 + C_5$ are cycles. (For those familiar with linear algebra, the process here is akin to forming a vector space from a vector basis.)

Example 7.2

Find all cycles of the graph in Fig. 7.1.

Solution.   Since there are seven cycles in the basis, there are $2^7 - 1 = 127$ nonzero combinations of these cycles. Some of these are shown in Table 7.2. Of the 127 combinations only 38 result in cycles.

TABLE 7.2

Combination	Cycle
C_1	C_1
$C_1 + C_2$	1 3 2 4 1
$C_1 + C_3$	1 2 4 3 1
$C_1 + C_4$	not a cycle
$C_1 + C_5$	not a cycle
$\vdots$	
$C^1 + C^2 + C^3$	1 3 4 1
$\vdots$	
$C_1 + C_2 + C_3 + C_4 + C_5 + C_6 + C_7$	not a cycle (sum of cycles 1 3 4 1, 2 4 6 5 2, 6 7 8 6, and 5 9 10 5)

We may represent each of these cycles by the corresponding binary (or decimal) numeral. Thus C_1 is represented by 1000000 (or 128), C_2 by 0100000 (or 64), $C_1 + C_2 + C_3$ by 1110000 (or 224), and so forth.

ANSWER.   The cycles are represented by 1, 2, 3, 4, 5, 6, 8, 9, 10, 13, 14, 16, 24, 25, 26, 29, 30, 32, 40, 41, 42, 45, 46, 48, 64, 80, 88, 89, 90, 93, 94, 96, 104, 105, 106, 109, 110, 112.

It can be shown that the same cycle space for G results, regardless of the cycle basis chosen. Several computer programs have been written to compute the cycle vector, with that by Gibbs [4] among the more efficient ones.

Example 7.3

Compute the cycle vector for the graph of Fig. 7.1.

Solution. The cycle vector merely counts the cycles in G, according to size. For the given graph there are six 3-cycles, five 4-cycles, and so forth.

ANSWER. $(6, 5, 4, 4, 4, 7, 6, 2)$.

Note that this graph has girth 3, and there are two Hamiltonian cycles.

EXERCISES

1. Verify the cycle vector in Example 7.3.

2. Write a program to generate a cycle basis for a graph, using as input the output from the spanning tree program of Section 4.

*3. (cont.) Write a program to generate the cycle vector of a graph, using as input the output from the cycle basis program.

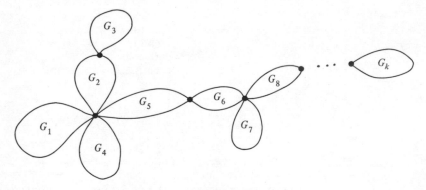

Figure 7.3

4. Let G have k components $G_1, G_2, ..., G_k$, and suppose that the cycle basis for G_i contains β_i cycles, $i = 1, 2, ..., k$. Let n be the number of cycles in a cycle basis for G.

(a) Determine n.
(b) Show that the maximum possible number of cycles in G is

$$N = \sum_{i=1}^{k} (2^{\beta_i} - 1).$$

(c) Compare N with $2^n - 1$.

5. (cont.) Let G be a separable graph with maximal nonseparable subgraphs $G_1, G_2, ..., G_k$,

and suppose that the cycle basis for G_i contains β_i cycles, $i = 1, 2, ..., k$. (The general schematic for G is given in Fig. 7.3.) Determine the maximum possible number of cycles in G.

Figure 7.4

6. (cont.) Let G be a 2-connected chain graph, as in Fig. 7.4, where $G_1, G_2, ..., G_k$ are maximal subgraphs of connectivity greater than 2. Suppose that the cycle basis for G_i contains β_i cycles, $i = 1, 2, ..., k$. Let n be the number of cycles in a cycle basis for G.

(a) Determine n.
(b) Determine a formula for the maximum possible number N of cycles in G in terms of the β_i.
(c) Compare N with $2^n - 1$.

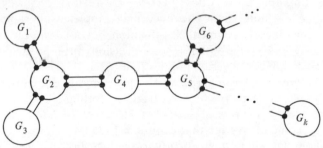

Figure 7.5

**7. (cont.) Extend the formula of Exercise 6 to cover 2-connected branching (Fig. 7.5) and ring (Fig. 7.6) graphs.

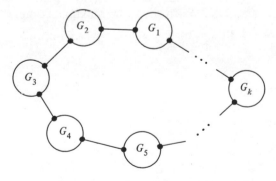

Figure 7.6

**8. (cont.) Use the results of Exercises 4–7 to improve your cycle vector program.

8. MAXIMAL COMPLETE SUBGRAPHS

One problem that arises in any classification process, whether determining biological species, or arranging a set of documents for efficient information retrieval, or some other purpose, is that of determining an appropriate grouping of the objects under consideration. Which animals form a species; which books pertain to the same subject? This question is answered by a group of techniques known as "cluster analysis," one of which we shall now describe.

We have already defined the maximal complete subgraph matrix S on p. 51. Recall that the rows of S represent vertices, while the columns represent maximal complete subgraphs of a given graph G. If we interpret an edge in G as representing a relation between its end points—they belong to the same species, or pertain to the same subject—then a maximal complete subgraph is a largest subgraph, all of whose vertices are related. A vertex may, of course, belong to more than one maximal complete subgraph. This is generally permissible for document classification, but not desirable for species differentation. Thus in the latter case a decision must be made on the proper classification of the vertex in question.

Assume the graph G has n vertices. We can compute the matrix $S = (s_{ij})$ from the adjacency matrix $A = (a_{ij})$. We begin by setting $s_{11} = 1$, and adding a row to S for each of the rows $2, \ldots, n$ of A. However, we add a new column only when we come to a vertex that cannot be related to any of the groups of vertices (complete subgraphs) already formed. Notice that in A, if $a_{ij} = 1$, then vertices i and j are related (joined by an edge). However, in S such a relationship is indicated differently. There, if in any column $k, s_{ik} = s_{jk} = 1$, the vertices i and j are related since they belong to the same maximal complete subgraph, the kth one. But if vertices i and j are not related (adjacent), they cannot belong to the same maximal complete subgraph. In this event $s_{ik} \cdot s_{jk} = 0$ for all k.

The algorithm for forming S is this.

1. $S_{1,1}, q = 1$
2. for $p = 2$ to n:
3.   for $k = 1$ to q: $S_{p,k} = 1$
4.   for $j = 1$ to $p-1 \ni a_{p,j} = 0$:
5.   for $k = 1$ to $q \ni S_{j,k} = 1$: $S_{p,k} = 0$
6.   for $j = 1$ to $p-1 \ni a_{p,j} = 0$:
7.   for $k = 1$ to $q \ni S_{j,k} S_{p,k} = 2$: $S_{j,k} = 0$
8.   for $j = 1$ to $p-1 \ni a_{p,j} = 1$:
9.   if $\exists_{k=1 \text{ to } q}(S_{j,k} S_{p,k} = 1)$:   reiterate

10.   if $\exists_{k=1 \text{ to } q}(S_{j,k}S_{p,k} = 2)$:
11. $S_{j,k} = 1$
12.   for $m = 1$ to $p-1 \ni S_{j,k}S_{m,k} = 2$ and $a_{j,m} = 0$: $S_{m,k} = 0$
13.   reiterate
14. $q = q+1$
15.   for $m = 1$ to $p-1$: $S_{m,q} = 2$
16. $S_{p,q}, S_{j,q} = 1$
17.   for $m = 1$ to $p \ni S_{m,q} = 2$:
18.   for $m1 = 1$ to $p \ni S_{m1,q} = 1$ and $a_{m,m1} = 0$: $S_{m,q} = 0$
19. for $k = 1$ to q:
20.   for $j = 1$ to $p \ni S_{j,k} = 2$:
21. $S_{j,k} = 1$
22.   for $m = 1$ to $p \ni S_{j,k}S_{m,k} = 2$ and $a_{j,m} = 0$: $S_{m,k} = 0$
23. $j = j-1$

In this algorithm statement 1 initializes the development of S, setting
both $S_{1,1}$ and q to 1. Statements 2 and 3 define each new row. Statements 4 and
5 change values in the new row to avoid including relations (edges) not in
the graph. This done for values of j such that ($\ni$) $a_{p,j} = 0$, and k such that
$S_{j,k} = 1$. In the developing array S, the value 2 is used to denote an entry
whose value is not yet determined. (Statement 15 initiates a new column with
undetermined entry values.) Statements 6 and 7 define some of the previously undetermined values to avoid improper edge inclusion.

Statements 8–18 are concerned with representing newly indicated edges in
S. Statement 9 determines if there exists ($\exists$) a column whose jth and pth
entries are both 1, thus indicating the edge (j,p). If the edge is already
indicated by default from statements 4–7, then "reiterate" in statement 9
passes control back to statement 8, to increment j and continue. Statements
10–13 utilize a previously undetermined value to include the new edge, and
also avoid any improper edge inclusion resulting from defining this value.
If there is no way to include the new edge in the matrix as it has developed
thus far, statements 14–18 initiate a new column, and define all entry values
that are forced by inclusion of the edge.

It is possible that the development of S to its maximal size still leaves
some undetermined entries. Statements 19–23 locate these, and define as
many as possible to have the value 1, thus including additional edges in
the maximal subgraphs. (To be complete, this procedure needs array
declarations, a termination statement, and so on.)

Example 8.1

Determine S for the graph G in Fig. 8.1.

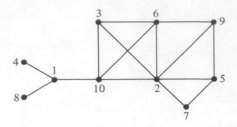

Figure 8.1

Solution.   First compute the adjacency matrix A.

$$
A = \begin{bmatrix}
0 & 0 & 0 & 1 & 0 & 0 & 0 & 1 & 0 & 1 \\
0 & 0 & 1 & 0 & 1 & 1 & 1 & 0 & 1 & 1 \\
0 & 1 & 0 & 0 & 0 & 1 & 0 & 0 & 0 & 1 \\
1 & 0 & 0 & 0 & 0 & 0 & 0 & 0 & 0 & 0 \\
0 & 1 & 0 & 0 & 0 & 0 & 1 & 0 & 1 & 0 \\
0 & 1 & 1 & 0 & 0 & 0 & 0 & 0 & 1 & 1 \\
0 & 1 & 0 & 0 & 1 & 0 & 0 & 0 & 0 & 0 \\
1 & 0 & 0 & 0 & 0 & 0 & 0 & 0 & 0 & 0 \\
0 & 1 & 0 & 0 & 1 & 1 & 0 & 0 & 0 & 0 \\
1 & 1 & 1 & 0 & 0 & 1 & 0 & 0 & 0 & 0
\end{bmatrix}.
$$

The arrays show the development of S after the algorithm step indicated in parentheses:

```
1    1    1    1    1    12   10   10   10   10   10   102  100  100  100
     1    0    0    0    01   01   01   01   01   01   011  011  011  011
               1    0    01   01   01   01   01   01   012  010  010  010
(1)  (3)  (5)  (3)  (5)  (16) (17) 11   10   10   10   102  100  100  100
                                   (3)  (5)  11   00   001  001  001  001
                                             (3)  (5)  (16) (17) 111  010
                                                                 (3)  (5)
```

```
100   100   100   100   1001   1001   1001   1001   10012   10010   10010
011   011   011   011   0112   0110   0110   0110   01101   01101   01101
010   010   010   010   0102   0100   0100   0100   01002   01000   01000
100   100   100   100   1002   1000   1000   1000   10002   10000   10000
001   001   001   001   0012   0010   0010   0010   00102   00102   00101
010   010   010   010   0102   0100   0100   0100   01002   01002   01002
111   001   001   001   0012   0010   0010   0010   00102   00100   00100
(3)   (5)   111   000   0001   0001   0001   0001   00012   00010   00010
                  (3)   (5)    (16)   (17)   1111   0000    00001   00001
                                             (3)    (5)     (16)    (17)    (11)
                                                     α
```

```
10010   100102   100100   100100   100100   1001001   1001001   1001001
01101   011012   011012   011012   011012   0110122   0110120   0110110
01000   010002   010000   010000   010000   0100002   0100000   0100000
10000   100002   100000   100000   100000   1000002   1000000   1000000
00101   001012   001010   001010   001010   0010102   0010100   0010100
01000   010001   010001   010001   010001   0100012   0100010   0100010
00100   001002   001000   001000   001000   0010002   0010000   0010000
00010   000102   000100   000100   000100   0001002   0001000   0001000
00001   000011   000011   000011   000011   0000112   0000110   0000110
(12)    (16)     (17)     111111   010000   0100001   0100001   0100001
          β                 (3)      (5)      (16)      (17)      (22)
```

In the first few additions of new columns, the undetermined values introduced in step 16 are all resolved in step 17. However, in the array marked α some undetermined values are introduced which are not resolved until the use of steps 11 and 12. Again, in the array β, one of the undetermined values survives step 17. In fact the value of this entry $(S_{2,6})$ is only determined in steps 18 through 22.

The graph G is covered by seven maximal complete subgraphs.

ANSWER.

$$
S = \begin{bmatrix}
1 & 0 & 0 & 1 & 0 & 0 & 1 \\
0 & 1 & 1 & 0 & 1 & 1 & 0 \\
0 & 1 & 0 & 0 & 0 & 0 & 0 \\
1 & 0 & 0 & 0 & 0 & 0 & 0 \\
0 & 0 & 1 & 0 & 1 & 0 & 0 \\
0 & 1 & 0 & 0 & 0 & 1 & 0 \\
0 & 0 & 1 & 0 & 0 & 0 & 0 \\
0 & 0 & 0 & 1 & 0 & 0 & 0 \\
0 & 0 & 0 & 0 & 1 & 1 & 0 \\
0 & 1 & 0 & 0 & 0 & 0 & 1
\end{bmatrix} .
$$

Note that had $S_{2,6}$ been assigned the value 0, G would still have been covered by complete subgraphs, but one of them would not have been maximal.

The entries of the matrix $A^* = S \times S^T$ have the following significance:

a_{ii}^*: the number of maximal complete subgraphs containing vertex i;

$a_{ij}^* \, (i \neq j)$: the number of maximal complete subgraphs containing the edge (i,j).

If we replace each a_{ii}^* with 0, and each $a_{ij}^* \neq 0 \ (i \neq j)$ with 1, we have returned to the adjacency matrix A.

We can also obtain the incidence matrix M from S by expanding each column of S with k ones, $k = 2, 3, ..., n$ into $\binom{k}{2} = k(k-1)/2$ columns with pairs of ones, and eliminating any duplicate columns that result. Thus a column with exactly two 1's remains intact, representing an edge; a column with three 1's expands into $\binom{3}{2} = 3$ columns, representing the three edges of the maximal complete subgraph, and so forth. For example, the second column of the matrix derived in Example 8.1 expands into $\binom{4}{2} = 6$ columns, representing the six edges of the complete subgraph on vertices 2, 3, 6, and 10.

Example 8.2

Compute A^* and an incidence matrix for the graph of Example 8.1.

Solution.   Begin with S. A^* is obtained by matrix multiplication.

$$
A^* = S \times S^T = \begin{bmatrix}
3 & 0 & 0 & 1 & 0 & 0 & 0 & 1 & 0 & 1 \\
0 & 4 & 1 & 0 & 2 & 2 & 1 & 0 & 2 & 1 \\
0 & 1 & 1 & 0 & 0 & 1 & 0 & 0 & 0 & 1 \\
1 & 0 & 0 & 1 & 0 & 0 & 0 & 0 & 0 & 0 \\
0 & 2 & 0 & 0 & 2 & 0 & 1 & 0 & 1 & 0 \\
0 & 2 & 1 & 0 & 0 & 2 & 0 & 0 & 1 & 1 \\
0 & 1 & 0 & 0 & 1 & 0 & 1 & 0 & 0 & 0 \\
1 & 0 & 0 & 0 & 0 & 0 & 0 & 1 & 0 & 0 \\
0 & 2 & 0 & 0 & 1 & 1 & 0 & 0 & 2 & 0 \\
1 & 1 & 1 & 0 & 0 & 1 & 0 & 0 & 0 & 2
\end{bmatrix}.
$$

(Compare this with the given adjacency matrix A.)
   To compute M, expand each column of S.

$$
M' = \begin{bmatrix}
1 & 000000 & 000 & 1 & 000 & 000 & 1 \\
0 & 111000 & 110 & 0 & 110 & 110 & 0 \\
0 & 100110 & 000 & 0 & 000 & 000 & 0 \\
1 & 000000 & 000 & 0 & 000 & 000 & 0 \\
0 & 000000 & 101 & 0 & 101 & 000 & 0 \\
0 & 010101 & 000 & 0 & 000 & 101 & 0 \\
0 & 000000 & 011 & 0 & 000 & 000 & 0 \\
0 & 000000 & 000 & 1 & 000 & 000 & 0 \\
0 & 000000 & 000 & 0 & 011 & 011 & 0 \\
0 & 001011 & 000 & 0 & 000 & 000 & 1 \\
\hline
1 & 2 & 3 & 4 & 5 & 6 & 7
\end{bmatrix}.
$$

(Under each group of columns in M' is shown the column in S from which they derive.) Now eliminate all duplicate columns.

$$
M = \begin{bmatrix}
1 & 0 & 0 & 0 & 0 & 0 & 0 & 0 & 0 & 0 & 1 & 0 & 0 & 0 & 1 \\
0 & 1 & 1 & 1 & 0 & 0 & 0 & 1 & 1 & 0 & 0 & 1 & 0 & 0 & 0 \\
0 & 1 & 0 & 0 & 1 & 1 & 0 & 0 & 0 & 0 & 0 & 0 & 0 & 0 & 0 \\
1 & 0 & 0 & 0 & 0 & 0 & 0 & 0 & 0 & 0 & 0 & 0 & 0 & 0 & 0 \\
0 & 0 & 0 & 0 & 0 & 0 & 0 & 1 & 0 & 1 & 0 & 0 & 1 & 0 & 0 \\
0 & 0 & 1 & 0 & 1 & 0 & 1 & 0 & 0 & 0 & 0 & 0 & 0 & 1 & 0 \\
0 & 0 & 0 & 0 & 0 & 0 & 0 & 0 & 1 & 1 & 0 & 0 & 0 & 0 & 0 \\
0 & 0 & 0 & 0 & 0 & 0 & 0 & 0 & 0 & 0 & 1 & 0 & 0 & 0 & 0 \\
0 & 0 & 0 & 0 & 0 & 0 & 0 & 0 & 0 & 0 & 0 & 1 & 1 & 1 & 0 \\
0 & 0 & 0 & 1 & 0 & 1 & 1 & 0 & 0 & 0 & 0 & 0 & 0 & 0 & 1 \\
\end{bmatrix}.
$$

EXERCISES

Compute S for the graph:

1. Fig. 8.2.

2. Fig. 8.3.

3. Fig. 8.4.

4. Fig. 8.5.

Figure 8.2

Figure 8.3

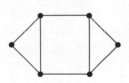

Figure 8.4

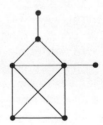

Figure 8.5

5. Program the algorithm to generate S, assuming $n \leqslant 20$, and test it on the graphs of Figs. 8.1–8.5.

6. Clearly if G is a complete graph on m vertices its maximal complete subgraph matrix is an $m \times 1$ matrix of ones. At the other extreme, if G is acyclic, its maximal complete subgraph matrix is its incidence matrix. Suppose that G is a connected graph with m vertices,

n edges, and k_i maximal complete subgraphs on i vertices, $i = 2, 3, ..., m$. Define the density of a matrix M of size $m \times n$ to be a/mn, where a is the number of nonzero entries in the matrix. Compute the density of the adjacency, incidence, and maximal complete subgraph matrices of G.

7. (cont.) Compare the densities of the three matrices for the graphs of Figs. 8.1–8.5.

8. (cont.) Considering the edges of the maximal complete subgraphs, show that

$$ n \leqslant \sum_{i=1}^{m} \binom{i}{2} k_i, $$

and that equality can hold.

*9. (cont.) Determine an upper bound, and if possible a least upper bound for $\sum \binom{i}{2} k_i$.

9. STORAGE MINIMIZATION FOR MATRICES

Suppose we wish to store the matrix

$$ M = \begin{bmatrix} 0 & 1 & 0 & 0 & -2 & 0 & 0 \\ 7 & 0 & 0 & 0 & 0 & 0 & 13 \\ 0 & -5 & 2 & 0 & 0 & 0 & 0 \\ 0 & 0 & 0 & 1 & 0 & 0 & 0 \\ 0 & 0 & 0 & 0 & 0 & -3 & 0 \end{bmatrix} $$

in a computer memory. Since it is a small matrix there is no real problem. But notice that it consists mostly of zeros. Less than 23 percent of the entries are nonzero. Such a matrix, containing mostly zeros, is called *sparse*. Now consider a sparse matrix M' of 4000 rows and 5000 columns, say one with 23 percent nonzero elements. (Matrices of comparable size occur in such application areas as information retrieval and linear programming.) To store M' in a straightforward manner requires 20,000,000 words of memory. Clearly this is wasteful since 15,400,000 of these words each contain zero. If we could find some way to store all of the nonzero entries and few, if any, zero entries, we would save a sizable amount of memory. We illustrate several such procedures with our small matrix M.

One technique is to replace M by a pair of matrices, M_1 containing the locations within each row of the nonzero entries, and M_2 containing the values of the entries:

$$ M_1 = \begin{bmatrix} 2 & 5 \\ 1 & 7 \\ 2 & 3 \\ 4 & 0 \\ 6 & 0 \end{bmatrix}, \qquad M_2 = \begin{bmatrix} 1 & -2 \\ 7 & 13 \\ -5 & 2 \\ 1 & 0 \\ -3 & 0 \end{bmatrix}. $$

For M, this reduces the storage requirement by 43 percent.

If M were a $0,1$—matrix, that is, if the nonzero entries were all ones, then we would not need M_2. In this case M can be interpreted as one of the graph matrices, say the bipartite matrix (see Section 4), and M_1 is called the *connection table* for the graph.

A similar technique uses a *bit map* to indicate the location of nonzero elements. Think of M as a linear array,

$$0\ 1\ 0\ 0\ -2\ 0\ 0\ 7\ 0\ 0\ 0\ \cdots\ 0\ 0\ -3\ 0.$$

(A program can correctly interpret this as a 5×7 matrix.) We now utilize a single bit to indicate the presence of each nonzero element, form the bit map

$$01001001000001011000000010000000010,$$

and accompany it by a vector of the nonzero values in corresponding order,

$$\langle 1, -2, 7, 13, -5, 2, 1, -3 \rangle.$$

If our computer has at least 35 bits per word, then this bit map will fit in a single computer word. In a 12- or 16-bit machine we would need three words for the bit map. But even in this case, the storage requirement is reduced by 69 per cent from that necessary to store M in the conventional way.

Another, less compact way to store the matrix is as a list of position–value pairs. The position may be indicated as the pair of row and column numbers

(1, 2),	1	(3, 2),	−5
(1, 5),	−2	(3, 3),	2
(2, 1),	7	(4, 4),	1
(2, 7),	13	(5, 6),	−3

or as the serial position in the linear array

2,	1	16, −5
5,	−2	17, 2
8,	7	25, 1
14,	13	34, −3

For most applications of matrices, it is not important that we keep the rows and columns in any particular order. Thus we are generally free to interchange rows and to interchange columns in an effort to reduce storage requirements. For example, if in M we write the rows $1, \ldots, 5$ in the order $1, 3, 4, 5, 2$, and the columns $1, \ldots, 7$ in the order $5, 2, 3, 4, 6, 1, 7$, we obtain the matrix

$$M_3 = \begin{bmatrix} -2 & 1 & 0 & 0 & 0 & 0 & 0 \\ 0 & -5 & 2 & 0 & 0 & 0 & 0 \\ 0 & 0 & 0 & 1 & 0 & 0 & 0 \\ 0 & 0 & 0 & 0 & -3 & 0 & 0 \\ 0 & 0 & 0 & 0 & 0 & 7 & 13 \end{bmatrix}.$$

Notice the large blocks of zeros in the corners. Thus we need only store the relatively narrow band containing nonzero elements as an array

$$M_4 = \begin{bmatrix} -2 & 1 & 0 \\ -5 & 2 & 0 \\ 0 & 1 & 0 \\ 0 & -3 & 0 \\ 0 & 7 & 13 \end{bmatrix},$$

remembering that it is really a diagonal band of M_3. This technique is called *banding* the matrix.

For M, this saves 57 percent on storage requirements. For larger matrices, the savings is often greater. The only restriction is one of symmetry. If the original matrix is symmetric ($m_{ij} = m_{ji}$), and we desire to preserve the symmetry, then for each row interchange there must be a corresponding column interchange. Of course if the matrix is symmetric, we need store only the upper (or lower) triangular half of it.

The value of this particular technique for storage reduction lies in the fact that it is also mathematically important. Numerical analysts have developed special procedures for computing with matrices all of whose nonzero elements lie in a narrow diagonal band. We shall explore this band concept further in Sections 10 and 11 of this chapter.

Another technique that is useful for certain large matrices is to permute rows and columns so that the nonzero elements are arranged in blocks which are quite dense (Fig. 9.1). Then only these blocks and some indication of their relationship to the original matrix must be stored.

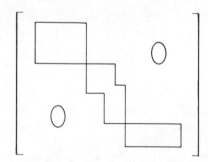

Figure 9.1

Finally, we discuss a technique that has been thoroughly developed by chemists. Consider the compound shown in Fig. 9.2. This diagram represents

Figure 9.2

a single molecule of an organic compound containing six atoms of carbon, five of hydrogen, one of chlorine, and two each of nitrogen and oxygen per molecule. The lines (or "bonds") and the positions of the letters within the figure describe the molecule structure to a chemist. The traditional chemical notation for this, $C_6H_5ClN_2O_2$, does not indicate the structure of the compound, and, in fact, is ambiguous, representing any of several related compounds. The compound could be represented as a matrix,

$$M = \begin{bmatrix}
C & 1 & 0 & 0 & 0 & 1 & 0 & 0 & 0 & 1 & 0 & 0 & 0 & 0 & 0 & 0 \\
1 & C & 1 & 0 & 0 & 0 & 1 & 0 & 0 & 0 & 0 & 0 & 0 & 0 & 0 & 0 \\
0 & 1 & C & 1 & 0 & 0 & 0 & 0 & 0 & 0 & 0 & 0 & 1 & 0 & 0 & 0 \\
0 & 0 & 1 & C & 1 & 0 & 0 & 0 & 0 & 0 & 0 & 0 & 0 & 0 & 0 & 1 \\
0 & 0 & 0 & 1 & C & 1 & 0 & 1 & 0 & 0 & 0 & 0 & 0 & 0 & 0 & 0 \\
1 & 0 & 0 & 0 & 1 & C & 0 & 0 & 1 & 0 & 0 & 0 & 0 & 0 & 0 & 0 \\
0 & 1 & 0 & 0 & 0 & 0 & H & 0 & 0 & 0 & 0 & 0 & 0 & 0 & 0 & 0 \\
0 & 0 & 0 & 0 & 1 & 0 & 0 & H & 0 & 0 & 0 & 0 & 0 & 0 & 0 & 0 \\
0 & 0 & 0 & 0 & 0 & 1 & 0 & 0 & H & 0 & 0 & 0 & 0 & 0 & 0 & 0 \\
1 & 0 & 0 & 0 & 0 & 0 & 0 & 0 & 0 & N & 1 & 1 & 0 & 0 & 0 & 0 \\
0 & 0 & 0 & 0 & 0 & 0 & 0 & 0 & 0 & 1 & H & 0 & 0 & 0 & 0 & 0 \\
0 & 0 & 0 & 0 & 0 & 0 & 0 & 0 & 0 & 1 & 0 & H & 0 & 0 & 0 & 0 \\
0 & 0 & 1 & 0 & 0 & 0 & 0 & 0 & 0 & 0 & 0 & 0 & N & 1 & 1 & 0 \\
0 & 0 & 0 & 0 & 0 & 0 & 0 & 0 & 0 & 0 & 0 & 0 & 1 & O & 0 & 0 \\
0 & 0 & 0 & 0 & 0 & 0 & 0 & 0 & 0 & 0 & 0 & 0 & 1 & 0 & O & 0 \\
0 & 0 & 0 & 1 & 0 & 0 & 0 & 0 & 0 & 0 & 0 & 0 & 0 & 0 & 0 & Cl
\end{bmatrix}$$

but this is wasteful of storage. (This matrix shows the bonds but not the type of bond (single or double), with the atoms listed on the diagonal.) As a result, chemists have developed linear notations that exactly represent any organic compound, and which are highly compact. The Wiswesser notation for the given compound,

ZR   DG   CNW,

is an example of these. This notation uses a code to represent each of the common substructures that a chemist might expect, and relative positions within the code string to describe the relationships among these structures. Such a linear notation reduces storage requirements greatly, but its utility is limited by the ease with which it can be used to determine substructures, or for some other application. (Wiswesser notation is quite good from this standpoint.) Since there are several hundred thousand organic compounds, compact linear notations are invaluable in chemical information storage and retrieval.

EXERCISES

Consider the matrices

$$A = \begin{bmatrix} 0 & 1 & 0 & 0 & 2 & 0 \\ 1 & -1 & 0 & 0 & 1 & 0 \\ 0 & 0 & 2 & 1 & 0 & 1 \\ 0 & 1 & 0 & 2 & 1 & 0 \end{bmatrix},$$

$$B = \begin{bmatrix} 0 & 0 & 0 & 1 & 0 & 0 & 0 & 0 \\ 1 & 0 & 2 & 0 & 0 & 1 & 0 & 0 \\ 0 & 2 & 0 & 0 & -1 & 0 & 2 & 0 \\ 1 & 0 & 0 & 1 & 0 & 0 & 0 & 2 \end{bmatrix},$$

$$C = \begin{bmatrix} -3 & 0 & 1 & 0 & 0 & 0 & 0 & 0 & 0 & 0 \\ 0 & 2 & 0 & 0 & 1 & 0 & 0 & 0 & 1 & 0 \\ 0 & 0 & 1 & 1 & 0 & 0 & 2 & 0 & 0 & -1 \\ 1 & 0 & 0 & 2 & 0 & 1 & 1 & 0 & 0 & 0 \\ -1 & 0 & 0 & 0 & 0 & 1 & 0 & -1 & 0 & 0 \\ 0 & 0 & 1 & 0 & 2 & 0 & 0 & 1 & -2 & 0 \end{bmatrix}.$$

For each matrix compute the storage saved by using the technique named for storage:

1. paired position and value matrices;

2. bit map and value vector;

3. position-value pair, with position as a coordinate pair;

4. position-value pair, with position as a single number;

5. banding the matrix.

6. Assume that m and n are fixed, and that k, the number of nonzero elements in an $m \times n$ array, is known. Estimate the storage required on a computer with word length w bits by each of the five storage methods discussed.

7. (cont.) Determine for a given quadruple $\langle m, n, k, w \rangle$ which storage method is best on the basis only of the storage required. (Other requirements such as accessibility might favor other storage methods.)

10. BANDWIDTH OF CUBIC GRAPHS

As indicated in the previous section, although banding a matrix may not reduce the storage requirements to a minimum, the technique is important because of the methods that numerical analysts have developed for computing with matrices of small bandwidth, that is, all of whose nonzero elements lie in a small diagonal band. Much of the work on banding matrices has centered on symmetric or zero-symmetric matrices since these are quite important in the applied mathematics of structural and network engineering. Since such matrices are readily interpreted as adjacency matrices for undirected graphs, it is natural to relate the banding of the matrix to operations on the associated graph.

If G is a graph on n vertices, let $\mathcal{N}$ be a *numbering* of G, that is, an assignment of the integers $1, 2, \ldots, n$ to the vertices of G. If i and j are the integers assigned to two adjacent vertices, then their adjacency is indicated by a 1 in the (i, j) position of the adjacency matrix for G. Note that this 1 is located a distance $|i-j|$ from the main diagonal of the matrix. Hence if we are interested in arranging all nonzero elements of the matrix in a small diagonal band, we are in effect interested in minimizing the maximum difference between numbers assigned to adjacent vertices of G. Thus if $\mathcal{B}_{\mathcal{N}}(G) = \max(|i-j|)$, where the maximum is taken over all adjacent pairs of vertices (i, j) we define the *bandwidth* of G to be $\mathcal{B}(G) = \min \mathcal{B}_{\mathcal{N}}(G)$, where the minimum is taken over all possible numberings of G. A numbering $\mathcal{N}$ for which $\mathcal{B}_{\mathcal{N}}(G) = \mathcal{B}(G)$ is called a *bandwidth numbering* of G.

Example 10.1

Determine the bandwidth of the graph G in Fig. 10.1.

Figure 10.1

Solution. One possible numbering $\mathcal{N}$ of G is shown in Fig. 10.2a. Since vertices 1 and 6 are adjacent, $\mathcal{B}_{\mathcal{N}}(G) = 5$ for this particular numbering. Observe, however, that the central vertex (numbered 2) is adjacent to all other vertices. Thus it must be adjacent, in any numbering, to one or both of the vertices numbered 1 or 6. Hence its number should be intermediate, namely either 3 or 4. Fig. 10.2b illustrates such a numbering $\mathcal{N}'$,

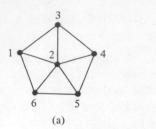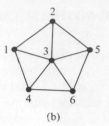

(a)   (b)

Figure 10.2

for which $\mathscr{B}_{\mathcal{N'}}(G) = 3$. Since numbering the central vertex 4 will not reduce this, $\mathcal{N'}$ is a bandwidth numbering of G.

ANSWER. $\mathscr{B}(G) = 3.$

If we were to embark on a general study of graph bandwidth, we would quickly find that it is difficult to discover patterns that will lead us to general statements about bandwidth. Thus we are motivated to look for a more restricted class of graphs, about whose bandwidth we may hope to make some enlightening statements. The cubic graphs provide such a class, simply because of the fact that each row and each column of the adjacency matrix for a cubic graph contains exactly three ones. (Why are the connected 2-regular graphs too narrow a class?)

Since a prime motivation for the study of graph bandwidth is to simplify calculations on sparse matrices, the study of cubic graphs is not very enlightening. Only rarely does one encounter a sparse zero-symmetric matrix that can properly be associated with a cubic graph. Nevertheless, the study does yield insight into the interplay of various graph parameters, and into characteristics of graphs which bear a general relationship to the bandwidth. In addition, the work that has been done on the bandwidth of cubic graphs illustrates nicely the use of computers in solving graph-theoretical problems.

We might conjecture, since each vertex is of degree 3, that the bandwidth of most cubic graphs should be quite small. How do we investigate this conjecture? One way is to generate some cubic graphs and determine their bandwidth.

The smallest cubic graph is K_4, which has bandwidth 3. See Fig. 10.3.

Figure 10.3

However, this is a very special case, since each vertex of the graph lies in three 3-cycles or triangles. We quickly note that if a cubic graph has six or more vertices, then each vertex either lies in two 3-cycles (Fig. 10.4a, vertices 1 and 2), in one 3-cycle (Fig. 10.4b, and vertices 3 and 4 in Fig. 10.4a), or in no 3-cycle (Fig. 10.4c). This provides a procedure for generating cubic

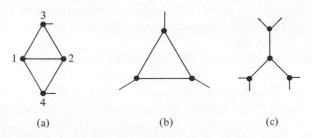

(a)   (b)   (c)

Figure 10.4

graphs by hand. Classify cubic graphs on n vertices by the number of vertices of type 1 (in two triangles), type 2 (in one triangle), and type 3 (in no triangle) they contain. These numbers are related by the formulas

$$T_1 = 2k_1, \qquad T_2 = T_1 + 3k_2, \qquad T_1 + T_2 + T_3 = n,$$

where T_i denotes the number of vertices of type $i, i = 1, 2, 3$, and k_1 and k_2 are nonnegative integers. For $n = 6$, the triples (T_1, T_2, T_3) are $(2, 2, 2)$, $(0, 6, 0)$, $(0, 3, 3)$, and $(0, 0, 6)$. Of these, $(2, 2, 2)$ and $(0, 3, 3)$ yield no graphs, while $(0, 6, 0)$ and $(0, 0, 6)$ each yield one graph. For $n = 8$, the triples are $(4, 4, 0)$, $(2, 5, 1)$, $(2, 2, 4)$, $(0, 6, 2)$, $(0, 3, 5)$, and $(0, 0, 8)$. There are no graphs of types $(2, 5, 1)$ and $(2, 2, 4)$, two graphs of type $(0, 0, 8)$, and one of each other type. For $n \geqslant 10$, there are graphs of every possible type. We find, for example, that there are exactly two 12-vertex graphs of type $(4, 4, 4)$, with two diamonds (paired 3-cycles) and no other 3-cycles (Fig. 10.5). (Recall also Figs. 3.3–3.5, pp. 43–44.)

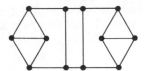 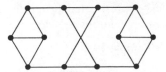

Figure 10.5

The hand generation of cubic graphs works well enough for graphs on at most 10 vertices, but even for graphs on 12 vertices it is tedious and subject to error. Hence we use the computer to generate cubic graphs of this size or larger, representing the graphs by their adjacency matrices. Now on the one hand, if we were to generate all symmetric $n \times n$ matrices with three ones in each row and column, we would generate for each graph on n vertices all of its $n!$ adjacency matrices. For example, for the two graphs on six vertices we would generate 1440 matrices (not all distinct). On the other hand, to specify a priori the numbers of diamonds and 3-cycles, and their positions within the graph, requires a sophisticated preanalysis for each possible type of graph.

Let us try a middle course. We can avoid a sophisticated preanalysis, and still write a computer program that generates a relatively small number of matrices. Let us think of constructing the graph as we build the matrix. The first vertex chosen can, without any loss of generality, be numbered 1, and attached to vertices numbered 2, 3, and 4. Similarly, without restricting the development of the graph, vertex 2 may be attached to the other three vertices already present (Fig. 10.6a), to two of them (Fig. 10.6b), or only to vertex 1 (Fig. 10.6c); and the numbering shown in each case is the only one

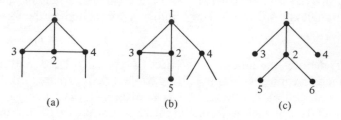

Figure 10.6

which need be considered. A computer program that develops only the distinct numberings at each stage will generate all of the desired adjacency matrices with a relatively small number of superfluous matrices. For example, such a program will generate only three 6×6 matrices (two corresponding to the planar cubic graph on six vertices and one for $K_{3,3}$), instead of the 1440 a less-sophisticated program would generate. (For larger graphs, the number of superfluous matrices is greater, although still relatively small.)

Example 10.2

Generate adjacency matrices for all cubic graphs on six vertices.

Solution. Let the first row of the matrix A be 011100. This attaches vertex 1 to vertices 2, 3, and 4, which restricts the number of matrices generated, without eliminating any graphs. Note that $a_{2,1} = 1$ and $a_{2,2} = 0$.

(Why?) Hence the possible second rows are

    101100,   101010,   101001,   100110,   100101,   100011.

Of these, only 101100, 101010, and 100011 represent distinct partial graphs. (The rows 101010 and 101001, for example, represent the same partial graph with different numberings, as in Fig. 10.7.)

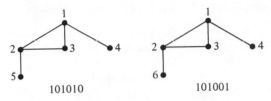

              101010   101001

Figure 10.7

Choose 101100 as the second row, resulting in the partial matrix

$$\begin{bmatrix} 0 & 1 & 1 & 1 & 0 & 0 \\ 1 & 0 & 1 & 1 & 0 & 0 \\ 1 & 1 & 0 & & & \\ 1 & 1 & & 0 & & \\ 0 & 0 & & & 0 & \\ 0 & 0 & & & & 0 \end{bmatrix}.$$

Since each row (and column) must contain exactly three ones, complete the fifth and sixth rows and columns:

$$\begin{bmatrix} 0 & 1 & 1 & 1 & 0 & 0 \\ 1 & 0 & 1 & 1 & 0 & 0 \\ 1 & 1 & 0 & & 1 & 1 \\ 1 & 1 & & 0 & 1 & 1 \\ 0 & 0 & 1 & 1 & 0 & 1 \\ 0 & 0 & 1 & 1 & 1 & 0 \end{bmatrix}.$$

This results in too many ones in the third and fourth rows. Hence the choice of 101100 yields no matrices.

    Now choose 101010 as the second row:

$$\begin{bmatrix} 0 & 1 & 1 & 1 & 0 & 0 \\ 1 & 0 & 1 & 0 & 1 & 0 \\ 1 & 1 & 0 & & & \\ 1 & 0 & & 0 & & \\ 0 & 1 & & & 0 & \\ 0 & 0 & & & & 0 \end{bmatrix}.$$

Again, complete the sixth row and column:

$$
\begin{bmatrix}
0 & 1 & 1 & 1 & 0 & 0 \\
1 & 0 & 1 & 0 & 1 & 0 \\
1 & 1 & 0 & & & 1 \\
1 & 0 & & 0 & & 1 \\
0 & 1 & & & 0 & 1 \\
0 & 0 & 1 & 1 & 1 & 0
\end{bmatrix}.
$$

This forces zeros into the remaining positions in the third row and column, which in turn forces completion of the remaining rows and columns:

$$
A_1 =
\begin{bmatrix}
0 & 1 & 1 & 1 & 0 & 0 \\
1 & 0 & 1 & 0 & 1 & 0 \\
1 & 1 & 0 & 0 & 0 & 1 \\
1 & 0 & 0 & 0 & 1 & 1 \\
0 & 1 & 0 & 1 & 0 & 1 \\
0 & 0 & 1 & 1 & 1 & 0
\end{bmatrix}.
$$

Now consider the remaining choice for row 2, 100011. The partial matrix is

$$
\begin{bmatrix}
0 & 1 & 1 & 1 & 0 & 0 \\
1 & 0 & 0 & 0 & 1 & 1 \\
1 & 0 & 0 & & & \\
1 & 0 & & 0 & & \\
0 & 1 & & & 0 & \\
0 & 1 & & & & 0
\end{bmatrix}.
$$

Of the three choices for the third row, 100110, 100101, and 100011, only two, namely 100110 and 100011, correspond to distinct partial graphs. In each case there is only one way to complete the adjacency matrix of a cubic graph:

$$
A_2 =
\begin{bmatrix}
0 & 1 & 1 & 1 & 0 & 0 \\
1 & 0 & 0 & 0 & 1 & 1 \\
1 & 0 & 0 & 1 & 1 & 0 \\
1 & 0 & 1 & 0 & 0 & 1 \\
0 & 1 & 1 & 0 & 0 & 1 \\
0 & 1 & 0 & 1 & 1 & 0
\end{bmatrix},
\qquad
A_3 =
\begin{bmatrix}
0 & 1 & 1 & 1 & 0 & 0 \\
1 & 0 & 0 & 0 & 1 & 1 \\
1 & 0 & 0 & 0 & 1 & 1 \\
1 & 0 & 0 & 0 & 1 & 1 \\
0 & 1 & 1 & 1 & 0 & 0 \\
0 & 1 & 1 & 1 & 0 & 0
\end{bmatrix}.
$$

ANSWER. A_1, A_2, and A_3.

Note that there are only two distinct cubic graphs on six vertices. Thus two of these matrices (which two?) represent different numberings of the same graph.

Having generated the graphs that we need, we are next faced with the problem of identifying and discarding any duplicate graphs that are generated. Without more information, this is a difficult task. Either we try, somehow, to interchange rows and columns of one matrix to transform it into another suspected of representing the same graph, or we resort to drawing the graphs, hoping that we can then determine which ones may be discarded. We may find a collection of graphs such as those in Fig. 10.8. It is not obvious that these graphs are all isomorphic.

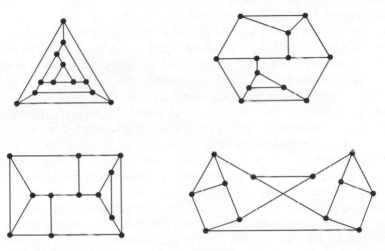

Figure 10.8

It is at this point that the invariants of the graph come into use. Since the graphs that we are studying are regular, the degree spectrum does not help; and since some other invariants are difficult to compute, the most useful invariant for our purposes is the cycle vector, which counts the number of cycles of various lengths. A number of programs exist for finding a spanning tree and from it a set of basic cycles. Having these, it is a simple task to determine the cycle space and count the cycles of various lengths. The utility of this invariant is indicated by the fact that in a study of over 700 distinct cubic graphs only three pairs of nonisomorphic graphs with the same cycle vector (metamorphic graphs) were found. Thus if two graphs have the same cycle vector, there is strong reason to suspect that they are isomorphic. (Different cycle vectors, of course, indicate nonisomorphic graphs.)

Let us now turn to the other aspect we wish to consider, the interplay among the graph parameters. In particular, we wish to relate the bandwidth of a graph to its other characteristics.

For each of the graphs that we have generated (some are shown in Figs. 3.2–3.5 of this chapter), we compute the bandwidth. If we classify these graphs by bandwidth, we might notice, for example, that each cubic graph of bandwidth 3 which we find contains a 3-cycle. Is this always the case?

In fact, if a cubic graph has bandwidth 3, then vertices 1 and 2 lie in one 3-cycle, and vertices n and $n-1$ lie in another 3-cycle. Observe Fig. 10.6: if vertices 1 and 2 do not lie in a 3-cycle, then vertex 2 is adjacent to a vertex numbered 6 or higher, and hence the bandwidth cannot be less than 4. A similar argument holds for vertices n and $n-1$. Thus a cubic graph of bandwidth 3 contains not one, but at least two, 3-cycles. This may be generalized: a cubic graph of bandwidth k has at least two cycles of length k or less. That is, its girth is at most k. This implies that the bandwidth of a cubic graph is not generally small. We can find a cubic graph of arbitrarily large bandwidth simply by finding one of sufficiently large girth. (It is known that the latter is always possible.)

We may observe another property of our graphs. Each graph of bandwidth 3 is planar. Is this always true? While the proof is somewhat more difficult, it is true that the bandwidth of a nonplanar graph must be at least 4. This may generalize to a relationship between crossing number and bandwidth, as the 3-cycle property generalized to a relationship between girth and bandwidth. Whether it does or not is not known.

Observations such as these can be used to refine our programs. For example, if the graph is nonplanar, the program to compute bandwidth should not try for a bandwidth of 3. If the graph has girth 7, the program should similarly recognize that the bandwidth cannot be less than this.

We see here the use of the computer in the classical loop of the experimental sciences. We use the computer to generate examples. Studying the examples, we make conjectures which lead to theoretical results. These results in turn enable us to write more efficient programs (refine our experimental technique) for generating and studying further examples.

EXERCISES

1. Observe the three generation processes for cubic graphs:

   P_1:   replace any one vertex by a triangle (Fig. 10.9);

Figure 10.9

P_2:   in the middle of an edge, insert a diamond (Fig. 10.10);

Figure 10.10

P_3:   join any two edges by a new edge (Fig. 10.11).

Figure 10.11

Write a program to take a cubic graph of at most 12 vertices, given as an adjacency matrix, and generate new graphs from it by these three processes.

2. Consider the graph fragment of Fig. 10.12. Let a denote numbering the triangle $1, 2, 3$, and b denote numbering it $1, 2, 4$. Further, if the last three numbers on the cross-links are k, l, and m with $k < l < m$, let a denote adding another cross-link to the verticals having vertices k and l, and let b denote adding another cross-link to the verticals having vertices k and m. In each case, the new vertex under k, l, or m is numbered $k+3$, $l+3$, or $m+3$, respectively. (If a b triangle heads the fragment, the first vertex under 1 is numbered 3.) Thus a sequence in a and b corresponds to a cubic graph, if we close the fragment off with a triangle. See Figs. 10.13–10.15. Show that sequences have no two adjacent b's correspond to graphs of bandwidth three, but that sequences with two or more adjacent b's correspond to graphs of higher bandwidth.

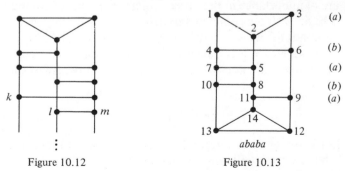

Figure 10.12   Figure 10.13

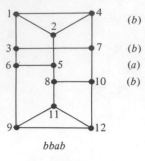

bbab

Figure 10.14

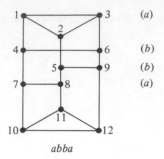

abba

Figure 10.15

3. (cont.) Show that the number of labeled cubic graphs of bandwidth 3 on $2n$ vertices, derived from this process, is f_n, the nth Fibonacci number. *Hint:* Either an a or a b follows an a, but only an a follows a b. Use this to establish the recurrence relation $f_n = f_{n-1} + f_{n-2}$.

4. Devise a procedure for reducing the bandwidth of a given symmetric matrix, that is, for finding a low $\mathscr{B}_N$ for a graph. You should be able to discover a procedure which yields good, but not generally minimal, results. *Hint:* Consider the implications which the degree of a vertex has for bandwidth.

5. (cont.) If you specialize your procedure to handle only cubic graphs, can you modify it to improve its performance?

11. BANDWIDTH OF BIPARTITE GRAPHS

The study of the bandwidth of cubic graphs exemplifies the use of the computer in mathematical studies which are not (because of the very special class of graphs) of great importance to computing. The study of the bandwidth of bipartite graphs presents the opposite situation. The conjectures leading to the theoretical results are readily made from a few hand-generated examples. The computer is not really needed for much of this work. However, since bipartite graphs can be related to arbitrary matrices, the theoretical results are of importance in the numerical applications of computers.

Let us begin with the complete bipartite graph $K_{m,n}$, $m \le n$. Observe that this corresponds to an $m \times n$ matrix with no zero entries (the bipartite matrix B of Section 4). Hence it is numerically uninteresting: no mere shuffling of rows or columns is going to save space. However, it is easy to show that the bandwidth is

$$B(K_{m,n}) = \left\lfloor \frac{n-1}{2} \right\rfloor + m,$$

where $\lfloor x \rfloor$ is the greatest integer less than or equal to x, called the *floor* of x. A bandwidth numbering is achieved by splitting the larger set of vertices, numbering them $1, 2, \ldots, \lfloor (n+1)/2 \rfloor$, $\lfloor (n+1)/2 \rfloor + m + 1, \ldots$,

$n+m-1$, $n+m$, with the smaller set numbered $[(n+1)/2]+1, ..., [(n+1)/2]$ $+m$. See Fig. 11.1.

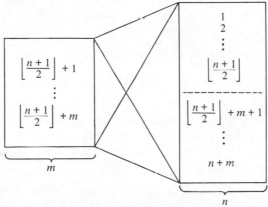

Figure 11.1

This type of numbering provides the clue for numbering incomplete bipartite graphs. In some way we want to break up the larger set of vertices and fit the vertices from the smaller set in between.

Let $B_{m,n,\langle p_1,p_2,...,p_m \rangle}$ denote a subgraph of $K_{m,n}$ with p_i edges missing from the ith vertex of the smaller set, $i=1,...,m$. For example, $B_{2,8,\langle 2,4 \rangle}$ denotes a subgraph of $K_{2,8}$ with two edges missing from one of the two ($m=2$) vertices, and four edges missing from the other. Two graphs fitting this description are shown in Fig. 11.2. (We shall be interested only in connected graphs.)

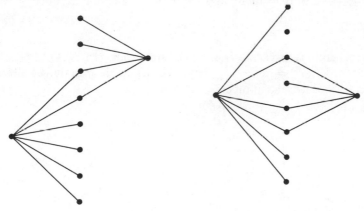

Figure 11.2

Now consider Fig. 11.3, representing $B_{2,n,\langle p_1,p_2 \rangle}$. The smaller set of vertices V is represented by the distinct vertices v_1 and v_2, while the larger set W is represented by the three subsets of vertices adjacent only to v_1 (shown as

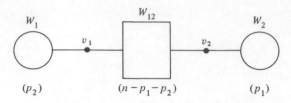

W_1

W_{12}

W_2

v_1

v_2

(p_2)

$(n-p_1-p_2)$

(p_1)

Figure 11.3

W_1), adjacent only to v_2 (shown as W_2), and adjacent to both v_1 and v_2 (shown as W_{12}). Vertex v_1 is adjacent to the p_2 vertices of W_1 and the $n-p_1-p_2$ vertices of W_{12}, while vertex v_2 is adjacent to all vertices of W_{12} and the p_1 vertices of W_2. Notice that there are three maximal complete bipartite subgraphs of this graph:

$$W_1 \cup \{v_1\} \cup W_{12} \qquad (K_{1,n-p_1}),$$
$$\{v_1\} \cup W_{12} \cup \{v_2\} \qquad (K_{2,n-p_1-p_2}), \qquad \text{and}$$
$$W_{12} \cup \{v_2\} \cup W_2 \qquad (K_{1,n-p_2}).$$

Hence the largest of the bandwidths for these three subgraphs provides a lower bound for the bandwidth of $B_{2,n,\langle p_1,p_2\rangle}$. That is, the bandwidth of the whole graph is at least this large. (Actually, the bandwidth of $B_{2,n,\langle p_1,p_2\rangle}$ either equals this lower bound or exceeds it by one, depending on the graph.)

A similar diagram for $B_{3,n,\langle p_1,p_2,p_3\rangle}$ is shown in Fig. 11.4. Here again the maximal complete bipartite subgraphs provide a good lower bound for the bandwidth of the graph.

In the general case, we let $W_{i,j,\ldots,k}$ be the set of vertices adjacent to $v_i, v_j, \ldots, v_k$, and only to these, and we let $w_{i,j,\ldots,k}$ be the number of vertices in $W_{i,j,\ldots,k}$. Then we let

$$p_{i,j,\ldots,k} = \sum w_{r,s,\ldots,t},$$

where the sum is over all $\{r,s,\ldots,t\}$ such that $\{i,j,\ldots,k\} \not\subset \{r,s,\ldots,t\}$. Finally we let $p^{(k)} = \min p_{i,j,\ldots}$, over all p's with exactly k subscripts, $k = 1, 2, \ldots, m$. Then the numbers

$$E_k = \begin{cases} \left\lfloor \dfrac{n-p^{(k)}-1}{2} \right\rfloor + k & \text{if} \quad n - p^{(k)} \geqslant k \\[2ex] \left\lfloor \dfrac{k-1}{2} \right\rfloor + n - p^{(k)} & \text{if} \quad 0 < n - p^{(k)} < k \\[2ex] 0 & \text{if} \quad n - p^{(k)} = 0 \end{cases}$$

$(k = 1, \ldots, m)$ are lower bounds for the bandwidth of any graph $B_{m,n,\langle p_1,\ldots,p_m\rangle}$.

Since the bandwidth concept for graphs is related to the bandwidth of the adjacency matrix, it is not clear how the bounds that we have calculated relate to the most compact form of the bipartite matrix corresponding to the graph. This is an area of active research. The best that can be said at present

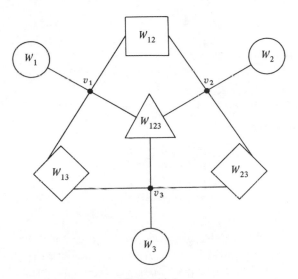

Figure 11.4

is that apparently if one determines a bandwidth numbering and orders the rows and columns of the bipartite matrix according to this numbering, the most compact arrangement of the matrix is achieved.

Example 11.1
Give the bandwidth analysis for the matrix

$$\begin{bmatrix} 0 & 1 & 0 & -1 & 0 & 0 & 1 \\ 1 & 0 & 0 & 2 & 1 & 0 & 0 \\ 0 & 3 & 0 & 0 & 1 & -1 & 0 \\ 0 & 0 & -2 & 0 & 0 & 0 & 2 \\ 1 & 0 & 2 & 0 & 0 & 0 & 0 \end{bmatrix}$$

and find the most compact arrangement of the matrix.

Solution. Since the matrix is 5×7 it corresponds to a subgraph of $K_{5,7}$. Thus an upper bound for the bandwidth is

$$B(K_{5,7}) = \left\lfloor \frac{6}{2} \right\rfloor + 5 = 8.$$

The actual graph, one of the family $B_{5,7,\langle 4,4,4,5,5 \rangle}$, is shown in Fig. 11.5. Compute the numbers $w_1, \ldots, w_{12345}$ and $p_1, \ldots, p_{12345}$:

$w_1 = 0,$ $w_2 = 0,$ $w_3 = 1,$ $w_4 = 0,$ $w_5 = 0,$

$w_{12} = 1,$ $w_{13} = 1,$ $w_{14} = 1,$ $w_{15} = 0,$ $w_{23} = 1,$

$w_{24} = 0,$ $w_{25} = 1,$ $w_{34} = 0,$ $w_{35} = 0,$ $w_{45} = 1,$

$w_{123} = 0,$ $w_{124} = 0,$ $w_{125} = 0,$ $w_{134} = 0,$ $w_{135} = 0,$

$w_{145} = 0,$ $w_{234} = 0,$ $w_{235} = 0,$ $w_{245} = 0,$ $w_{345} = 0,$

$w_{1234} = 0,$ $w_{1235} = 0,$ $w_{1245} = 0,$ $w_{1345} = 0,$ $w_{2345} = 0,$ $w_{12345} = 0;$

$p_1 = 4,$ $p_2 = 4,$ $p_3 = 4,$ $p_4 = 5,$ $p_5 = 5,$

$p_{12} = 6,$ $p_{13} = 6,$ $p_{14} = 6,$ $p_{15} = 7,$ $p_{23} = 6,$

$p_{24} = 7,$ $p_{25} = 6,$ $p_{34} = 7,$ $p_{35} = 7,$ $p_{45} = 6,$

$p_{123} = 7,$ $p_{124} = 7,$ $p_{125} = 7,$ $p_{134} = 7,$ $p_{135} = 7,$

$p_{145} = 7,$ $p_{234} = 7,$ $p_{235} = 7,$ $p_{245} = 7,$ $p_{345} = 7,$

$p_{1234} = 7,$ $p_{1235} = 7,$ $p_{1245} = 7,$ $p_{1345} = 7,$ $p_{2345} = 7,$ $p_{12345} = 7.$

Hence, with $n = 7$,

$$
\begin{aligned}
p^{(1)} &= 4 & E_1 &= 2 \\
p^{(2)} &= 6 & E_2 &= 1 \\
p^{(3)} &= 7 & E_3 &= 0 \\
p^{(4)} &= 7 & E_4 &= 0 \\
p^{(5)} &= 7 & E_5 &= 0.
\end{aligned}
$$

Thus the bandwidth of the graph is at least 2, the largest E_i, and at most 8. The actual bandwidth of the graph is 3, achieved by the numbering shown in Fig. 11.6. The corresponding rearrangement of the matrix is

	(1)	(4)	(5)	(7)	(9)	(10)	(12)
(2)	-2	2	0	0	0	0	0
(3)	2	0	1	0	0	0	0
(6)	0	1	0	-1	1	0	0
(8)	0	0	1	2	0	1	0
(11)	0	0	0	0	3	1	-1

where the numbers for the rows and columns are the vertex numbers of the graph.

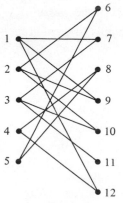

Figure 11.5

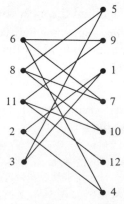

Figure 11.6

EXERCISES

For the matrix, determine the corresponding bipartite graph, find its bandwidth, and reorder the matrix according to a bandwidth numbering. Compute the saving in storage:

1.
$$\begin{bmatrix} 0 & 1 & 0 & 2 & 0 & 3 & 0 & 0 \\ 1 & 0 & 0 & 1 & 0 & 0 & -1 & 0 \\ 0 & 0 & 2 & 0 & 1 & 3 & 0 & -1 \\ 2 & 0 & 1 & 0 & 2 & 0 & 0 & 0 \end{bmatrix}.$$

2.
$$\begin{bmatrix} 1 & 0 & -1 & 0 & 0 & 2 & 0 & 0 & 0 \\ 0 & 2 & 0 & 1 & 3 & 0 & 0 & 0 & 0 \\ 0 & 0 & 1 & 0 & 0 & 0 & -2 & 0 & 1 \\ 2 & 0 & 0 & -3 & 0 & 1 & 0 & 1 & 0 \\ 0 & 3 & 0 & 0 & 1 & 0 & 0 & 2 & 0 \\ 1 & 0 & 2 & 0 & 0 & 0 & -1 & 0 & 2 \\ 0 & 1 & 0 & 0 & 0 & 1 & 0 & 0 & 0 \end{bmatrix}.$$

3. Write a program to compute upper and lower bandwidth bounds for the bipartite graph corresponding to arbitrary $m \times n$ matrix ($m, n \leqslant 20$). Test your program on Exercises 1 and 2.

4. Devise a procedure for finding a low $\mathcal{B}_N$ for bipartite graphs. (This may be a modification of your procedure for Exercise 4, Section 10.)

12. PLANAR GRAPHS AND THE FOUR COLOR CONJECTURE

"As far as I see at this moment, if four ultimate compartments have each boundary line in common with one of the others, three of them enclose the fourth, and prevent any fifth from connexion with it. If this be true, four colours will colour any possible map, without any necessity for colour meeting colour except at a point." (Portion of a letter from Augustus DeMorgan to Sir William Rowan Hamilton, October 23, 1852, quoted in Ore [5].)

This letter is the earliest known reference to the four color conjecture, apparently posed as a question by one of DeMorgan's students. As we stated earlier, the conjecture, that four colors suffice to color any planar map, remains unproven today. In this section we discuss a characteristic of planar graphs, and give a brief glimpse into the four color conjecture.

Suppose G is a planar graph, drawn with no false crossings in a plane (Fig. 12.1). Observe that certain cycles (for example, $v_4 v_5 v_6 v_8 v_7 v_4$) naturally bound plane regions containing no edge of the graph (the "ultimate compartments" of DeMorgan), while others (for example, $v_2 v_3 v_4 v_5 v_{11} v_2$) do not, at least in this representation. Note that the infinite region, "outside" the graph is an "ultimate compartment." Observe that the number of vertices

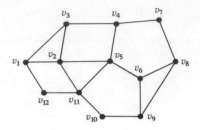

Figure 12.1

(v), edges (e), and regions (r) are related by the formula $v - e + r = 2$. This, the *Euler formula*, characterizes planar graphs. Since the cyclomatic number is $e - v + 1$, we observe that the number of cycles in a cycle basis for a planar graph is $r - 1$, the number of finite regions.

By algebraic manipulation of the Euler formula it is possible to show that for any planar graph with at least two edges,

$$\frac{3}{2}r \leqslant e \leqslant 3v - 6.$$

This is sometimes useful as a planarity test.

The four color problem was originally stated in terms of coloring regions of a map. It can be transformed into a graph coloring problem by representing regions by vertices, and placing an edge joining two vertices whenever the corresponding regions have a common boundary line (Fig. 12.2). The

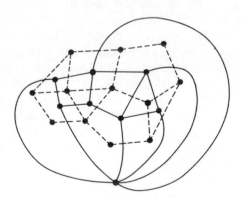

Figure 12.2

problem becomes: can one always color the vertices of a planar graph in such a way that no two adjacent vertices have the same color, and using at most four colors? That is, if G is planar, is its chromatic number, $\chi(G)$, at most four?

Two further transformations may be made in the problem. First, any vertex of degree 4 or more may be replaced by a set of vertices of degree 3, surrounding a new region (Fig. 12.3). Obviously the one color used for the original

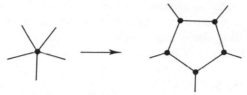

Figure 12.3

vertex cannot be used for all of the new vertices, but it can easily be shown that the resulting cycle of vertices can be colored in at most three colors, matching properly with the colors already used for other vertices in the graph. Similarly, any vertex of degree 2 can be replaced by a properly colored diamond (Fig. 12.4). Thus the four color problem need only be concerned with cubic graphs.

Figure 12.4

The problem may be further reduced to an edge coloring problem. The vertices of a graph may be properly 4-colored if and only if the edges of the graph may be properly 3-colored. (That is, no two edges of the same color have a vertex in common.) The correspondence between pairs of colors for the vertices and colors for the edges is given in Table 12.1. For example,

TABLE 12.1

Edge color	Vertex pair colors
a	1–2 or 3–4
b	1–3 or 2–4
c	1–4 or 2–3

each edge with vertices colored 1 and 2, or 3 and 4, is assigned color a. Thus we need only examine edge colorings of planar cubic graphs.

We leave as an exercise the proof that any Hamiltonian cubic graph may be 3-edge colored. Returning to bandwidth (Section 10), it can be shown that any connected cubic graph of bandwidth three is either separable (has connectivity 1), or is Hamiltonian. Thus for nonseparable cubic graphs of bandwidth 3, the four color conjecture holds.

(This illustrates both the fact that work on one concept, in this case bandwidth, may lead to results quite unanticipated at the beginning of the work, and also Ore's comment that "writers on graph theory...nearly always seem to have some thoughts in mind regarding...application [of their work] to the four-color problem." [5, p. vii].)

EXERCISES

1. Construct a planar graph whose vertices cannot be properly 3-colored.

2. Show that the transformation into a cubic graph does not affect colorability. That is, the resulting cubic graph is 4-colorable if and only if the original graph is.

3. Prove that any Hamiltonian cubic graph can be 3-edge colored.

4. Find a planar Hamiltonian cubic graph which can be 3-edge colored, but which does not have bandwidth 3.

REFERENCES

1. N. E. Gibbs and P.-S. Mei, A planarity algorithm based on the Kuratowski theorem, *Proc. 1970 Spring Joint Computer Conf.*, pp. 11–94. AFIPS Press, Montvale, New Jersey, 1970.
2. R. E. Tarjan, An efficient planarity algorithm, STAN-CS-244-71. Computer Science Department, Stanford University, Palo Alto, California, 1971.
3. D. J. A. Welch and M. B. Powell, An upper bound for the chromatic number of a graph and its application to timetabling problems, *Computer J.* **10** (1967) 85–86.
4. N. E. Gibbs, A cycle generation algorithm for finite undirected linear graphs, *J. Assoc. Comput. Mach.* **16** (1969) 564–568.
5. O. Ore, "The Four-Color Problem." Academic Press, New York, 1967.

CHAPTER 3

Gorn Trees

1. INTRODUCTION

The concept of a tree was introduced in Section 3 of the previous chapter. There we presented the classical definition of a tree as a connected acyclic graph. As we use trees to establish search patterns, parse a sentence, delineate a proof, or indicate a hierarchical relationship among a group of objects, we find that this bare concept is not enough. For virtually all applications, we find that there is a need to label the vertices, and often the edges, of a tree. For example, a genealogical tree which describes the relationships within a family is useless unless we know who in the family fits where in the tree. Hence each vertex of the tree carries a label identifying a particular member of the family. Often the label is more complex, giving information about sex, birth and death dates, and so forth. From the computational point of view, we generally want to label the vertex with the name of a file that contains all of the pertinent information.

Edge labeling of a tree is particularly important for decision trees. In this application, the content (label) of each vertex is a question. Depending on the answer to the question at one vertex, we pass to another vertex

101

chosen from a set of possible successors. By labeling the edges of the decision tree we identify the possible answers to each question, and their relationship to a chosen path through the tree. This type of labeling is common in flowcharts (which may or may not be trees), where the various exits from each decision point are labeled with the decision response which causes the program to follow that edge.

At times it is also necessary to establish a direction within a tree, or an order among the branches of a tree. Consider, for example, the expressions $a - b$ and $b - a$. The parsing trees for these expressions are shown in Fig. 1.1.

$(a - b)$ $(b - a)$

Figure 1.1

From the classical point of view, these are equivalent labeled trees. However, with the usual algebraic interpretation $a - b$ is quite a different expression from $b - a$. Thus in the parsing trees we must be able to distinguish between the left and the right branches. While this is easy enough visually, we need some notation or convention that will enable us to convey this information to a computer program. That is, there are situations for which we must be able to indicate to the computer an order or sequencing among the vertices or edges of a tree.

Indeed, as we store a tree (or any set or graph) in a computer, the memory device used for storage, for example, magnetic tape, drum, or disk, imposes an order on the tree. If we recognize this order, we can at times use it to advantage in computations on the tree. Consider, for example, the tree in Fig. 1.2. Since there are five vertices in the tree, the symbols

Figure 1.2

at these vertices can be stored on magnetic tape in any one of $5! = 120$ orders. Depending on the computations to be performed on the tree, it may be sensible to store the tree elements by levels within the tree ($+ * cab$ or $ab * c+$), or by tracing down the branches in some order ($+ * abc$ or $+ c * ba$). It would probably not be sensible to store the tree elements with complete disregard for the tree structure ($a+bc*$ or $ca+b*$). We generally need to know relationships between parts of the tree which are obscured by the mode of storage. In the example which we have been using, c is an operand of $+$. This is not clear from the storage string $+*abc$. The relationship may be explicitly shown by adding appropriate pointers ("The operands for the symbol at location n are found at locations $n+1$ and $n+4$."), or it may be implicitly used in designing the program that manipulates the tree.

The need for an address structure for a tree is made evident whenever we wish to substitute for portions of a tree. The ordinary distributive law $a(b+c) = ab+ac$, for example, can be expressed as an equivalence between trees (Fig. 1.3). Now consider the tree in Fig. 1.4. The distributive law

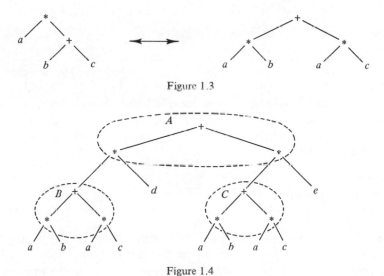

Figure 1.3

Figure 1.4

allows substitutions that simplify this tree at the three locations marked A, B, and C. If we make the substitution at A first, we find that we need make only one more substitution (Fig. 1.5), whereas substituting for

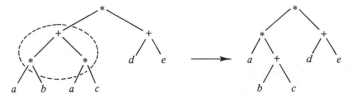

Figure 1.5

B and C first requires a total of three substitutions (Fig. 1.6). Thus we would prefer, as a matter of economy, to make the substitution at A first. To do so, we need some uniform scheme (since location A may be anywhere in the tree) of addressing the various vertices of the tree.

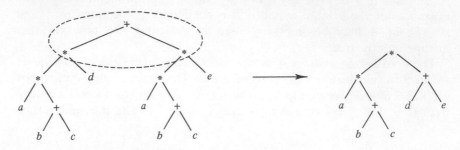

Figure 1.6

Such considerations have led Saul Gorn to redefine the concept of a tree so that it relates more directly to computing [1, 2]. Basically, Gorn defined an infinite treelike domain in which we embed trees as labeled, directed structures. Our objective in this chapter is to discuss Gorn's concept and its relationship to three computer problems: searching, subroutines, and theorem proving.

2. TREE DOMAINS

Let $\mathcal{N}$ be a numeral system for natural numbers, and let . be a symbol not in $\mathcal{N}$. For example, $\mathcal{N}$ might be the decimal numerals (which we shall use), binary or hexadecimal numerals, or even Roman numerals. The *universal address set* $\mathcal{U}$ is a set of symbols constructed from $\mathcal{N} \cup \{.\}$ as follows.

1. $0 \in \mathcal{N}$ is the *root address* of $\mathcal{U}$. ($0 \in \mathcal{U}$.) If $a \in \mathcal{U}$, then $a.0 = 0.a = a$.
2.   If $a \in \mathcal{U}$, and $i \in \mathcal{N}$, then $a.i \in \mathcal{U}$. Moreover, if $i \neq 0$, $a.i$ is an *immediate successor* of a, and a is the *immediate predecessor* of $a.i$.
3.   An address b is a *successor* of an address a if there is an integer $i \in \mathcal{N}$ and an address $c \in \mathcal{U}$ (possibly $c = 0$) such that $b = a.i.c$. If b is a successor of a, then a is a *predecessor* of b.

We define a *depth function* d on $\mathcal{U}$ by

1. $d(0) = 0$,
2.   for $a \in \mathcal{U}$, $i \neq 0 \in \mathcal{N}$, $d(a.i) = d(a)+1$;

and a *partial order* $\leqslant$ by $a_1 \leqslant a_2$ if and only if there is an address $a_3 \in \mathscr{U}$, such that $a_2 = a_1 . a_3$. As usual, we write $a_1 < a_2$ if $a_1 \leqslant a_2$ and $a_1 \neq a_2$. In addition, there is a *lexicographic order* $<_{\mathscr{L}}$ defined by $a_1 <_{\mathscr{L}} a_2$ if and only if either

(1) $a_1 < a_2$, or
(2) $a_1 - b.i.c_1, a_2 = b.j.c_2$, and i is numerically less than j. (We permit $b = 0$, as well as $c_1 = 0$ or $c_2 = 0$.)

We may think of the universal address set as the infinite tree structure represented in Fig. 2.1. The depth function corresponds to the layers in

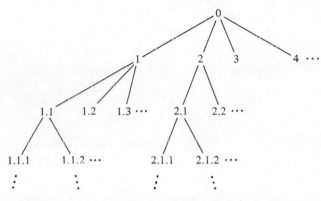

Figure 2.1

the tree. Thus the top layer (0) is at depth 0, the second layer $(1, 2, 3, \ldots)$ at depth 1, and so forth. The partial order $<$ is shown by the lines in the figure, with an address being less than ($<$) all addresses below and connected to it. Thus 2.1 is less than ($<$) 2.1.1, 2.1.2, 2.1.1.2, and 2.1.3.1.4, among others. The lexicographic order $<_{\mathscr{L}}$ imposes a left-to-right order in addition to the top-to-bottom one. Since this is a highly complex order for the entire universal address set, and since we will use it only with trees and tree domains (for which it is much simpler), we shall postpone description of it momentarily.

A set of addresses $A_1 \subset \mathscr{U}$ is an *independent set* if no address in A_1 is the successor of any other address in A_1.

A *complete end point set* is an independent set of addresses A_1 such that if $a.i \in A_1$ or if $a.i$ is a predecessor of some element in A_1, and if $0 < j < i$ in $\mathscr{N}$, then either $a.j$ or some successor of $a.j$ is also in A_1.

Example 2.1

If S is a complete end point set and $2.3.2 \in S$, what other elements of $\mathscr{U}$ are in S?

Solution.   Since $2.3.2 \in S$ and $0 < 1 < 2$, the address 2.3.1 or a successor of this address is in S. Since 2.3 is a predecessor of an element of S, the addresses 2.1 and 2.2 or successors of these must be in S. Finally, since 2 is a predecessor of an element of S, the address 1 or some successor is in S. See Fig. 2.2.

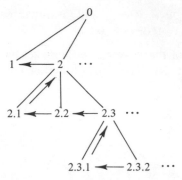

Figure 2.2

ANSWER.   The addresses $1, 2.1, 2.2$, and $2.3.1$ or successors of these are in S.

Note that if $1.4.10.2$ is an element of the set S of the example (as a successor of 1), this implies that another group of addresses is in S, namely, 1.1, 1.2, 1.3, $1.4.1$, ..., $1.4.9$, and $1.4.10.1$, or successors of these. Note also that from the given information it is impossible to determine whether the addresses $3, 4, ...$, or any successors of these are in S.

A finite set of addresses $D \subset \mathcal{U}$ is a *tree domain* if and only if

(1)   whenever $a_1 \in D$ and $a_2 < a_1$, then $a_2 \in D$; and
(2)   whenever $a.i \in D$ and $j < i$ in N, then $a.j \in D$.

A *maximal address* of a set S of addresses is an address $a \in S$ such there is no address $b \in S$ with the property that $a < b$. A finite set of maximal addresses determines a minimal tree domain (Exercise 2).

Example 2.2

Consider the set of addresses $A_1 = \{1.1.1, 1.1.2, 1.1.3, 1.2.1.1, 1.2.1.2, 1.2.2, 2.1, 3.1.1, 3.1.2, 3.2\}$. Show that A_1 is a complete end point set. Describe the tree domain D generated by A_1, and the orders $<$ and $<_{\mathscr{L}}$ on D.

Solution. Note first that A_1 is an independent set of addresses; that is, no successor of an address in A_1 is in A_1. To show that it forms a complete end point set, check the condition of the definition for each element of A_1 and each predecessor of the elements of A_1. The predecessors of elements of A_1 are 0, 1, 1.1, 1.2, 1.2.1, 2, 3, and 3.1. Because of the condition $0 < j < i$, the address 0 and the addresses of the form $a.1$ need not be checked. For the remaining addresses, Table 2.1 shows the required

TABLE 2.1

| | Required addresses | |
Test address	in A_1	Predecessors to A_1
1.1.2	1.1.1	
1.1.3	1.1.1, 1.1.2	
1.2		1.1
1.2.1.2	1.2.1.1	
1.2.2	1.2.1.1 or 1.2.1.2	
2		1
3		1, 2
3.1.2	3.1.1	
3.2	3.1.1 or 3.1.2	

addresses. Since each required address is either in A_1 or predecessor to some address in A_1, the set A_1 is a complete end point set. Hence the set A_1 together with all predecessors of addresses in A_1 constitute a tree domain (Fig. 2.3).

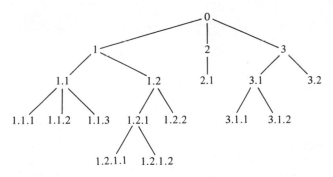

Figure 2.3

The partial order $<$ describes all paths from 0 to the end point set:

$0 <1 <1.1 <1.1.1$
$0 <1 <1.1 <1.1.2$
$0 <1 <1.1 <1.1.3$
$0 <1 <1.2 <1.2.1 <1.2.1.1$
$0 <1 <1.2 <1.2.1 <1.2.1.2$
$0 <1 <1.2 <1.2.2$
$0 <2 <2.1$
$0 <3 <3.1 <3.1.1$
$0 <3 <3.1 <3.1.2$
$0 <3 <3.2$

The lexicographic order $<_\mathscr{L}$ completely orders the tree domain. Note that the leftmost subdomains at each level come first in this order. Thus the tree (in the sense of Chapter 2) at 1 precedes the tree at 2; that at 1.1 precedes that at 1.2; and so forth. The complete lexicographic order is

$0 <_\mathscr{L} 1 <_\mathscr{L} 1.1 <_\mathscr{L} 1.1.1 <_\mathscr{L} 1.1.2 <_\mathscr{L} 1.1.3 <_\mathscr{L} 1.2 <_\mathscr{L} 1.2.1$
$<_\mathscr{L} 1.2.1.1 <_\mathscr{L} 1.2.1.2 <_\mathscr{L} 1.2.2 <_\mathscr{L} 2 <_\mathscr{L} 2.1 <_\mathscr{L} 3 <_\mathscr{L} 3.1$
$<_\mathscr{L} 3.1.1 <_\mathscr{L} 3.1.2 <_\mathscr{L} 3.2.$

We see from this example that the lexicographic order extends down the leftmost branch of the tree domain, then the next branch to the right, then the second branch to the right, and so on. (Imagine this for the universal address set.) We shall encounter this order again in Section 4 and during our discussion of the Polish notation in Chapter 9.

The effect of this definition of a tree domain is to embed any tree in the universal address set as far left as possible. Note however that this does not autonatically make the leftmost branch the longest. Thus Figs. 2.4a and b are both legitimate tree domains.

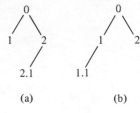

(a)   (b)

Figure 2.4

EXERCISES

1. Show that the maximal addresses of a tree domain form a finite complete end point set; conversely, a finite complete end point set and all its predecessors form a tree domain.

2. Let M_0 be a finite set of maximal addresses. Show that M_0 defines a tree domain D_0 such that if M is any complete end point set containing M_0, then the tree domain D defined by M contains D_0.

For each complete end point set, describe the generated tree domain D and the orders $<$ and $<_{\mathscr{L}}$ on D:

3. $A = \{1.1.1.1, 1.1.2, 1.2.1.1.1, 1.2.2.1, 1.3.1, 2.1\}$.

4. $A = \{1, 2.1, 3.1.1, 4.1.1.1, 5.1.1.1.1\}$.

5. $A = \{1.1.1.1.1, 1.1.1.2, 1.1.2.1.1, 1.2.1.1, 1.2.2, 1.3, 2.1, 2.2.1, 3.1.1, \\ 3.2.1.1, 3.2.1.2.1\}$.

6. Show that $\leqslant$ is a partial order on the universal address set.

7. Show that $<_{\mathscr{L}}$ is a linear order on the universal address set.

8. Write a program to accept a set of at most 50 addresses, having at most 20 characters each, and determine if the set is a complete end point set.

9. (cont.) For each complete end point set, extend the program to generate its tree domain.

10. Write a program to determine $<$ and $<_{\mathscr{L}}$ on an address set (same specifications as in Exercise 8).

3. TREES

What we have now is a "labeled tree" in the classical sense of the term. However, we wish to view it as an address set onto which we will fit a labeled tree.

Let us consider the language of ordinary algebra. Within the language we have numerical constants, variables, and symbols for the various arithmetic operations such as $+$ and $\times$. Whenever we see an algebraic expression, we know that we should expect to find two operands associated with each occurrence of $+$ or $\times$, and that variables and constants, not being operators, have no "operands" associated with them. The minus sign causes a little trouble since it is used in two ways: to indicate negative numbers (for example, -3.1), for which there is one operand; and to indicate subtraction $(5-3.1)$, for which we expect two operands. However, we can, if we desire, agree to always use the minus sign with only one operand, and to indicate subtraction by $5+(-3.1)$. Then each symbol in our language will have a fixed number of operands (0, 1, or 2) associated with it.

We say that the ordered pair $\langle S, f \rangle$ is a *simply stratified alphabet* if S is a finite or denumerable set and f is a single-valued function, called the *stratification function*, from S into the natural numbers. In effect f assigns to each operator the number of operands associated with it.

Thus the language of ordinary algebra, if we assume only a denumerable set of symbols and adopt the unary convention for the minus sign, becomes a simply stratified alphabet by the mapping

$$f: S_1 \to 2, \qquad f: S_2 \to 1, \qquad f: S_3 \to 0,$$

where $S_1 \in \{+, \times, \div,$ exponentiation $(x^y)\}$, $S_2 \in \{-\}$, $S_3 \in \{$numerical constants and variables$\}$.

We ignore the slight difficulty caused by the parentheses necessary in ordinary algebra. As we shall see in Section 4, a change in notation eliminates the need for parentheses.

If we allow the dual use of the minus sign, or introduce a summation symbol $\sum$ with an arbitrary number of operands, then this language is not simple stratified under a function f that assigns to each symbol the number of operands it has.

In $a \in D$, a tree domain, the *ramification of a in D* is the number $r(a; D)$ of immediate successors of a in D. Let $\mathscr{A} = \langle S, f \rangle$ be a simply stratified alphabet. A *(labeled) tree over* $\mathscr{A}$ is any mapping $t: D \to \mathscr{A}$, where D is a tree domain, such that for each $a \in D, r(a; D) = f(t(a))$. Thus a labeled tree over $\mathscr{A}$ is an assignment of labels to the addresses of a tree domain in such a way that the stratification of the assigned label matches the ramification of the address. Simply put, the structure of the tree must match the structure of the alphabet. If $+$ is a binary operator in the alphabet, then each occurrence of $+$ in a tree must have exactly two immediate successors, representing its operands.

Example 3.1

Determine a tree domain for the expression $(4x + y) \cdot (3 + xy)^2$.

Solution. Use an arrow ($\uparrow$) to denote exponentiation, and an asterisk ($*$) to denote multiplication. The conventional tree for the expression is then Fig. 3.1. The root address in the tree domain is 0, corresponding to the

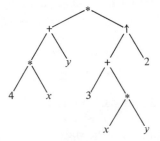

Figure 3.1

major operator ($*$). The operators $+$ and $\uparrow$ at the next level have addresses 1 and 2, respectively. Proceed in this manner until the tree domain is fully analyzed.

ANSWER. See Fig. 3.2.

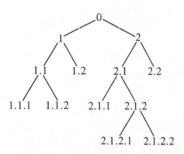

Figure 3.2

In order to discuss trees and the operations on them we shall need several terms. These will be defined for tree domains, with the corresponding definitions for trees left as an exercise.

A *scope set* is an address set of the form $a.D$, where D is a tree domain. If D is a tree domain and $a \in D$, then the *scope of a in D*, $S a D$, consists of a and all its successors in D. Clearly $S a D$ is a scope set. Note that if $a \notin D$, we define $S a D = \emptyset$; also $S 0 D = D$. An address $a \in D$ is an *end point of D* if and only if $S a D = \{a\}$.

Example 3.2

Find the scope of each of the addresses 1.2, 2.1, and 3 in the tree domain of Fig. 2.3.

Solution. Each scope set consists of the given address and all of its successors in D. Since 2.1 has no successors in D, $S 2.1 \; D = \{2.1\}$. Thus, 2.1 is an end point.

ANSWER. $S 1.2 D = \{1.2, 1.2.1,$
$1.2.1.1, 1.2.1.2,$
$1.2.2\};$
$S 2.1 D = \{2.1\};$
$S 3 D = \{3, 3.1, 3.1.1,$
$3.1.2, 3.2\}.$

If D is a tree domain and $a \in D$, then the *height of a in D*, $h(a;D)$, is defined:

(1) if a is an end point of D, then $h(a;D) = 0$;

(2) if the immediate successors in D of a are $a.1, a.2, ..., a.n$, then $h(a; D) = 1 + \max_{1 \le i \le n} h(a.i; D)$.

The *depth of a in D* is the same $d(a)$ that was previously defined (p. 104). The *depth of D* is $d(D) = \max d(a)$. If $S a D = a.D_1$, the *depth of the scope $S a D$* is $d(D_1)$.

Example 3.3

Find the height of each address in the tree domain of Fig. 2.3. Also find the depth of the scopes $S 1.2 D$, $S 2.1 D$, and $S 3 D$ of this tree domain.

Solution.   The simplest way to determine the height of all addresses is to work backward from the end points. The results of this procedure are displayed in Fig. 3.3, where each node is labeled with its height in D.

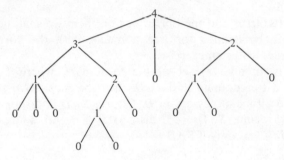

Figure 3.3

The depths of the scope sets are determined by examining the subdomains based at 1.2, 2.1, and 3.

$$\text{ANSWER.}\quad d(S 1.2 D) = 2,$$
$$d(S 2.1 D) = 0,$$
$$d(S 3 D) = 2.$$

Example 3.4

Find the height, depth, and scope of the exponentiation in $(4x + y)(3 + xy)^2$.

Solution.   In the tree domain for this expression the address of the exponentiation ($\uparrow$) is 2. Hence its depth is 1, and its scope consists of all symbols whose address in the domain is $2.a$ for some address a. Thus its scope includes all symbols at the addresses 2, 2.1, 2.1.1, 2.1.2, 2.1.2.1, 2.1.2.2, and 2.2. In terms of the original expression, the scope is $(3 + xy)^2$.

The height of $\uparrow$ in this tree T is determined by working backward from the end points 2.1.1, 2.1.2.1, 2.1.2.2, and 2.2. From this, $h(\uparrow; T) = 3$. See Fig. 3.4.

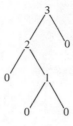

Figure 3.4

ANSWER. $h(\uparrow; T) = 3,$
$$d(\uparrow) = 2,$$
$$S \uparrow T = (3 + xy)^2.$$

The height of a symbol (address) indicates the number of levels of analysis required before the effect of the symbol can be determined. Since $h(\uparrow; T) = 3$, three levels of analysis (variable values, multiplication, addition) are required before the exponentiation can be analyzed.

EXERCISES

1. Show that if $S a_1 D = a_1 . D_1$, then $S a_1 a_2 D = a_1 . S a_2 D_1$.

Determine the tree domain for the given expression:

2. $3(x + 4) + (x^2 - 3x + 7)(y + x)$.   3. $x^3 + 3x^2 + 3x + 1$.

4. $x(x(x + 3) + 3) + 1$.   5. $(x + 1)^3$.

6. For the tree domain D shown in Fig. 3.5, find the scopes of the addresses 1.2, 3.1.1, 4.2, and 4.4 in D.

7. Find the height of each address in the tree domain of Fig. 3.5.

8. Find the depth of the scopes $S 1.2 D$, $S 3.1.1 D$, $S 4.2 D$, and $S 4.4 D$ in the tree domain of Fig. 3.5.

9. Let α be an element in a tree T. Define the scope of α in T, the height of α in T, and the depth of the scope $S \alpha T$.

Let M be a set of at most 50 addresses, having at most 20 characters each. Write a program, and test it on the tree domain of Fig. 3.5:

10. To determine the scope of each address in M.

11. To determine the height of each address in M.

12. To determine the depth of each address in M.

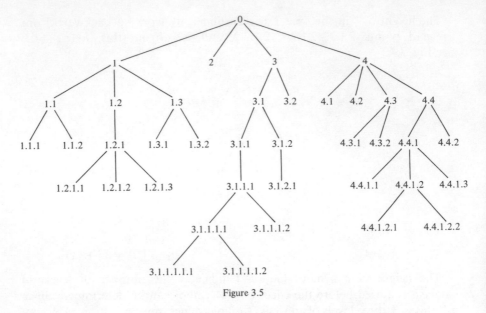

Figure 3.5

4. PREFIX REPRESENTATION AND TREE FORMS

It is often convenient to represent a tree by a linear notation, rather than by the diagrams that we have used heretofore for the tree domains. This is most conveniently done in the *prefix representation*, obtained by listing all symbol occurrences in the tree in the lexicographic order of the underlying tree domain.

Example 4.1

Develop the prefix representation for the tree of Fig. 4.1.

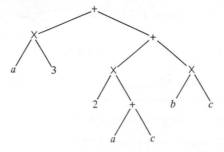

Figure 4.1

Solution. The underlying tree domain is shown in Fig. 4.2. Its lexicographic order is $0 <_{\mathscr{L}} 1 <_{\mathscr{L}} 1.1 <_{\mathscr{L}} 1.2 <_{\mathscr{L}} 2 <_{\mathscr{L}} 2.1 <_{\mathscr{L}} 2.1.1 <_{\mathscr{L}} 2.1.2$

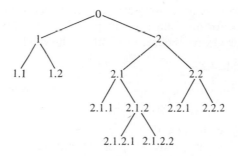

Figure 4.2

$<_{\mathscr{L}} 2.1.2.1 <_{\mathscr{L}} 2.1.2.2 <_{\mathscr{L}} 2.2 <_{\mathscr{L}} 2.2.1 <_{\mathscr{L}} 2.2.2$. The prefix representation is obtained by listing the symbol occurrences in the address order thus given: $+ \times a3 + \times 2 + ac \times bc$. Note that this amounts to scanning the tree as shown in Fig. 4.3.

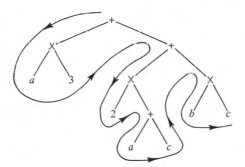

Figure 4.3

ANSWER. $+ \times a3 + \times 2 + ac \times bc.$

Prefix representation (or the equivalent postfix representation, giving a right-to-left scan) is used by many compilers for the internal representation of expressions because of the simple scanning and evaluation algorithms it permits. In scanning an expression we often find that we must lay aside a partially analyzed piece of the expression, and analyze another portion before we can complete the analysis of the first piece. For example, if we discover an *n*-ary operator at address *a*, we know that its operands are at addresses $a.1, a.2, \dots, a.n$. We may then set aside addresses $a.2, \dots, a.n$ in order to analyze address $a.1$. Similarly, we must in turn complete the analysis of addresses $a.2, \ldots, a.n$ before we can finish the analysis of the first operator. We may think of placing the pieces that we have set aside on a stack from which we retrieve them as needed. This image is particularly apt for an expression in prefix notation since the piece to be analyzed next is always the last one put on the stack.

Example 4.2

Find the tree corresponding to the prefix expression
$+ \times 3 + a \times bc + \times 7a \times b + 3c.$

Solution.   Use the fact that $+$ and $\times$ are binary operators, and the other symbols are operands. In prefix notation the symbols are in lexicographic order. Thus, from the left, $+$ is at the root address 0. It has two operands, at addresses 1 and 2. Stack the second address temporarily. The second symbol $\times$ is at address 1 and has operands at addresses 1.1 and 1.2. Again stack the second address. The third symbol 3, whose address is 1.1, is a constant and hence is an end point of the tree. Retrieve the last stacked address 1.2. It is the address of the fourth symbol. Continue in this way, producing the addresses in Table 4.1. From the reconstructed address set, draw the tree.

ANSWER.   See Fig. 4.4.

TABLE 4.1

Symbol	Address	Stack condition
$+$	0	
$\times$	1	Stack 2
3	1.1	Stack 1.2
$+$	1.2	Retrieved
a	1.2.1	Stack 1.2.2
$\times$	1.2.2	Retrieved
b	1.2.2.1	Stack 1.2.2.2
c	1.2.2.2	Retrieved
$+$	2	Retrieved, stack empty
$\times$	2.1	Stack 2.2
7	2.1.1	Stack 2.1.2
a	2.1.2	Retrieved
$\times$	2.2	Retrieved, stack empty
b	2.2.1	Stack 2.2.2
$+$	2.2.2	Retrieved, stack empty
3	2.2.2.1	Stack 2.2.2.2
c	2.2.2.2	Retrieved, stack empty

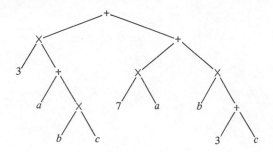

Figure 4.4

Let $\mathscr{A}_c = \langle S, f \rangle$ be a simply stratified alphabet, $\mathscr{A}_v = T$ be an unstratified alphabet. That is, no stratification function is assigned for $\mathscr{A}_v$. Form the new stratified alphabet $\mathscr{A} = \langle \mathscr{A}_c \cup \mathscr{A}_v, f' \rangle$, where $f' | \mathscr{A}_c = f$, and for $a \in \mathscr{A}_v$, $f'(a) = 0$. Any tree over $\mathscr{A}$ is a *tree form over* $\mathscr{A}_c$. Note that the elements of $\mathscr{A}_v$ can occur only at the end points of the tree. Thus we may think of a tree form over $\mathscr{A}_c$ as representing a set of trees over $\mathscr{A}_c$ with the operands (constants) of $\mathscr{A}_c$ replaced by new symbols from $\mathscr{A}_v$ (variables). A simple example is the use of the tree form in Fig. 4.5 to represent the trees in Fig. 4.6 and others.

Figure 4.5   Figure 4.6

A tree form is *normal* if its prefix representation has the property: if α_j occurs in the form, $j > 1$, then α_{j-1} also occurs, and the first occurrence of α_{j-1} precedes the first occurrence of α_j. Thus if c is a binary operator, α_1 and $c\alpha_1 c\alpha_2 \alpha_1$ are normal, whereas α_3 and $c\alpha_3 c\alpha_1 \alpha_3$ are not.

A *simple form* is a normal form of depth zero or one, all of whose end points are in $\mathscr{A}_v$. Thus the only simple forms are α_1 and forms of the type $c\alpha_1 \cdots \alpha_n$ (prefix), where $c \in \mathscr{A}_c$ has stratification n.

EXERCISES

Develop the prefix representation for the tree:

1. Fig. 4.7.   2. Fig. 4.8.

3. Fig. 4.9.   4. Fig. 4.10.

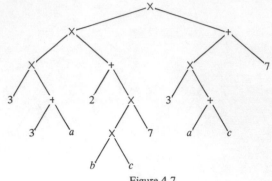

Figure 4.7

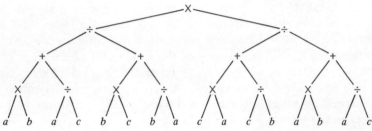

Figure 4.8

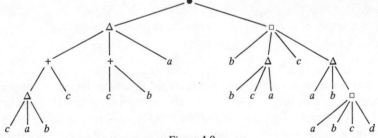

Figure 4.9

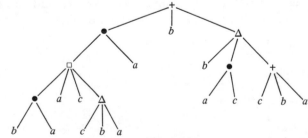

Figure 4.10

Construct the tree corresponding to the prefix expression:

5. $+ + \times 3 \times 4ac + + 7 + \times abc2$

6. $\times + \times + \times + 1a2b3c + \times + \times + \times c3b2a1a$

7. $\square \times 3 + \triangle ab + \square ac2 \times aba2abc$
    stratification   0: $2, 3, a, b, c$
                   1: none
                   2: $+, \times$
                   3: $\triangle$
                   4: $\square$

8. $\cdot + \triangle \times \square ab \triangle cda \cdot bca + acd$
    stratification   0: a, b, c, d
                   1: $\cdot$
                   2: $+, \times$
                   3: $\triangle$
                   4: $\square$

9.   Define the lexicographic order $<_{\mathscr{P}}$ corresponding to a postfix notation.

10. Write a program to accept as input

    (a) an alphabet of at most 20 symbols,
    (b) a stratification function with at most five levels,
    (c) a prefix expression of at most 50 symbols,

    and as output, produce a list of ⟨address, symbol⟩ pairs which may be used to draw the tree. Test your program on the expressions of Exercises 5–8.

*11. (cont.) Modify your program to produce as output a drawing of the tree. Test the program on the expressions of Exercises 5–8.

5. EXPLICIT DEFINITIONS

One of the more common features of large programs is the subroutine. Many of the subroutines we use are built-in "library" routines, such as the input and output routines of a language, or routines for the standard mathematical functions such as sine and square root. Others are defined by the individual programmer to suit his particular problem. But whether we use library routines or tailored subroutines, we are in effect doing the same thing. We are substituting for a perhaps large block of code with undefined ("dummy") variables, a very simple expression. This expression merely gives the name we have chosen for the block of code, and the values or names to be assigned to the dummy variables in that instance. We expect the computer to make the inverse substitution. Upon encountering the expression SQRT.(3*x) in a program, the compiler or interpreter should locate the code labeled SQRT., and prepare to substitute at run time the current value (or for some subroutines the name) of $3*x$ for the dummy variable in the code. This type of substitution, effectively equating a simple expression with a more complex program structure, is an instance of an explicit definition. An *explicit definition* of d is an equation of the form $\phi_1 =_{df} \phi_2$, where

    (1) ϕ_1 is a simple form, called the *definiendum*, whose root is d,
    (2) ϕ_2 is a form, called the *definiens*, each of whose variables occurs in ϕ_1.

For example, Figs. 5.1 and 5.2 show explicit definitions of the symbols d and k. In prefix form, these equations are $d\alpha_1\alpha_2 =_{\text{df}} +\alpha_1 +\alpha_2\alpha_1$, and $k\alpha_1\alpha_2\alpha_3 =_{\text{df}} *+\alpha_3\alpha_1 *\alpha_1\alpha_2$.

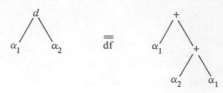

Figure 5.1

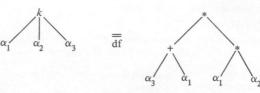

Figure 5.2

Associated with the concept of explicit definition are two operations introducing and removing defined symbols from trees. Any tree that can be obtained from a given tree by a series of introductions or removals of a defined symbol d is called d-*equivalent* to the given tree. If d is an explicitly defined symbol, then d-*elimination* is an operation replacing an occurrence of d in a tree by the definiens. If the definiendum is $d\alpha_1\cdots\alpha_n$, and the d-occurrence to be eliminated is at address a, then the symbols $\alpha_1, ..., \alpha_n$ denote the trees or tree forms rooted at $a.1, ..., a.n$, respectively.

Example 5.1

Using d as defined in Fig. 5.1, apply d-elimination to the tree F_0, shown in Fig. 5.3, obtaining d-equivalent trees.

Figure 5.3

Solution.   Apply d-elimination either at address 0 or at address 1. In the first case, obtain the tree F_1. See Fig. 5.4. There is still a choice for the next elimination. Eliminating the d at 1 results in tree F_2; eliminating the occurrence at address 2.2 results in tree F_3. See Figs. 5.5 and 5.6.

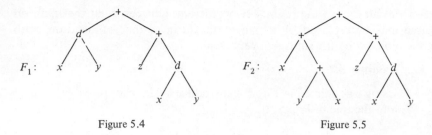

Figure 5.4   Figure 5.5

Choosing initially to eliminate the d at 1 rather than at 0 results in tree F_4. See Fig. 5.7.

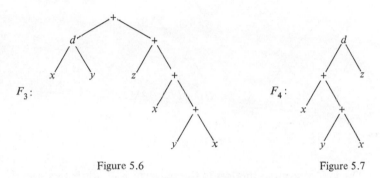

Figure 5.6   Figure 5.7

Finally, eliminating the one d remaining in F_2, F_3, or F_4 results in tree F_5. See Fig. 5.8.

This whole process may be represented by the structure shown in Fig. 5.9, where F_i is below, and joined to, F_j if and only if F_i obtained from F_j by a single d-elimination.

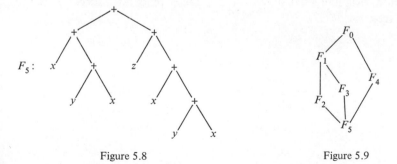

Figure 5.8   Figure 5.9

The inverse process to d-elimination is d-*introduction*. If the definiens of an explicit definition can be matched to a subtree of a given tree in

such a way that no α_i represents two different subtrees, then the matched
subtree may be replaced by a tree in the form of the definiendum, with
each α_i replaced by the subtree it represents.

Example 5.2

Using k as defined in Fig. 5.2, introduce k wherever possible
into the tree of Fig. 5.10.

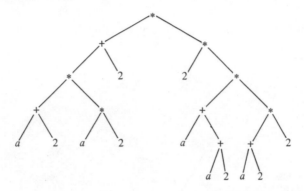

Figure 5.10

Solution.   The key to k-introduction is the pattern: $*$ at addresses a and
$a.2$, $+$ at address $a.1$, and the same tree at addresses $a.1.2$ and $a.2.1$. There
are two such patterns in the tree, with $a = 0$ and $a = 2.2$. The pattern with
$a = 1.1$ matches the $+$ and $*$ criteria, but the trees at $1.1.1.2$ and $1.1.2.1$
are not the same. Hence k may be introduced at addresses 0 and 2.2 only.
For this particular problem, the order of introduction does not matter.
Note, however, that if the first introduction is at 0, then the second
introduction, originally scheduled for 2.2, takes place at address 2 in the
modified tree.

ANSWER.   See Fig. 5.11.

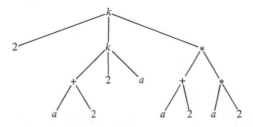

Figure 5.11

The process of d-introduction leads to genuine difficulties—in particular a fractionation of the language. The cause of this is the existence of "dominoes"—overlapping tree forms for which a d-introduction may be made in two or more ways.

Example 5.3

Discuss d-introduction for the tree F'_0 in Fig. 5.12.

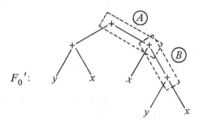

Figure 5.12

Solution. Introducing d for A yields F'_1, in Fig. 5.13. Or introducing d for B yields F'_2, in Fig. 5.14. But then it is not evident to the casual observer of the results that F'_1 and F'_2 are d-equivalent. In fact, F'_1 and F'_2 are nonisomorphic trees (even as unlabeled trees), and no sequence of solely d-eliminations or solely d-introductions will transform one into the other.

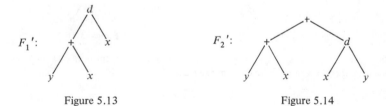

Figure 5.13   Figure 5.14

This domino phenomenon has been interpreted as implying that through simple explicit definitions, two dialects of a programming language (for example, Fortran) may develop enough differences that the relationship between them is obscured to the point that only a formal linguist would recognize it.

EXERCISES

1. Write the tree form of the explicit definition

$$h\alpha_1 \alpha_2 \alpha_3 =_{df} * + \alpha_1 * \alpha_2 \alpha_3 + * \alpha_1 \alpha_2 * \alpha_1 \alpha_3.$$

Use the definitions of d and k in Figs. 5.1 and 5.2, and the definition of h in Exercise 1 to eliminate d, h, and k from the given tree. Diagram the relationships between the generated trees:

2. Fig. 5.15.

3. Fig. 5.16.

4. Fig. 5.17.

5. Fig. 5.18.

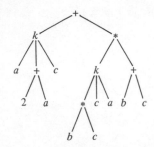

Figure 5.15

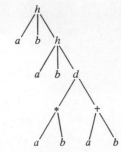

Figure 5.16

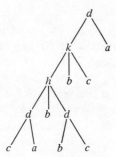

Figure 5.17

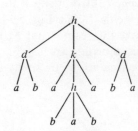

Figure 5.18

Use the definitions of d and k in Figs. 5.1 and 5.2, and the definition of h in Exercise 1 to introduce d, h, and k wherever possible in the tree:

6. Fig. 5.19.

7. Fig. 5.20.

8. Fig. 5.21.

9. *hkdabdcabdkacbhabchdbckacbhcab.*

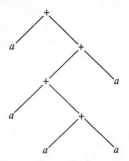

Figure 5.19

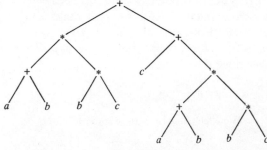

Figure 5.20

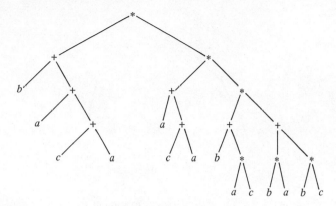

Figure 5.21

10. Show that d-equivalence is an equivalence relation for any one symbol d.

11. Show by a sequence of d-eliminations and d-introductions that the trees in Figs. 5.13 and 5.14 are equivalent—that is, that each can be transformed into the other by explicit definitions.

Show by a sequence of eliminations and introductions that the trees are equivalent. Use the definitions of d, h, and k given in this section:

12. Figs. 5.22 and 5.23.   13. Figs. 5.24 and 5.25.

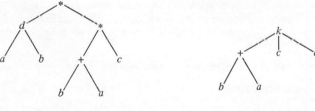

Figure 5.22   Figure 5.23

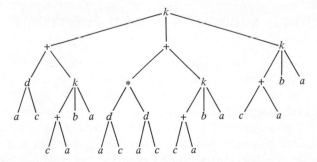

Figure 5.24

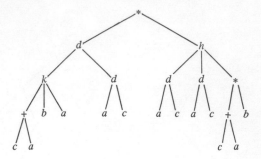

Figure 5.25

14. Show that there exist trees equivalent to the tree in Fig. 5.26 which are not obtained by pure *d*-elimination. Find all such trees.

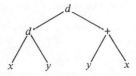

Figure 5.26

Suppose $\mathscr{A}_c$ contains at most 20 symbols, and $\mathscr{A}_v$ contains at most 10 symbols; the stratification function has at most five levels, an input prefix expression has at most 50 symbols, and an explicit definition is given as an ordered pair ⟨definiendum, definens⟩ of prefix expressions. Write the required program, assuming that no more than three explicit definitions are used at any one time:

15. To eliminate all defined symbols. Test on Exercises 2–5.

*16. To introduce defined symbols wherever possible. Test on Exercises 6–9.

*17. To determine if two given expressions are equivalent. Test on Exercises 11–13.

6. SEARCHING, SUBROUTINES, AND THEOREM PROVING

The importance of Gorn's conception of a tree lies in the clarity with which it exposes the tree structure, the underlying addressing scheme, and the contents of the tree vertices. Also of significance in understanding program and subroutine relationships is the concept of explicit definition.

Suppose we have a set of objects to search, to determine whether a given object is included. For example, the set may be a collection of documents, each having certain descriptors, and we wish to determine if the collection contains a document with certain given descriptors. This is a basic information retrieval problem. We could extract a document at random from the set, examine it, and either keep it as satisfying the search query, or reject it and look for another. However, if the set did not contain the required document we would have to search the entire set to determine that fact.

The only way to improve materially the time required for a search is to structure the set in some way, and use this structure in devising a search strategy. Not all structures are helpful, however. For example, books and other documents are often given "accession numbers" when acquired for a collection. Thus it is a simple matter to sequentially order a collection of books by accession number. Yet this generally does not aid a searcher since he will have no idea of the accession number of the document he requires. The best he can hope for is a sequential search of the file.

However, an alphabetically sequenced file of names will materially reduce the search time required to determine whether a given name is on file since massive portions of the file may be quickly eliminated. For example, if the desired name begins with K, then no names beginning with any other letter need be searched. A common example of this is the telephone directory. The directory could be sequenced by telephone number, but this would be useless in searching to find the number of a known person. Yet we make use of the alphabetic order of the directory to quickly search for a desired phone number.

Utilizing a more complex order on the set of documents will further reduce the required search effort. A hierarchical or tree structure is often the basis for this type of order, as in classification schemes used within libraries. The tree structure that we have defined in this chapter illustrates the roles which the addresses and contents of the tree play in such a search. The addressing structure, that is, the underlying tree domain, is essentially the classification scheme which is used in organizing the document set. These addresses are independent of the contents of each node. In another example, a business or industry may be organized in a certain hierarchical structure, with a president at the top, several vice presidents under him, and further administrative personnel lower in the tree structure. The addressing scheme in this case consists of the titles of the company officers. The contents, that is, the names of the people who actually are the officers, have nothing to do with the organizational structure per se.

Returning to the document search, the lexicographic order that we have discussed earlier in the chapter is not of significant value here. Following such a structure would impose a linear order on the entire tree and hence reduce our search to the previous pattern. Rather, we should think of each node of the tree as a decision point. Arriving at a node, we compare

our given descriptors with the descriptors in that node and then take one
or another branch of the tree, depending upon the relationship between
our descriptors and those that we find in the tree.

The strictly hierarchical or treelike structure has limitations in the
organization of data sets. For example, it does not allow for cross references
between various branches of the tree. Nevertheless, many data sets are
organized basically upon such a tree structure, with additional cross-links
added if this is necessary.

We have referred in Section 5 to the relationship between explicit
definition and the computer subroutine. While a subroutine may or may
not have a treelike structure, the concept of both naming the subroutine, and
supplying dummy variables for it, is precisely the concept of explicit
definition. In defining a subroutine we take an arbitrary structure, which
may be a tree, and replace it by a simple tree form, having a name at
the root, and the dummy variables listed following the name in prefix
notation. If we are, in fact, naming a tree by this process, then the process
of d-elimination causes no problem, although as we have seen, d-introduction
may lead to divergent forms for the same tree. If the structure being
defined by explicit definition is more complex than a tree, then the
difficulties that d-introduction causes could also arise through d-elimination.

One other problem that generally assumes a treelike form is the problem
of theorem proving. While we shall consider this in more detail in
Chapter 9, here we should note that the proof of a theorem corresponds
to the development of a tree from the end points toward the root. In the
schematic shown in Fig. 6.1, from statement E we may prove statement

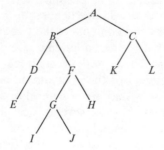

Figure 6.1

D. Similarly, from statements I and J we may prove statement G, and from
that and H derive statement F. Then having both D and F we may derive
statement B. On the other branch, statements K and L enable us to deduce
statement C, and from that and statement B we can finally prove statement
A. Recognition of this tree structure is at times the key to the develop-

ment of efficient theorem proving programs. Of particular interest is the case when each statement is a characterization of a linguistic structure. In this case the "theorem" at A is "S is a sentence," where S is the sequence of words at the end points of the tree ($EIJHKL$). For example, if statement I were an article and J were an adjective, then G would be an adjectival phrase. If then statement H were a noun, F would be a noun phrase. This particular interpretation of the tree structure is called a *parsing tree*. The terms *top-down* and *bottom-up parsing* are used to indicate an approach to the parsing problem from either the root or the end points of the tree, respectively.

EXERCISE

1. Consider a binary search of words, that is, a search in which each question has only a yes or no answer. Show that while a single letter of the alphabet requires at most five questions to determine, and sequences of two or three letters require at most 10 or 15 questions, respectively, a sequence of four letters requires only 19 questions (not 20) to determine.

REFERENCES

1. S. Gorn, Explicit definitions and linguistic dominoes, *in* "Systems and Computer Science" (J. F. Hart and S. Takasu, eds.), pp. 77–115. Univ. of Toronto Press, Toronto, Canada, 1967.
2. S. Gorn. Handling the growth by definition of mechanical languages. *Proc. 1967 Spring Joint Computer Conf.*, pp. 213–224, Thompson Books, Washington, D.C.

CHAPTER 4

Directed Graphs

1. INTRODUCTION

In Chapter 2 we developed many of the basic concepts of the theory of undirected graphs, and examined some of the applications of these to computing, as well as the use of computers in solving some problems in undirected graphs. In the discussion of Gorn trees in Chapter 3 we imposed, through the addressing scheme and its associated orders, an orientation or direction upon the basic tree structure which we had defined in the previous chapter. In the present chapter we wish to look at more general directed graphs, discuss the associated concepts, and examine some of the uses of these graphs.

2. BASIC DEFINITIONS

To facilitate comparison with undirected graphs, the organization of this section will be similar to that of Section 2 in Chapter 2. Let $G = \langle V, A, f \rangle$, where $V = \{v_i\}$ is a set of vertices of G, $A = \{a_j\}$ is a set of *arcs* of G, and

$f: A \rightarrow V \otimes V$ is a function mapping arcs into *ordered* pairs of vertices. Then G is called a *directed graph* or *digraph*. For $a \in A$, if $f(a) = \langle v, w \rangle$, we say that a is *incident from* v and *incident to* w. Note that if for $b \in A$, $f(b) = \langle w, v \rangle$, then b is a different arc from a, since the direction of b is opposite that of a. That is, the order in which the vertices are written in a pair is significant in the definition of directed graphs. We generally draw directed graphs with arrows on the arcs, indicating the direction of the arcs.

Example 2.1

Draw the directed graph with vertices $V = \{v_1, ..., v_7\}$, arcs $A = \{a_1, ..., a_9\}$, and f defined by Table 2.1.

TABLE 2.1

a	$f(a)$
a_1	$\langle v_1, v_2 \rangle$
a_2	$\langle v_2, v_3 \rangle$
a_3	$\langle v_3, v_2 \rangle$
a_4	$\langle v_3, v_4 \rangle$
a_5	$\langle v_5, v_4 \rangle$
a_6	$\langle v_4, v_6 \rangle$
a_7	$\langle v_5, v_7 \rangle$
a_8	$\langle v_6, v_7 \rangle$
a_9	$\langle v_7, v_2 \rangle$

ANSWER.   See Fig. 2.1.

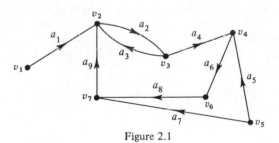

Figure 2.1

While in the study of undirected graphs multiple edges have little significance, in work with directed graphs they are often quite significant. Multiple arcs joining two vertices may be either *parallel*, that is, directed in the same direction, or *antiparallel*, that is, directed in opposite direction. Since the directions on the arcs are often used to indicate a flow of information or material, it is important to permit the use of multiple arcs in directed graphs. Note that the graph in Fig. 2.1 has a pair of antiparallel arcs joining v_2 and v_3.

In a directed graph a *loop* is an arc beginning and terminating at the same vertex $\langle v, v \rangle$. As with undirected graphs, we shall assume that directed graphs are loop-free.

The concepts of *walk*, *trail*, *cycle*, and *path* are defined as for undirected graphs, with the additional proviso that the arcs involved be consistently directed. Thus, in Fig. 2.1, the vertex sequence $v_2, v_3, v_4, v_6, v_7, v_2$ constitutes a cycle. However, the sequence v_5, v_4, v_6, v_7, v_5 does not constitute a cycle since the arc a_7 is directed in the wrong direction. Similarly, in this particular graph there is no path from v_7 to v_1 because of the adverse direction of arc a_1.

The concept of connectivity, while simple for undirected graphs, becomes more complex with directed graphs. For example, in some sense the graph of Fig. 2.1 is connected. However, as we just stated, there does not exist a path from v_7 to v_1, so that we cannot use this as a criterion of connectedness. Connectedness in the sense exemplified by the graph of Fig. 2.1 is termed "weak." A directed graph is *weakly connected* if the same graph as an undirected graph, that is, ignoring the directions of the arcs, is a connected graph. A directed graph is *unilaterally connected* if, given any two vertices of the graph, there exists a path from one vertex to the other, although a reverse path does not necessarily exist. Finally, a directed graph is *strongly connected* if given any two vertices of the graph v and w, there exists a path from v to w, and also a path from w to v. Thus, while the graph of Fig. 2.1 is only weakly connected, that of Fig. 2.2 is unilaterally connected, and the graph of Fig. 2.3 is strongly connected. Note that in a strongly connected graph, the path from v to w together with the path from w to v forms a circuit which may be a cycle, but which may also contain repeated vertices or edges. Thus the graph of Fig. 2.4 is strongly connected, although the circuit from v to w and back is not a cycle in the strict sense of the word.

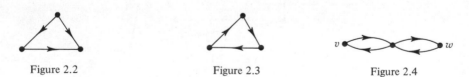

Figure 2.2   Figure 2.3   Figure 2.4

The *degree* of a vertex is defined in the same manner as it is for undirected graphs. However, because of the fact that arcs may be directed into or out from a given vertex, it is possible to split the degree of a vertex into two numbers called the *indegree*, which is the number of arcs incident to a vertex, and the *outdegree*, which is the number of arcs incident from a vertex. In particular, a vertex is called a *source* if its indegree is 0, that is, if all arcs emanate from it; or a *sink* if its outdegree is 0, that is, all arcs enter it. The concepts of source and sink are important in the study of networks (Section 6).

EXERCISES

Determine if the graph is weakly, unilaterally, or strongly connected:

1. Fig. 2.5.   2. Fig. 2.6.

3. Fig. 2.7.   4. Fig. 2.8.

5. Fig. 2.9.   6. Fig. 2.10.

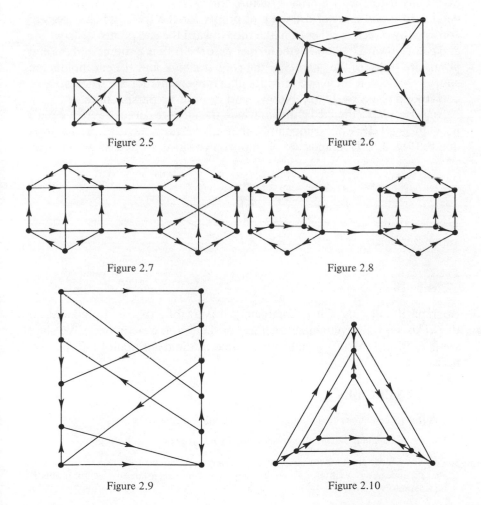

Figure 2.5   Figure 2.6

Figure 2.7   Figure 2.8

Figure 2.9   Figure 2.10

7. Let G be a digraph of at most 100 arcs, specified by an $n \times 2$ array, $A = (a_{i,j})$, with $\langle a_{1,j}, a_{2,j} \rangle$ representing the arc from vertex $a_{1,j}$ to vertex $a_{2,j}$. Write a program to determine the type of connectivity of G. Test the program on Exercises 1–6.

3. SPECIAL CLASSES OF GRAPHS

Although regular, complete, and bipartite directed graphs can be defined, they are not so important for our purposes as are the corresponding types of undirected graphs. However, directed trees, called *arborescences*, are important, as are the graphs known as isographs. An arborescence is essentially a tree with Gorn's structure imposed on it. That is, there is a defined root and a set of defined end points for the tree, and the arcs are consistently directed, either from the root toward the end points, or from the end points toward the root. In the former case the root is a source and the end points are sinks; in the latter case the root is a sink and the end points are sources. Note that these two directional arrangements for an arborescences correspond respectively to top-down and bottom-up parsing trees.

An *isograph* is a directed graph in which the indegree of each vertex equals its outdegree. This does not imply that all vertices have the same total degree (Fig. 3.1). Isographs are of importance since a directed graph is an

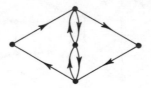

Figure 3.1

isograph if and only if it is completely traversable, that is, if and only if there exists a walk containing each arc of the graph exactly once. We shall see in Section 11 that isographs play a role in the definition of information networks.

EXERCISES

1. Under what conditions does a digraph have a spanning arborescence?

2. Show that the undirected graph corresponding to any isograph is an Euler graph.

3. Given any Euler graph on n vertices, show that any orientation of the arcs in such a way that the indegree of each of $n-1$ vertices equals the outdegree for that vertex results in an isograph.

4. Let G be an isograph such that the degree of each vertex is at least 4. For any two vertices v and w, do there always exist paths from v to w and from w to v which are independent, that is, have no vertices in common other than v and w? If not, what conditions are required on G so that such paths exist? (This problem relates to information network design.)

4. MATRIX REPRESENTATION OF DIRECTED GRAPHS

Because of the direction associated with the arcs of a digraph, the matrix representation of digraphs is somewhat different from that of undirected graphs. Assuming that we have no parallel arcs within our graph, we may use $+1$ together with -1 to indicate the direction of the arcs involved. For example, the *incidence matrix* for a directed graph G with m vertices and n arcs is an $m \times n$ matrix $M = (m_{ij})$ such that

$$M_{ij} = \begin{cases} 1 & \text{if } a_j \text{ is incident to } v_i, \\ -1 & \text{if } a_j \text{ is incident from } v_i, \\ 0 & \text{otherwise.} \end{cases}$$

A *(vertex) adjacency matrix* for a directed graph G is an $m \times m$ matrix $A = (a_{ij})$ such that

$$a_{ij} = \begin{cases} 1 & \text{if } \langle v_i, v_j \rangle \text{ is an arc of } G, \\ 0 & \text{otherwise.} \end{cases}$$

Note that the adjacency matrix allows representation of antiparallel, but not parallel, arcs. For example, matrices for the graph of Fig. 2.1 are

$$M = \begin{bmatrix} -1 & 0 & 0 & 0 & 0 & 0 & 0 & 0 & 0 \\ 1 & -1 & 1 & 0 & 0 & 0 & 0 & 0 & 1 \\ 0 & 1 & -1 & -1 & 0 & 0 & 0 & 0 & 0 \\ 0 & 0 & 0 & 1 & 1 & -1 & 0 & 0 & 0 \\ 0 & 0 & 0 & 0 & -1 & 0 & -1 & 0 & 0 \\ 0 & 0 & 0 & 0 & 0 & 1 & 0 & -1 & 0 \\ 0 & 0 & 0 & 0 & 0 & 0 & 1 & 1 & -1 \end{bmatrix},$$

$$A = \begin{bmatrix} 0 & 1 & 0 & 0 & 0 & 0 & 0 \\ 0 & 0 & 1 & 0 & 0 & 0 & 0 \\ 0 & 1 & 0 & 1 & 0 & 0 & 0 \\ 0 & 0 & 0 & 0 & 0 & 1 & 0 \\ 0 & 0 & 0 & 1 & 0 & 0 & 1 \\ 0 & 0 & 0 & 0 & 0 & 0 & 1 \\ 0 & 1 & 0 & 0 & 0 & 0 & 0 \end{bmatrix}.$$

Observe that for digraphs, the column sums of the incidence matrix are all zero and the row sums are the difference between the indegree and the outdegree, a rather useless statistic. In the adjacency matrix, the symmetry that is present for undirected graphs is absent here. However, in compensation, we find that the row sums are the outdegrees of the vertices, while the column sums are the indegrees.

EXERCISES

1. How are sources and sinks characterized in the incidence and adjacency matrices?

2. Let A be the adjacency matrix for a directed graph. If the graph is an arborescence, what can be said about A and its powers? Do these properties characterize an arborescence? That is, if A and its powers have the properties you describe, is the graph necessarily an arborescence?

3. (cont.) Answer the same questions for a directed graph that has no cycles but is not an arborescence.

4. Let A be the adjacency matrix of an isograph on n vertices with row sums $\{r_i\}$ and and column sums $\{c_i\}$. Show by example that even though $r_i = c_i, i = 1, ..., n$, the matrix A is generally not symmetric.

   Let G be a finite digraph of at most 20 vertices and 100 arcs. Write the required program, assuming the input to be either an incidence or an adjacency matrix for G, a code to indicate which matrix is supplied, and the size of the matrix:

5. To convert from the given matrix to one of the other form (adjacency or incidence).

6. To locate the sources and sinks of G.

7. To determine if G is an arborescence.

8. To determine if G is an isograph.

*9. If G is an isograph, to find independent paths from v to w and w to v, if they exist, for all vertex pairs (v, w).

10. To find a spanning arborescence of G, if one exists.

11. Show that in a directed acyclic graph it is always possible to number the nodes so that v precedes w if there is an arc from v to w. What does this imply about the adjacency matrix for such a graph?

5. FLOWCHARTS

For the computer programmer, the most commonly seen directed graphs are the flowcharts of his programs. In virtually all cases the flowchart is a graph with a single source and one or more sinks. That is, there is generally a single entry point for a program, and usually, but not always, a single exit statement for the program. This directed graph is labeled on both its vertices and arcs. The vertex labels consist of both addresses and contents.

In terms of the flowchart, the addresses are the line numbers for the statements within the program, and possibly specific statement labels. The contents of a node consist of the one or more program statements that are designated by that particular box in a flowchart, be they control statements, assignment statements, or some other type of statement allowed by the particular programming language. The arc labels, when they exist, are related to branch points within the flowchart, and define the conditions under which program flow follows a particular path.

The labeling of a flowchart is of utmost importance. Consider, for example, Fig. 5.1. If the vertices labeled A and B denote Fortran DO statements, and vertices C and D denote the ends of the respective loops, then the structure shown is an illegal one. However, if A and B do not denote DO statements, and C and D are conditional transfers of control back to A and B, the structure shown is perfectly legal. The same is true in Algol and many other languages. Thus, while the unlabeled digraph underlying a flowchart contains some information, the labels are vital for proper interpretation of the graph.

It is entirely reasonable to consider a hierarchy of directed graphs as a representation of a program, each successive level of the hierarchy being a more detailed representation of the program. For example, a vertex in a level 1 flowchart for a program might contain the statement, "Compute the variance of x." At level 2, this statement is the name of a digraph (flowchart) representing an algorithm for computing the variance of x. Each vertex at this level is a step in the algorithm, perhaps an arithmetic assignment statement. Level 3 digraphs would show the analysis of these statements into machine or assembly language constructs; level 4 might involve microprogramming,

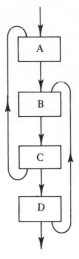

Figure 5.1

and so forth. By this hierarchical arrangement it is possible to suppress details when they need not be considered, or to concentrate on a particular detail, effectively blotting out the larger context in which it occurs. Note the analogy between this hierarchical leveling and the process of explicit definition.

Code optimization involves a series of transformations on the flowchart of a program. For example, statements not involving program loop variables may be moved outside of loops to increase program efficiency. Common subexpressions may be recognized, either within a statement (thus affecting the next hierarchical level) or between two or more statements. Decisions must then be made either to store computed values for these subexpressions, or to recompute them as needed. This recognition and decision process is much akin to the *d*-introduction process, and suffers from the same pitfalls.

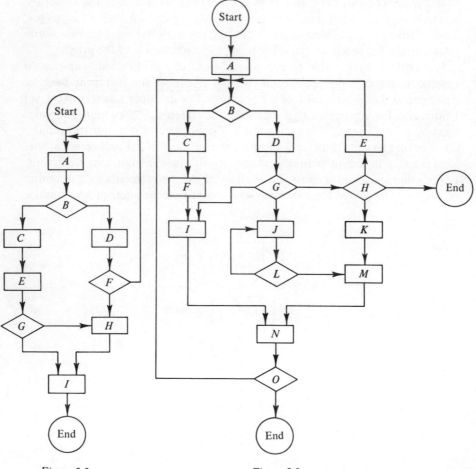

Figure 5.2   Figure 5.3

EXERCISES

One problem in debugging a program is to determine if all possible paths have been examined. For the flowchart in the figure determine all possible paths from START to END. Assume that each loop is either bypassed or traversed only once each time it is encountered:

1. Fig. 5.2.   2. Fig. 5.3.

6. NETWORKS

As is the case with most terms in graph theory, the word *network* has been used in many ways by various authors, to denote everything from an arbitrary graph to a directed graph with several constraints upon it. We shall define a *network* to be a finite directed graph, without loops, having assigned to each edge a numerical label, called the *capacity*. Intuitively, each vertex or node of the graph represents a location or event, such as a warehouse, a library, a computer program step, the completion of a certain phase of construction, or the like. The arcs in a network represent channels of flow for information, control, cars, water, manufactured goods, money, and so on. In many applications, the vertices may be considered to represent events, while the arcs represent activities leading from one event to another. For certain network problems, it is important that there be a single source and a single sink; for other problems, it is equally important that there be no source and no sink. (If there are multiple sources and sinks, these can be artificially combined to single sources and sinks.)

The capacity assigned to each arc of a network is related to the concept of flow within the network. Specifically, the capacity along a given arc is considered to be an upper bound on the permissible flow along that arc. While there is no requirement that the capacities be integers, for many applications integer capacities are sufficient. A typical problem is to maximize the total flow through a network subject to the condition that the flow along any arc not exceed that arc's capacity. Algorithms to solve this problem are generally iterative. Hence the use of noninteger capacities may result in accumulation of sufficient round-off error that the program will not halt. However, it can be shown that use of integer capacities results in an integer-valued maximal flow. In this case algorithms for maximal flow can be forced to satisfactory termination.

Consider the network shown in Fig. 6.1. This network has a single source s and a single sink t. The integers labeling the arcs of the graph are the capacities for each individual arc. For example, the arc sA has a capacity of 6 units. This is interpreted to mean that 6 units of information, money, machinery, or whatever, may flow along that particular arc. While there is no requirement that the capacities of the arcs entering a node total the

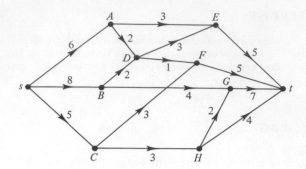

Figure 6.1. Network capacities.

same as the capacities of the arcs leaving the node, there is a requirement that the actual flow entering a node be the same as the amount of flow leaving the node. Thus, for example, since the capacities of the arcs leaving node A are 3 and 2, totaling to 5, even though the capacity of the arc sA entering A is 6, at most 5 units can flow into A during any actual flow since at most 5 units can leave A.

Let N be a network with source s and sink t, and let $c(i,j)$ be the capacity of arc $\langle i,j \rangle$ in the network. A *flow* in the network is an assignment of nonnegative real numbers $f(i,j)$ to each arc of the network, satisfying the conditions

(1) for each arc $\langle i,j \rangle$, $0 \leqslant f(i,j) \leqslant c(i,j)$;
(2) there is a real number v such that for each vertex i of the network

$$\sum f(k,i) - \sum f(i,j) = \begin{cases} -v & \text{if } i = s \\ 0 & \text{if } i \neq s, t \\ v & \text{if } i = t, \end{cases}$$

where the sums are over all arcs $\langle k,i \rangle$ and $\langle i,j \rangle$ of the network.

The number v is called the *value* of the flow. A flow in a network is *maximal* if there is no other flow in the network with a higher value.

A *cut* in a network is a set of arcs $\langle x,y \rangle, x \in X, y \in Y$, where the sets X and Y partition the vertex set of the network. Typically, if the network has a single source and a single sink, we are interested in cuts that include the source in one of the sets X and the sink in the other set Y. Figure 6.2 illustrates a cut dividing the nodes of the network into the sets $X = \{s, A, B, F\}$ and $Y = \{C, D, E, G, H, t\}$. The *cut capacity* is defined to be the sum of the capacities on arcs leading from set X to set Y. Thus in this example, the cut capacity is the sum of the capacities of the arcs AE, AD,

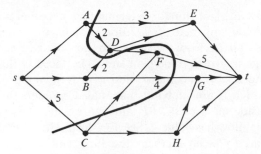

Figure 6.2. Cut of capacity 16.

BD, BG, Ft, and sC, or a total of 21. Note that the capacities of the arcs DF and CF are not used since these arcs lead from set Y to set X. That is, arcs DF and CF are not elements of this cut.

Of fundamental importance in the study of single source, single sink networks is the *max-flow min-cut theorem*. This theorem states that the value of the maximal flow from the source s to the sink t in any such network is equal to the minimal cut capacity of the network for all cuts separating s and t. For the network of our example, Fig. 6.3 illustrates a cut having capacity 15, and Fig. 6.4 illustrates a flow of value 15 (arc capacities omitted).

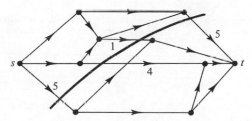

Figure 6.3. Minimal capacity cut.

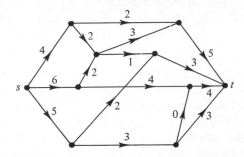

Figure 6.4. Maximal value flow.

We present now an algorithm for determining a maximal flow in an acyclic single source, single sink network. Unfortunately, the algorithm depends upon examining all paths from the source to the sink, and hence may be impractical for large networks. One problem to be faced is that we may find ourselves chasing a false maximum; that is, a flow value which is locally maximal, but which is not globally maximal. Consider, for example, Fig. 6.5a. Before considering arcs sC, BE, and ET we may have decided that a tentative maximal flow has value 4. See Fig. 6.5b. However, by considering the remaining arcs we can increase the flow value to 5. See Fig. 6.5c.

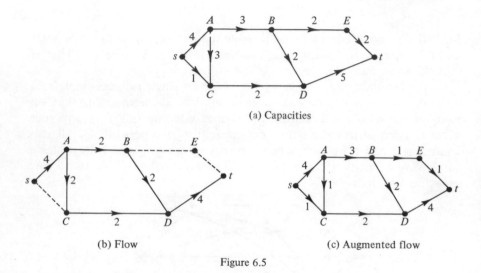

(a) Capacities

(b) Flow   (c) Augmented flow

Figure 6.5

Note that this requires an increase in flow along AB, but a decrease along AC. Thus we must have a way to "backtrack"—to retreat and pursue an alternative development.

We define two length functions d and d' by

$$d(a) = \begin{cases} 0 & \text{if } f(a) < c(a) \\ \infty & \text{if } f(a) = c(a), \end{cases} \qquad d'(a) = \begin{cases} 0 & \text{if } f(a) > 0 \\ \infty & \text{if } f(a) = 0. \end{cases}$$

We examine all paths from source to sink, ignoring for the moment the individual arc directions. The length of any path from source to sink is the sum of its arc lengths, using $d'(a)$ if the direction of a is inconsistent with the path direction. Thus any path has either zero length or infinite length. We wish to develop a new flow from an existing one by choosing a path and modifying the flow along it, adding or subtracting some flow. A path length of zero indicates that this is possible; an infinite path length indicates that there is no way to modify along that particular path. Either some arc in the path is saturated, $f(a) = c(a)$, so that the flow along the path cannot be increased; or some arc is empty, $f(a) = 0$, so that the flow along the path cannot be decreased.

For example, consider the network of Fig. 6.6, with the capacities shown in Fig. 6.6a, and a particular flow shown in Fig. 6.6b. We may increase this

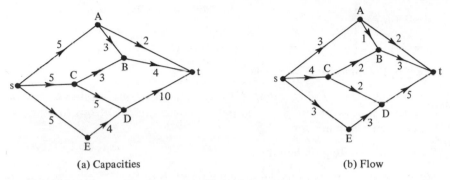

(a) Capacities   (b) Flow

Figure 6.6

flow by constructing an additional flow of two units along the path $sABCDt$. Note that this path includes the arc BC—the reversal of an arc which is actually in the network. Thus adding two units of flow along this arc amounts to subtracting two units from the arc CB, which is possible since the given flow has two units flowing along arc CB. The result is the flow shown in Fig. 6.7. In this new flow, since CB carries zero flow, $d'(CB) = \infty$. That is, we cannot subtract any further flow from CB.

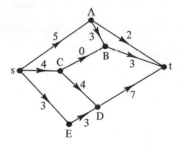

Figure 6.7

The maximum modification that can be achieved in any step is determined by the minimum residual capacity—that is, the excess of capacity over assigned flow, and the minimum flow. The former defines the flow which can be added, the latter the flow which can be subtracted from any arc in the path. The algorithm for determining a maximal flow is this. We assume that initially $f(a) = 0$ in all arcs.

Step 1.   Choose any minimal length path p from the source to the sink.

Step 2. For each arc a_i in the chosen path let

$$t(a_i) = \begin{cases} c(a_i) - f(a_i) & \text{if } \text{the direction of } a_i \text{ agrees with} \\ & \qquad \text{the path direction,} \\ f(a_i) & \text{if } \text{the direction of } a_i \text{ opposes the} \\ & \qquad \text{path direction.} \end{cases}$$

Let $t(p) = \min_{a_i} t(a_i)$.

Step 3. Augment the flow along p by the amount $t(p)$, noting that if the reversal of a network arc is in the path, we subtract $t(p)$ from the flow in the corresponding network arc.

Step 4. Recompute path lengths.

Step 5. If any paths of finite length remain, return to Step 1; otherwise go to Step 6.

Step 6. The maximal flow in the given network is then the sum of the path flows which have been computed.

Note that the above algorithm allows for a choice of paths in Step 1. By systematically choosing all possible paths, we may determine all possible maximal flow patterns within the network.

Example 6.1

Determine the maximal flow in the network of Fig. 6.6.

Solution.   There are nine paths from s to t, with lengths as indicated in Table 6.1.

TABLE 6.1

	path	length
p_1:	sAt	0
p_2:	$sABt$	0
p_3:	$sABCDt$	∞
p_4:	$sCBAt$	∞
p_5:	$sCBt$	0
p_6:	$sCDt$	0
p_7:	$sEDCBAt$	∞
p_8:	$sEDCBt$	∞
p_9:	$sEDt$	0

Choose any path of length zero, say p_1. Then $t(p_1) = 2$; increment the flow along p_1 (initially 0) by this amount (Fig. 6.8a). (Along each arc the numeral pair is (flow, residual capacity).) Recompute the path lengths: p_1 now has infinite length.

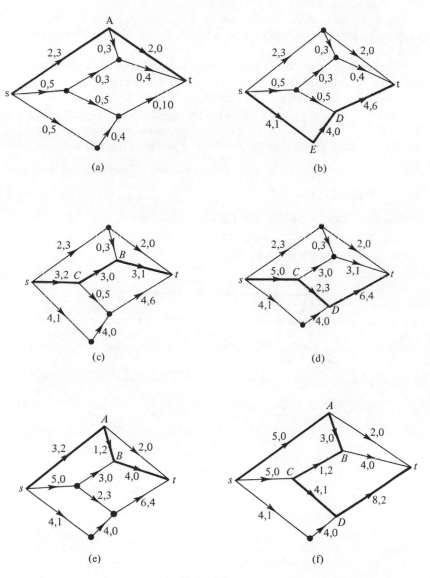

Figure 6.8

Choose another path of length zero, say p_9. A flow increment of 4 ($t(p_9) = 4$) may be added here (Fig. 6.8b). Now the length of p_9 is infinite.

For the third path, choose p_5, with $t(p_5) = 3$. Adding this flow, at this stage there is a total flow value of 9. See Fig. 6.8c. Now the length of p_5 is infinite, but the length of p_3 has become finite, since p_3 involves the reversal of an arc now having a positive flow.

There are now three finite paths, p_2, p_3, and p_6. The p_6 flow may be augmented by $t(p_6) = 2$ units (Fig. 6.8d); and the p_2 flow by $t(p_2) = 1$ unit (Fig. 6.8e).

Now the only path of finite length is p_3, $sABCDt$, for which $t(p_3) = 2$. Augment the flow by this amount, noting that in the process two units of flow are subtracted from the arc BC. (See Fig. 6.8f.)

There is now a total flow of value 14, equal to the capacity of the cut $X = \{s, E\}$, $Y = \{A, B, C, D, t\}$. Hence by the max-flow min-cut theorem, this flow is maximal. (Note that the residual capacity of this cut is zero.)

ANSWER.   The maximal flow is 14.

EXERCISES

1. Determine all flows in the network in Fig. 6.6 having a value 14.

2. Using the algorithm, determine the maximal flow for the network of Fig. 6.9. Also determine a cut having capacities equal to the maximal flow value.

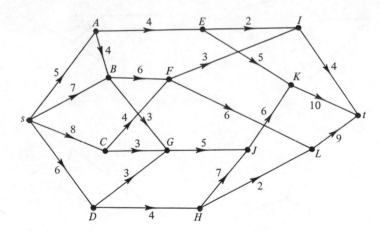

Figure 6.9

3. Write a computer program implementing the algorithm of this section, and test it on the networks of Figs. 6.6 and 6.9. Assume a network of at most 20 nodes and 200 arcs, given as an adjacency matrix, with the matrix entries being the arc capacities.

4. What is the effect of permitting cycles in the network?

7. MINIMAL COST FLOWS

When several flows in a network have the same maximal value, then it is of interest to determine a flow with minimal cost, whether this cost is measured in dollars, time, or some other unit. The algorithm that was developed in Section 6 applies equally well for determining minimal cost networks, with a change in length function. Wherever the length function for a path has the value 0 for the algorithm as we have defined it above, for use with minimal cost problems the value of the length function is the sum of the cost of traversing the arc for each direct arc included in the path, and the negative of this cost for each reverse arc.

Example 7.1

Using the given algorithm, determine a minimal cost flow through the network of Fig. 7.1. The triples are (flow, residual capacity, cost).

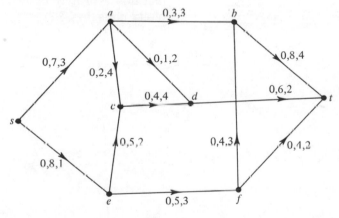

Figure 7.1

Solution. By inspection there is a cut of capacity 13 (where?), so this is an upper bound on the flow. The procedure is the same as that used in Example 6.1, except for the use of a different length function. The paths, their lengths, and $t(p_i)$ are given in Table 7.1. (If $t(p) = 0$, there is no need to compute the path length.)

At each stage choose a path of minimal length from among those with $t(p_i) \neq 0$. Thus, first choose p_{16}, and institute a flow of 4 along it. Following the algorithm, choose in sequence flows of 1 along p_4, 4 along p_{15}, 3 along p_1, and 1 along p_9. The total flow is then 13, equal to the cut capacity which

TABLE 7.1

		length	$t(p_i)$			length	$t(p_i)$
p_1:	sabt	10	3	p_{10}:	secabt		0
p_2:	sabft		0	p_{11}:	secabft		0
p_3:	sabfecdt		0	p_{12}:	secadt		0
p_4:	sadt	7	1	p_{13}:	secdabt		0
p_5:	sadceft		0	p_{14}:	secdabft		0
p_6:	sadcefbt		0	p_{15}:	secdt	9	4
p_7:	sacdt	13	2	p_{16}:	seft	6	4
p_8:	saceft		0	p_{17}:	sefbt	11	4
p_9:	sacefbt		0	p_{18}:	sefbadt		0
				p_{19}:	sefbacdt		0

was noted. The cost is found by multiplying the flow along each arc by
the cost for that arc, and totaling.

ANSWER.   The minimal cost maximal flow is a flow of 13 at a cost
of 112 units (Fig. 7.2).

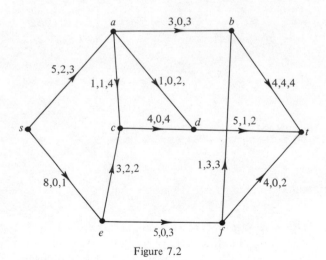

Figure 7.2

EXERCISES

Compute the minimal cost of a maximal flow in the network:

1. Fig. 7.3.   2. Fig. 7.4.

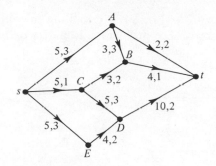

Figure 7.3. Capacity, cost.

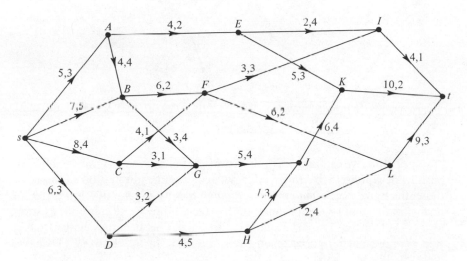

Figure 7.4. Capacity, cost.

3. Modify the program written for Exercise 3 of Section 6 to solve minimal cost problems, and test it on the networks of Figs. 7.1 and 7.3. Assume an additional adjacency matrix, whose entries are the arc costs.

8. PRUNING BRANCHES TO FIND THE SHORTEST PATH

In the previous sections we have considered problems associated with a total flow through a network, that is, a flow involving several paths through the network. In this and the next section we turn our attention to a single path through the network.

Consider the network of Fig. 8.1, interpreting the numbers as the length or cost of a network arc. The problem is to find a path from s to t whose total length (cost) is minimal.

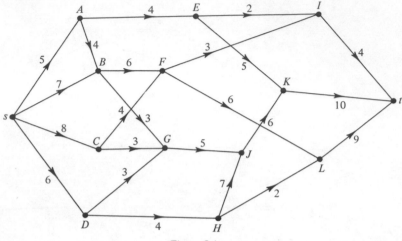

Figure 8.1

In principle, we examine all paths from s to t and choose a minimal length path from among them. In practice, we can avoid examination of many of the paths by a *branch and bound technique*. Essentially we develop a tree of partial paths, and prune a branch of the tree ("bound" the solution) as soon as it is evident that its partial path is too long. This becomes evident when two or more partial paths reach the same node in the network. Then we discard all but the shortest path.

The algorithm we use, due to R. E. Nance, utilizes both numerical and string-valued variables. The former represent lengths (costs), and the latter represent the partial paths through the network. Some algorithms for this problem do not explicitly keep track of the minimal length path as it is found, and hence need to reconstruct the path is a final step. Keeping track of the paths by means of string-valued variables eliminates the need for this reconstruction. We use s and t to denote the source and sink, respectively.

For each vertex v in the network, we let l'_v denote a tentative minimal path length from s to v, l_v denote the previously determined minimal path length from s to v, and p_v denote a particular path from s to v whose length is l_v. Initially, for each vertex v other than the source, l_v is set to the maximum positive value allowed on the computer (effectively infinity), and p_v is undefined. Thus l'_v and l_v have numerical values, while p_v has a string value, $sAB \cdots Kv$, listing the vertices along a particular path from

s to v, having path length l_v. For any two vertices v and w such that there is an arc from v to w, we let l_{vw} denote the length of that arc, and p_{vw} denote the path $sAB \cdots Kvw$, where $p_v = sAB \cdots Kv$. We say that p_v has been *finally determined* if all paths leading into v have been examined. The algorithm is this.

Step 1. Set $p_s = s$, $l_s = 0$, $l_v = \infty$ for all v other than s.

Step 2. Let v be a vertex other than t for which p_v has been finally determined, but no p_{vw} has been defined. For each w such that there is an arc vw in the network, let $l'_w = \min(l_w, l_v + l_{vw})$.

Step 3. If $l'_w = l_w (<\infty)$, then p_w has been previously defined. Let it stand. If $l'_w < l_w$, then set $p_w = p_{vw}$, and replace l_w with l'_w.

Step 4. Return to Step 2 as long as there are vertices v satisfying the condition; otherwise go to Step 5.

Step 5. The shortest path is p_t, with length l_t.

Example 8.1

Find the shortest path from s to t in the network of Fig. 8.1.

Solution. Let $l_s = 0$, $p_s = s$. The immediate successors of s in the network are A, B, C, and D. Compute the tentative lengths:

$$l'_A = \min(\infty, 0+5) = 5, \qquad l'_B = \min(\infty, 0+7) = 7,$$
$$l'_C = \min(\infty, 0+8) = 8, \qquad l'_D = \min(\infty, 0+6) = 6.$$

In each case $l' < l$. Hence $p_A = sA$, $p_B = sB$, $p_C = sC$, and $p_D = sD$; $l_A = 5$, $l_B = 7$, $l_C = 8$, and $l_D = 6$. The paths p_A, p_C, and p_D, but not p_B, have been finally determined.

Choose vertex A. Its successors are B and E.

$$l'_B = \min(7, 5+4) = 7, \qquad l'_E = \min(\infty, 5+4) = 9.$$

Since $l'_B = l_B$, leave $p_B = sB$. (The path p_B is now finally determined.) Since $l'_E < l_E$, set $p_E = p_{AE} = sAE$, and $l_E = 9$.

Similarly, proceed through the network.

From B: $l'_F = \min(\infty, 7+6) = 13$,
$\qquad\qquad l'_G = \min(\infty, 7+3) = 10$.
$\qquad\therefore p_F = sBF$, $l_F = 13$; $p_G = sBG$, $l_G = 10$.

From C: $l'_F = \min(13, 8+4) = 12$,
$\qquad\qquad l'_G = \min(10, 8+3) = 10$.
$\qquad\therefore p_F = sCF$, $l_F = 12$; $p_G = sBG$, $l_G = 10$.

From D: $l'_G = \min(10, 6+3) = 9,$
$\qquad\quad l'_H = \min(\infty, 6+4) = 10.$
$\qquad \therefore p_G = sDG, \quad l_G = 9; \quad p_H = sDH, \quad l_H = 10.$

From E: $l'_I = \min(\infty, 9+2) = 11,$
$\qquad\quad l'_K = \min(\infty, 9+5) = 14.$
$\qquad \therefore p_I = sAEI, \quad l_I = 11; \quad p_K = sAEK, \quad l_K = 14.$

From F: $l'_I = \min(11, 12+3) = 11,$
$\qquad\quad l'_L = \min(\infty, 12+6) = 18.$
$\qquad \therefore p_I = sAEI, \quad l_I = 11; \quad p_L = sCFL, \quad l_L = 18.$

From G: $l_J = \min(\infty, 9+5) = 14.$
$\qquad \therefore p_J = sDGJ, \quad l_J = 14.$

From H: $l'_J = \min(14, 10+7) = 14,$
$\qquad\quad l'_L = \min(18, 10+2) = 12.$
$\qquad \therefore p_J = sDGJ, \quad l_J = 14; \quad p_L = sDHL, \quad l_L = 12.$

From I: $l'_t = \min(\infty, 11+4) = 15.$
$\qquad \therefore p_t = sAEIt, \quad l_t = 15.$

From J: $l'_K = \min(14, 14+6) = 14.$
$\qquad \therefore p_K = sAEK, \quad l_K = 14.$

From K: $l'_t = \min(15, 14+10) = 15.$
$\qquad \therefore p_t = sAEIt, \quad l_t = 15.$

From L: $l'_t = \min(15, 12+9) = 15.$
$\qquad \therefore p_t = sAEIt, \quad l_t = 15.$

ANSWER. $p_t = sAEIt, \quad l_t = 15.$

This particular network has 14 paths through it from s to t. By using the algorithm we discard many paths as soon as their partial development indicates that they cannot be minimal. For example, when we examine vertex C, we discard all paths starting $sBF\ldots$, since sCF is a shorter path to F. This pruning of the branches of the path tree of the network reduces the number of steps needed to locate the minimal path (Fig. 8.2).

The condition that p_v be finally determined is easily met if the nodes of the network are ordered in such a way that v precedes w in the order whenever there is an arc from v to w. Since the network is acyclic, such an ordering is always possible. (See Exercise 11, Section 4.) The algorithm will also work for directed graphs containing cycles.

The algorithm which we have discussed "floods" across a network, examining paths in parallel. A similar algorithm by E. W. Dijkstra examines paths more serially, always extending one minimal length partial path [1]. While

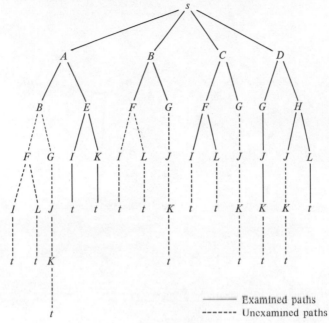

Figure 8.2

—————— Examined paths
- - - - - - Unexamined paths

the Dijkstra algorithm is slightly more efficient than the Nance algorithm, it cannot easily be converted to a longest path algorithm. As we shall see in the next section, such a conversion is simple for the Nance algorithm.

EXERCISES

Using the algorithm find a minimal length path through the network:

1. Fig. 8.3.   2. Fig. 8.4.
3. Fig. 8.5.   4. Fig. 8.6.

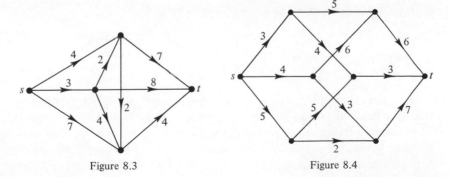

Figure 8.3   Figure 8.4

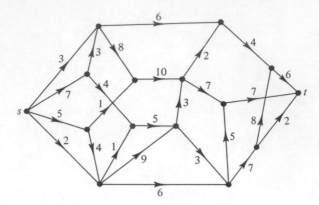

Figure 8.5

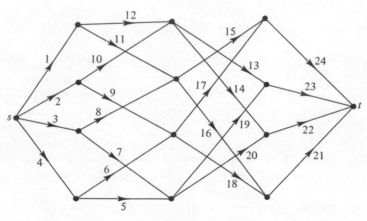

Figure 8.6

5. Write a program to implement the algorithm and test it on the networks of Example 8.1 and Exercise 3. Assume a network of at most 20 nodes, given as an adjacency matrix of arc lengths.

6. Determine the effect of cycles on the shortest path algorithm.

9. CRITICAL PATHS

One network problem that has been widely implemented on computers is that of critical path scheduling. Consider the construction of a large building. This feat involves numerous tasks, some of which may be carried out in parallel, while others must be sequenced. One must purchase the building site, purchase materials, prepare the foundation, build the frame, install wiring and plumbing, hang the walls, paint, and so on. It may be possible to install wiring and plumbing simultaneously, but the foundation must be laid before construction of the building frame can begin. Each of these tasks requires a certain minimum length of time to complete. The entire process may be represented by a network whose source represents initiation of the process, and whose other nodes each represent the completion of a specific task, with the sink representing completion of the entire project. The arcs in the network then represent the execution of these tasks, and may be labeled with associated variables such as cost and time.

This same model applies to other types of problems as well. For example, work toward a college degree may be modeled in this way. The tasks in this case are the completion of courses; the sequencing is determined by the prerequisites for each course; and two courses may be taken in parallel if neither is a prerequisite for the other.

The problem for each of these examples is to determine the minimum length of time from initiation of the project to its completion. This time is determined by the longest path from the source to the sink in the network. For example, if a course that is required for a college degree has nine prerequisites which must be taken in sequence, then it is impossible to complete the degree in less than 10 terms. Paths of maximal length in the network are called *critical paths*. Computer programs to determine these paths are known by such acronyms as CPM (critical path method), CPS (critical path scheduling), and PERT (program evaluation review technique).

The procedure for determining the longest path is identical to that for determining the shortest path, except that at each stage a maximal finite length partial path is selected instead of a minimal length one.

Example 9.1

Find a critical path through the network of Fig. 8.1.

Solution. Apply the shortest path algorithm, choosing a maximal finite length partial path at each stage.

$$p_s = s, \quad l_s = 0.$$

From s: $\quad l'_A = 5, \quad l'_B = 7, \quad l'_C = 8, \quad l'_D = 6.$
$\quad \therefore p_A = sA, \quad l_A = 5; \quad p_B = sB, \quad l_B = 7; \quad p_C = sC,$
$\quad l_C = 8; \quad p_D = sD, \quad l_D = 6.$

From A: $\quad l'_B = 9, \quad l'_E = 9.$
$\quad \therefore p_B = sAB, \quad l_B = 9; \quad p_E = sAE, \quad l_E = 9.$

From B: $\quad l'_F = 15, \quad l'_G = 12.$
$\quad \therefore p_F = sABF, \quad l_F = 15; \quad p_G = sABG, \quad l_G = 12.$

From C: $l'_F = 12,$ $l'_G = 11.$
      $\therefore p_F = sABF,$ $l_F = 15;$ $p_G = sABG,$ $l_G = 12.$

From D: $l'_G = 9,$ $l'_H = 10.$
      $\therefore p_G = sABG,$ $l_G = 12;$ $p_H = sDH,$ $l_H = 10.$

From E: $l'_I = 11,$ $l'_K = 14.$
      $\therefore p_I = sAEI,$ $l_I = 11;$ $p_K = sAEK,$ $l_K = 14.$

From F: $l'_I = 18,$ $l'_L = 21.$
      $\therefore p_I = sABFI,$ $l_I = 18;$ $p_L = sABFL,$ $l_L = 21.$

From G: $l'_J = 17.$
      $\therefore p_J = sABGJ,$ $l_J = 17.$

From H: $l'_J = 17,$ $l'_L = 12.$
      $\therefore p_J = sABGJ,$ $l_J = 17;$ $p_L = sABFL,$ $l_L = 21.$

From I: $l'_t = 22.$
      $\therefore p_t = sABFIt,$ $l_t = 22.$

From J: $l'_K = 23.$
      $\therefore p_K = sABGJK,$ $l_K = 23.$

From K: $l'_t = 33.$
      $\therefore p_t = sABGJKt,$ $l_t = 33.$

From L: $l'_t = 30.$
      $\therefore p_t = sABGJKt,$ $l_t = 33.$

ANSWER. $p_t = sABGJKt,$ $l_t = 33.$

Note that this is still a branch and bound algorithm.

The examination of all paths in a network is essential to finding critical path. If we always selected only the longest partial path to extend, then for the graph in Fig. 9.1 the sequence would be:

$$sC \quad \text{(length 2)}, \qquad sCt \quad \text{(length 4)}.$$

Since sA has length 1 it would be rejected, and the longer path $sABt$ (length 6) would be missed.

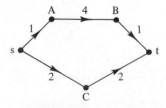

Figure 9.1

Critical path problems are generally far more complex than merely finding the longest path through a network. Suppose, for example, that the critical path is so long that a vital deadline cannot be met. Perhaps there is even a severe penalty associated with missing the deadline. Then one must search for ways to reduce the critical path length. Yet shortening the length of one path (for example, by doubling the work force for a particular task) may perturb path lengths elsewhere in the network, even increasing critical path length for the network. In general, efforts to modify path lengths involve a great deal of computation and recomputation, impossible to complete without a computer.

EXERCISES

Use the algorithm to determine a critical path in the network:

1. Fig. 8.3.   2. Fig. 8.4.

3. Fig. 8.5.   4. Fig. 8.6.

5. Modify the program for the shortest path algorithm (Exercise 5, Section 8) to compute critical paths, and test it on the networks of Example 9.1 and Exercise 3. *Hint:* Initially define all path lengths to be zero.

10. GRAPHS OF MULTIPROCESSING SYSTEMS

With the development of multiprocessing computer systems has come the need to systematically study the relationships between concurrent or sequential processes. In this type of computer environment we typically have several different kinds of resources such as CPU's, disk drives, and printers, with one or more copies of each type of resource available to the users. However, since several processes are being carried out simultaneously, they must compete for the limited set of resources. Many times two different processes will compete for a single piece of hardware or software. In addition, as each process proceeds through its various stages toward completion, the initiation of any stage depends on the completion of the immediately preceding stages, generally in one of two ways. Either *all* immediately preceding stages of the process must be completed, or *any one* of the immediately preceding stages must be completed.

This type of environment is easily represented by a labeled acyclic directed graph, where the label of each vertex denotes the particular resource used, the process using it, and how many times the process has used that particular resource, and labels on edges or sets of edges denote the conditions for initiation of a process stage. Thus the power of the theory of digraphs is brought to bear on significant problems in computing. The graphs are networks, so the scheduling, cost, and flow algorithms that have been developed for networks can be used to analyze the performance of a large computer system.

11. INFORMATION NETWORKS

Information networks are networks without sources and sinks. The underlying concept is that a user at any node of the network must be able to direct an information request to any other node of the network and receive a reply to the request. This means that there must be paths from any node in the network to any other node. That is, an information network is a strongly connected digraph. The most common examples of information networks are management information systems, library information networks, and computer networks. While any strongly connected digraph can be considered an information network, it is instructive to define several basic types for study. Despite the fact that such information transfer systems as the mail and telephone lines are bidirectional, we assume that the information transfer channels of our networks are unidirectional. This allows us to include television, for example, as an information transfer channel. (Most television receiving stations do not have the ability to originate a television signal.) We can, of course, easily model bidirectional channels by antiparallel arcs.

The simplest information network, called a *cyclic network*, consists of a single directed cycle (Fig. 11.1). In such a network an information request

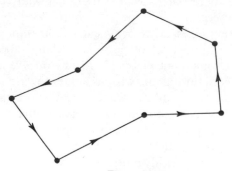

Figure 11.1

will flow along one path of the cycle, and the reply to the request will come back along the complementary path. This type of network, while obviously simple and relatively inexpensive to construct and maintain, is inefficient if the cycle is long, and highly susceptible to failure in the event that one of the information transfer links fails.

A second type of information network, the *hierarchical network*, is perhaps more characteristic of existing networks (Fig. 11.2). Such a network is a

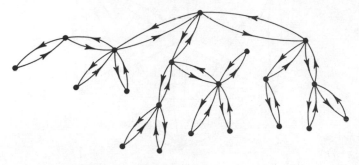

Figure 11.2

treelike structure in which information flows in both directions along each branch of the tree. This type of network is less susceptible to failure since the failure of any single transmission link will isolate a portion of the network, without affecting operations within either that portion of the network, or the other portion of the network. Moreover, this type of network provides excellent control of information flow in situations where that is required. A hierarchical network in which all end nodes are connected directly to one central node is called a *star network* (Fig. 11.3). Such a network typifies a central computer with attached terminals.

Figure 11.3

At the other extreme of the spectrum of information networks lies the *decentralized network* (Fig. 11.4). This network, a complete digraph, allows

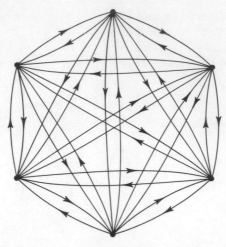

Figure 11.4

the user at any node immediate and direct access to information at any other node. Since there are many paths from one node to another, the network is highly tolerant of failure in transmission links, but at the same time very expensive to construct and maintain.

For some networks such as the ARPA computer network, one of the criteria in design of the network is that there be at least two alternative paths from any one node to any other. It can be shown that a minimum requirements for such a network is that it be a 2-regular 3-connected isograph (Fig. 11.5). The purpose of this criterion is reliability. While a

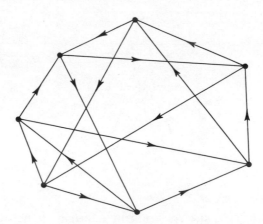

Figure 11.5

2-regular 3-connected isograph does not have the high reliability of the decentralized network, it is much less expensive than the latter, and is not susceptible to a single transfer channel failure. Thus if the probability of two or more simultaneous channel failures is sufficiently low, this type of network is highly reliable.

While many factors must be considered in assessing the merits of a network, a basic graph-theoretic parameter is the *flexibility* F defined by

$$F = \frac{Q - N}{N(N - 2)},$$

where Q is the number of arcs in the network, and N is the number of nodes. The flexibility is a measure of the alternative paths available for information transfer. Its values for the four basic network types are given in Table 11.1. The values of this parameter are normalized between 0 and 1,

TABLE 11.1

Network type	Flexibility
Cyclic	0
Hierarchical	$1/N$
Two-regular isograph	$1/(N - 2)$
Decentralized	1

with 1 representing the maximum freedom in choosing an information transfer channel, and 0 representing no freedom of choice among channels. Observe also that to maintain constant flexibility as a network grows (if that is desirable) the number of arcs must vary as the square of the number of nodes.

Another aspect of this type of information network model is being intensively studied. This is the effect of the probability of use of each information transfer channel. For example, while the network of Fig. 11.6a is decentralized, if the values given represent probability of use of each channel, then the network acts very much like the cyclic network of Fig. 11.6b. The study of this aspect relates strongly to the modification of existing networks. Thus we might consider modification of a hierarchical network by the introduction of cross-links, in hopes of improving information flow. The actual effect of such an introduction depends on the probability of use of the cross-links, which in turn depends on the incentives for using them. Hence we can begin to model and study quantitatively the effects of different types of information transfer channels, and of the costs and benefits of their use.

It is important to note that the use of various network arcs is often determined solely by the network management, and is of little interest to the

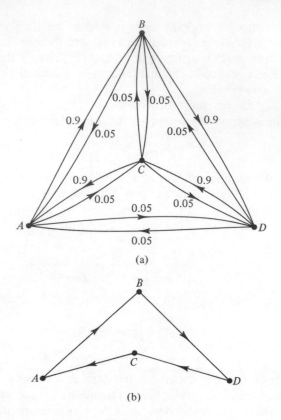

(a)

(b)

Figure 11.6

user. It is important to the user that his message from New York reach the proper destination in Chicago quickly, accurately, and inexpensively; but he generally does not care what particular route the message follows through the network. Thus the incentives for adding, modifying, or deleting a network link are most often related to management's concept of providing a level of service which will attract and hold network users.

The probability information for a network is easily handled by substituting the arc probabilities for the 1's in an adjacency matrix of the network. We call the matrix thus derived a *probabilistic adjacency matrix* of the network.

Example 11.1

Determine a probabilistic adjacency matrix of the network in Fig. 11.6a.

Solution. The adjacency matrix has ones everywhere but on the main diagonal. Use the vertex order *A, B, C, D*.

ANSWER.

$$\begin{bmatrix} 0 & 0.9 & 0.05 & 0.05 \\ 0.05 & 0 & 0.05 & 0.9 \\ 0.9 & 0.05 & 0 & 0.05 \\ 0.05 & 0.05 & 0.9 & 0 \end{bmatrix}.$$

Example 11.2

In the network of Fig. 11.6a, what is the probability of reaching B from A in at most three steps?

Solution. Consider all five paths from A to B having length three or less. Along any one path the probabilities of using the arcs multiply to yield the probability of using that path. The total probability of reaching B in three or fewer steps is the sum of the probabilities on the paths:

AB	0.9	= 0.9
ACB	(0.05)(0.05)	= 0.0025
ADB	(0.05)(0.05)	= 0.0025
$ACDB$	(0.05)(0.05)(0.05)	= 0.000125
$ADCB$	(0.05)(0.9)(0.05)	= 0.00225
		0.907375

ANSWER.

The probability is 0.907375.

Note that this probability is very close to 0.9. In fact, if the given probabilities are accurate to only one significant digit, then the answer is also, hence is 0.9. Thus there is virtually no network traffic from A to B except along the direct path.

The results of a network analysis must be carefully interpreted in terms of the effect of various actions on each node. Thus while the deletion of arcs BC, CB, CD, and DB will have little effect on the traffic at node A, it will affect 10 percent of the incoming traffic and 5 percent of the outgoing traffic at node B. See Fig. 11.7. The affected traffic may disappear from the network, or may shift to other arcs in the network.

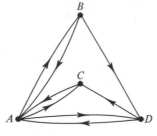

Figure 11.7

EXERCISES

1. For the network of Fig. 11.6a compute the probabilities of reaching C and D from A in three or fewer steps.

2. Write a program to compute the flexibility of an information network of at most 20 nodes, given an adjacency matrix as input. Test it on the networks of Fig. 11.8.

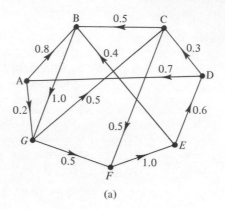

(a)

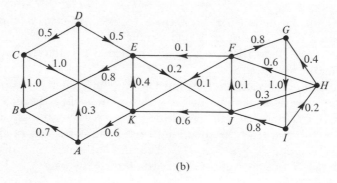

(b)

Figure 11.8

3. Given as input a probabilistic adjacency matrix of an information network of at most 20 nodes, two specified nodes v and w, and the number k of steps, write a program to compute the probability of reaching w from v in at most k steps. Test your program on the networks of Figs. 11.6a and 11.8. *Hint:* Modify the shortest path algorithm of Section 8.

4. (cont.) Suppose that you do not know the two nodes v and w. Write a program to compute for *all* pairs of nodes the probability of reaching one from the other in at most k steps. *Hint:* You do not need to trace paths through the network.

REFERENCE

1. E. W. Dijkstra, A note on two problems in connection with graphs. *Numerische Matematik* **1** (1959) 269–271.

CHAPTER 5

Formal and
Natural Languages

1. INTRODUCTION

We now turn from graph theoretic ideas to concepts that are more closely related to those of ordinary arithmetic. In this and the succeeding three chapters we shall consider a range of rather simple algebraic systems which are related to work with computers. These are by no means all of the existing algebraic systems, and indeed, some rather important ones are not discussed.

An *algebra* or *algebraic system* is a set S together with a set F of finitary operators defined on S. By *finitary* we mean that each operator in F maps an n-tuple from $S \otimes \cdots \otimes S$ into an element of S, for some finite n. The usual arithmetic operators, such as addition and multiplication, are finitary; in particular, they are all binary, except for the unary use of the negative sign. Study of an algebra or an algebraic system centers on the properties and relations that hold for its operators.

We begin with an examination of the most simply defined algebraic system, the semigroup. Because of the simplicity of definition, semigroups form a large and complex class of objects with little structure. For many

166

years, algebraic research was concentrated on more fully structured algebraic systems. However, since semigroups provide models for languages, recent interest in logic, computing, and mathematical linguistics has stimulated interest in the study of semigroups.

2. SEMIGROUPS

Among the simplest of algebraic systems are those having only a single binary operator. If we consider the ordinary arithmetic operators, we note that the properties of these operators are different, and depend upon the set on which they are defined. For example, if we restrict our attention to the set of nonnegative integers, then addition and multiplication are always defined, while subtraction and division may be undefined for certain pairs of numbers. If we change the set to be the real numbers, then all four arithmetic operators are defined for all pairs of real numbers, with one exception. Division by 0 is undefined. Among other properties, we note that addition is commutative for either set, that is, $a+b = b+a$, whereas subtraction is not.

A *semigroup* is an ordered pair $\langle S, * \rangle$ where S is a set, $*$ is a binary operator on S, and the following properties hold:

(*Closure*)   For all $a, b \in S, a * b \in S$.
(*Associativity*)   For all $a, b, c \in S, (a * b) * c = a * (b * c)$.

By convention, the operator $*$ is usually considered to be "multiplication" or "concatenation" so that the associative law is often written $a(bc) = (ab)c$. Note that the operator in a semigroup may or may not have these additional properties:

(*Identity*)   There exists a special element $e \in S$ such that for any
                   $a \in S, ea = ae = a$.
(*Commutativity*) For any $a, b \in S, ab = ba$.

The identity element is especially important in linguistic studies, but commutativity generally does not hold (*dog* is not the same as *god*).

The positive integers form a commutative semigroup under either addition or multiplication, with 1 acting as the identity of the multiplicative semigroup. However, this set does not form a semigroup under either subtraction or division because closure and associativity do not hold. Expanding the set to include all integers, positive and negative, provides closure for subtraction, but associativity is still lacking. For example, $5-(3-2) \neq (5-3)-2$.

We may use the English alphabet as the basis for a semigroup under the operation of concatenation. That is, the elements of the semigroup are formed by writing finite sequences of letters. From elements of the semigroup, say *dog* and *house*, we form new elements by concatenation, as *doghouse*, *housedog*, and *doghousedog*. Since not all letter combinations (*sqtxz*) form English words, we must impose additional rules if we wish to model the English language. In linguistic applications such as this, the semigroup identity is generally taken to be the null string, containing no letters at all.

EXERCISES

Show that $\langle S, * \rangle$ is a semigroup. Determine its identity, if there is one, and whether the semigroup is commutative:

1. S is the set of positive integers; $x * y$ is $x + y + xy$.

2. S is the set of all subsets of a set T; $x * y$ is $x \cup y$.

3. S is the set of all subsets of a set T; $x * y$ is $x \cap y$.

4. Show that the set of positive integers is not a semigroup if $x * y$ is x^y.

5. Let T be an infinite set. Do the answers to Exercises 2 and 3 change if S is the set of all finite subsets of T? the set of all infinite subsets of T?

6. Show that for any semigroup the generalized associative law holds. That is, given $a_1, a_2, ..., a_k$, the semigroup element $a_1 a_2 ... a_k$ is the same, regardless of the order of computation of the subproducts. For example, $((ab)c)d = (ab)(cd) = (a(bc))d = a((bc)d) = a(b(cd)) = abcd$.

3. FORMAL LANGUAGES

A formal language may be specified in terms of an *alphabet* and a *grammar*. The alphabet is a finite set of symbols. Using the operation of concatenation, we construct from the alphabet A the *free semigroup A^* on A*, consisting of all finite strings of elements of A including the null string Λ. (This is called "free" because there are no relations given other than the identity; that is, two strings are equal if and only if they consist of exactly the same arrangement of symbols from A.) We are not interested in calling A^* a language any more than we call all strings of letters from the English alphabet a language. The grammar for a language specifies which strings of letters are of interest.

A useful, if mildly inaccurate, distinction between formal languages and natural languages such as English or French is this. A natural language develops and grows through daily use by many people. A grammar for a natural language is subsequently developed to codify the rules for the language, as they have developed. Thus the language comes first, and the grammar follows it. However, for a formal language the grammar is developed first and firmly fixed. Thus the formal language grows out of its grammar, rather than the grammar arising from the already developed language. As a result of this difference in development, the structure of a formal language is usually very simple, involving few rules and components, whereas that of a natural language is highly complex.

The terminology used with formal languages mirrors the terminology of natural languages. Thus we call the elements of a formal language *words*, and combine these by the grammatical rules of the language to form *sentences*. The grammar for a language specifies its *syntax*, the structure of the strings that are to be considered sentences. The *semantics* of the language, or the meaning of its sentences, is determined not by the grammar alone, but rather by external references. Thus it is possible to determine that the Algol statement

$$xbar := xbar + x[i] \uparrow 2;$$

is syntactically correct, simply from knowing the grammar of Algol. However, its meaning must be determined not solely from the rules of Algol, but from the meanings assigned to the variables by the programmer.

It is even possible to have syntactically correct sentences, such as "Rainy telephones think," which have no semantic content or meaning. Note, however, that there is an interplay between syntax and semantics. If the syntax of a language were to specify that weather adjectives cannot modify nouns referring to communications media, then "Rainy telephones think," would be a nonsentence, rather than a sentence without meaning.

The formal languages that we shall examine are called *phrase structure languages* since each properly formed string or sentence in the language consists of one or more substrings or *phrases*. Each of these in turn is broken into subphrases and, ultimately, into the elements or words of the language. Thus for each sentence in the language we have *parsing tree* describing this structure (Figs. 3.1 and 3.2).

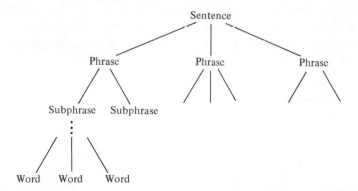

Figure 3.1

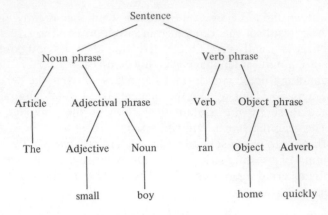

Figure 3.2

A grammar for a language may be either *generative*, starting with the concept "sentence" and generating particular sentences, or *descriptive*, starting with a particular string of words and describing exactly how it can (or cannot) be deemed a sentence. Because of the conventional manner of drawing a parsing tree, the action of a generative grammar is called *top-down parsing* while that of a descriptive grammar is *bottom-up parsing*. We shall work with generative grammars, mentioning only briefly the corresponding descriptive grammars.

A generative grammar begins with a single initial symbol, denoting the concept "sentence." By the *transformation rules* or *productions* of the grammar this symbol is transformed into a longer symbol string, involving other symbols from the alphabet of the language. Ultimately, repetition of this transformation process should result in a string consisting only of words of the language to which no further transformations apply. Since the words cannot be further changed, they are *terminal* symbols of the language. Those symbols that may be further transformed are called *nonterminals*.

A *phrase structure grammar* or *Type 0 grammar* is a quadruple $G = \langle S, T, \Pi, s \rangle$ where

1. S is an alphabet;
2. $T \subset S$; the elements of T are called *terminals* or *terminal words*, while the elements of $S - T$ are the *nonterminals*;
3. Π is a set of ordered pairs $\langle p, q \rangle$, such that $p \in S^*$ and includes at least one nonterminal, and $q \in S^*$;
4. $s \in S - T$.

The nonterminal s that is particularly singled out is thought of as denoting the concept "sentence," while the other nonterminals are essentially names for various phrases and grammatical categories. The ordered pairs $\langle p, q \rangle$ in Π are the *productions* and are generally written $p \to q$.

A production $p \to q$ permits replacement of the string p by the string q in any of its occurrences. That is, if rpt is a string and $p \to q$ is a production, then we can generate the string rqt. Note in particular that q may be Λ, that is, we may simply erase p. Using x to denote the string rpt and y to denote the string rqt we write $x \vdash y$ to indicate this generation process. More generally, we write $x \vDash y$ if there exist strings $z_0, z_1, z_2, ..., z_n, n \geqslant 0$, such that $x = z_0 \vdash z_1 \vdash z_2 \cdots \vdash z_n = y$, that is, if y can be generated by applying a finite sequence of productions to x. (Note that $x \vDash x$.)

Corresponding to each phrase structure grammar $G = \langle S, T, \Pi, s \rangle$ there is a *phrase structure language* or *type 0 language*

$$\mathscr{L} = \mathscr{L}(G) = \{x \in T^* \mid s \vDash x\}.$$

Thus the language consists of all strings of terminals generated from s.

For a descriptive grammar, the above structure is reversed. The strings of terminal words of the generative grammar become a special set of initial words, replacing s in the fourth clause of the definition. The symbol s in turn becomes the sole terminal word. All productions are reversed, so that they may be used to analyze a given symbol string. A successful analysis will terminate with just the terminal word s, while an unsuccessful analysis will result in more than one terminal word, or will leave a string containing nonterminals also. Note that successful termination in a generative grammar results is a string of terminals, whereas a successful termination in a descriptive grammar results always in the single terminal.

Example 3.1

Find the language generated by the grammar $G = \langle S, T, \Pi, s \rangle$, where

$$S = \{s, a, b, c, d\}, \qquad T = \{c, d\},$$
$$\Pi = \{s \to ab, s \to bbc, a \to cbd, b \to dcc\}.$$

Find also the descriptive grammar G' for this language.

Solution. $\mathscr{L}(G)$ consists of all terminal strings generated from s. Begin the generation:

$$
\begin{aligned}
s &\vdash \underline{a}b \\
&\vdash c\underline{b}db \\
&\vdash cdcc\underline{d}b \\
&\vdash cdccddcc. \\
s &\vdash \underline{b}bc \\
&\vdash dcc\underline{b}c \\
&\vdash dcdccc.
\end{aligned}
$$

(The underlined symbol is altered by the next production application.) Thus there are only two terminal strings produced. Hence $\mathscr{L}(G) = \{cdccddcc, dccdccc\}$.

The descriptive grammar for this language is $G' = \langle S, T', \Pi', \Sigma \rangle$, where

$$S = \{s, a, b, c, d\}, \qquad T' = \{s\},$$

$$\Pi' = \{ab \to s, bbc \to s, cbd \to a, dcc \to b\}, \qquad \Sigma = \{c, d\}^*.$$

Thus $cdccddcc \vDash s$, while $cdcc \vDash cb$, and $cdccddccdccdccc \vDash ss$. Only the first of these is a successful termination.

ANSWER.

$$\mathscr{L}(G) = \{cdccddcc, dccdccc\}.$$

Note that a string, such as $cbdcc$, may have two or more analyses in G'.

Just as the string $cbdcc$ has two analyses in the descriptive grammar G', it is possible for a string to have two or more derivations in a generative grammar. For example, if a grammar contains productions $a \to aa$ and $b \to ab$, then the string aab has two distinct derivations from the string ab. Such a grammar is called *ambiguous*. If it develops that a grammar G is ambiguous, then it is of interest to find an alternative unambiguous grammar for $\mathscr{L}(G)$. Often this is possible; if it is not, $\mathscr{L}(G)$ is called *inherently ambiguous*. If an alternative grammar $\tilde{G}$ can be found, then it is *equivalent* to G in the sense that $\mathscr{L}(G) = \mathscr{L}(\tilde{G})$. It is easy to see that this is an equivalence relation on grammars.

By restricting the rules of a phrase structure grammar we obtain a hierarchy of language types. In the theory of automata it is shown that these language types correspond to different kinds of theoretical models for a computer, with differing capabilities.

A *context-sensitive* or *Type 1* grammar is a phrase structure grammar in which all productions are of one of the forms $rpt \to rqt$, $rpt \to rtq$, or $rpt \to qrt$, with $p \in S - T$, $\{r, t\} \subset S^*$, and $q \in S^* - \{\Lambda\}$. Note that while either r or t may be the null string, p and q cannot be null. It can be shown that only productions of the form $rpt \to rqt$ are necessary. See Exercise 9.

A *context-free* or *Type 2* grammar is a context-sensitive grammar for which all contexts r and t are null.

A *regular* or *Type 3* grammar is a context-free grammar with the restriction that the nonterminal p is replaced either by a single terminal, or by a single terminal followed by a single nonterminal.

Since these types of grammars are formed by adding restrictions, each grammar of Type i is automatically a grammar of Type $i-1$, for $i = 1$,

2, and 3. In each case there is a real restriction: there exist grammars (and languages) of Type $i-1$ which are not of Type i.

While it may seem that a context-free language has fewer restrictions than a context-sensitive one since the context does not matter, in fact the restrictions on the descriptive power of context-free grammar are very real. Intuitively the situation is this. While the production

$$\langle \text{noun phrase} \rangle \rightarrow \langle \text{adjective} \rangle \langle \text{noun} \rangle$$

is clearly context-free, it permits such constructions as "paranoid seaweed" and "green ideas." We might avoid such constructions by minutely classifying nouns and adjectives and replacing the given production with a set of productions

$$\langle \text{noun phrase} \rangle \rightarrow \langle A\text{-adjective} \rangle \langle A\text{-noun} \rangle$$

$$\langle \text{noun phrase} \rangle \rightarrow \langle B\text{-adjective} \rangle \langle B\text{-noun} \rangle$$

$$\vdots$$

However, it is conceivable that complexity of the language would not permit us to adequately classify all names and adjectives, and that we would ultimately be forced to use a context-sensitive production. The language in Example 3.2 can be proven to be context-sensitive and not context-free.

Example 3.2

Develop a context-sensitive grammar G for $\mathscr{L}(G) = \{a^n b^n a^n \mid n \geqslant 1\}$.

Solution. Let $G = \langle S, T, \Pi, s \rangle$. Then S and Π are presently unspecified, while $T = \{a, b\}$. An obvious way to proceed is to generate a string of the form $(abc)^n$, for example, *abcabcabc*, sort this into the order $a^n b^n c^n$, and convert the c's to a's. However, there are two difficulties. First, there must be a way of assuring that the sorting is complete before conversion begins, and second, there is no way within the rules for a context-sensitive grammar to convert *ba* to *ab* since both strings consist of terminals only. An alternative approach is to generate pairs consisting of a and a nonterminal t. Since these can be formed into strings $a^n t^n$, part of the sorting problem is eliminated. The remaining portion can be easily handled. In fact, if you can generate new pairs in the middle of a string, $at \vdash a(at)t \vdash aa(at)tt \vdash \cdots$, the generation of $a^n t^n$ proceeds without any sorting. Thus you need a seed u to generate the string at, together with a means of stopping the process, and a nonterminal v, which initiates the formation of the b's. But a specific nonterminal symbol must be available to start the process.

Let $S = \{a, b, s, t, u, v\}$ and let

$$\Pi = \{s \to auv$$
$$s \to av$$
$$u \to aut$$
$$u \to at$$
$$v \to ba$$
$$tb \to bt$$
$$ta \to baa\}.$$

The first four productions produce strings $a^n t^{n-1} v$, for $n \geq 1$. The second and fifth productions are used to generate the string aba, while the fourth production halts the at production for the longer strings. The last three productions convert these strings to the desired form. Note that the context plays a role only in the last two productions, which describe what happens to t. Note also that the productions are used in any arbitrary order. For example, the following derivation is perfectly legitimate.

$$\underline{s} \vdash a\underline{uv} \vdash aa\underline{ut}v \vdash aautb\underline{a} \vdash aaaut\underline{tb}a \vdash$$

$$aaaut\underline{bt}a \vdash aaau\underline{tb}baa \vdash aaaat\underline{tb}baa \vdash$$

$$aaaa\underline{tb}tbaa \vdash aaaabt\underline{tb}aa \vdash aaaabt\underline{bt}aa \vdash$$

$$aaaab\underline{tb}baaa \vdash aaaabb\underline{tb}aaa \vdash aaaabbbt\underline{aa} \vdash$$

$$aaaabbbbaaaa.$$

A context-free grammar can be designed to produce $\{a^n b^n a^m \mid n, m \geq 1\}$. While this contains the language of Example 3.2, there is no context-free way to separate the strings for which $m = n$.

This derivation process is analogous to d-elimination in Gorn trees. However, since we do not necessarily have a unique "definition" of each nonterminal (for example, $u \to aut$ and $u \to at$), and since context plays a role, we cannot guarantee that the process will terminate with the same string, regardless of the order of choice of the productions.

While context-free languages are particularly simple to study, it is difficult to find a realistic language that is context-free. For example, in Fortran IV, IF may be the name of an integer variable, or it may be the first word in a conditional statement. The compiler must examine the context in which IF occurs in order to decide which meaning it has.

Intuitively, the correspondence between language types and computer model types is this. Consider a "computer" that can read an input tape one symbol at a time, do some "computation," and produce an output. Type 3 or regular languages correspond to machines with a fixed amount of memory. Because the memory size is fixed and an input tape can be made arbitrarily long, it is not generally possible for such a machine to remember the entire input tape. This, of course, limits its computational power. Such a machine is called a *finite state machine* or *finite automaton*.

Type 2 or context-free languages correspond roughly to machines with a memory size that is linearly proportional to the length of the input string. For example, with each input string one might provide a "scratch tape" that is twice as long as the input string. Thus such machines can remember the entire input string, but have only a limited amount of memory available for computation. Because of the increased memory, these *linear bounded automata* are more powerful than are finite state machines.

Type 1 or context-sensitive languages correspond to computers with an unlimited "stack" memory in addition to a small "core." That is, they can store unlimited amounts of data, but can only access the last datum which was stored and has not yet been retrieved. These *push-down automata*, with unlimited memory, present a further increase in computational power.

Finally, Type 0 languages correspond to *Turing machines*, which have infinite memory and the ability to search for and retrieve any datum from memory. These are the most powerful computational devices known, and theoretical work over the last half century has led some to believe that it is impossible to design a more powerful computer model.

How do these models correspond to reality? If one ignores economics, then large digital computers are essentially equivalent to Turing machines. Memory is not infinite, but we can make it as large as desired simply by mounting another tape or disk pack. However, the supply of tapes and disk packs is not unlimited, and most computing centers place tight restrictions on the time and memory available to a user. Hence for the average user, a large digital computer is more powerful than a linear bounded automaton, but does not have the infinite memory required to make it a push-down automaton.

EXERCISES

Let $G = \langle S, T, \Pi, s \rangle$, where $S = \{s, a, b, c, d\}$, $T = \{a, b\}$, and

$$\Pi = \{s \to acda$$
$$c \to bacdb$$
$$d \to cab$$
$$ac \to baa$$
$$bdb \to abab\} :$$

1. Show that $baabbabaaabbaba \in \mathscr{L}(G)$.

2. Show that $\mathscr{L}(G)$ contains infinitely many elements (words).

3. Show that if Σ is any string generated from s by Π and including some nonterminal symbols, then it is always possible to extend the generation from Σ to a string involving only terminals. That is, any generation process in G can be forced to terminate.

4. Devise a grammar G' with a generation process that cannot be forced to terminate.

Devise a grammar whose language is the given set:

5. $\{a^{n^2} \mid n \geqslant 1\}$.

6. $\{a^n b^n a^m \mid n, m \geqslant 1\}$.

7. $\{ww \mid w \in \{a, b\}^*\}$. (That is, w is any word in a and b.)

8. $\{ww^R \mid w \in \{a, b\}^*\}$, where w^R is the reversal of w, that is, the string formed by writing the symbols of w in reverse order.

9. Show that if t is a single nonterminal symbol, the production $rpt \to rtq$ can be replaced by a sequence of three productions of the form $uxw \to uyw$.

10. (cont.) Hence, show that in a context-sensitive language a production of the form $rpt \to rtq$ can be replaced by a sequence of productions of the form $uxw \to uyw$, for any string t.

11. Suppose $G = \langle S, T, \Pi, s \rangle$ is a generative grammar such that S contains at most 10 symbols, T contains at most three symbols, and Π contains at most five productions, expressed as ordered pairs of strings. Suppose also that no string in Π contains more than five symbol occurrences. Write a program to accept G as so described and generate all strings in $\mathscr{L}(G)$ having at most 20 symbol occurrences.

12. (cont.) Write a program to accept a string in S^* of length 30 or less, and determine whether it is in $\mathscr{L}(G)$.

4. BACKUS NAUR FORM AND ALGOL-LIKE LANGUAGES

Backus Naur form, abbreviated BNF and also called *Backus normal form*, is a widely used convention for defining programming languages. The elements of the notation are individual symbols, a separator ($|$), a definition symbol ($::=$), and class names, enclosed in angular brackets, such as $\langle \text{digit} \rangle$. We illustrate the use of this notation with a fragment from Algol-60, defining concept "number" [1].

$$\langle \text{digit} \rangle ::= 0 \mid 1 \mid 2 \mid 3 \mid 4 \mid 5 \mid 6 \mid 7 \mid 8 \mid 9$$

$$\langle \text{unsigned integer} \rangle ::= \langle \text{digit} \rangle \mid \langle \text{unsigned integer} \rangle \langle \text{digit} \rangle$$

$$\langle \text{integer} \rangle ::= \langle \text{unsigned integer} \rangle \mid + \langle \text{unsigned integer} \rangle \mid$$
$$- \langle \text{unsigned integer} \rangle$$

$$\langle \text{decimal fraction} \rangle ::= . \langle \text{unsigned integer} \rangle$$

$$\langle \text{exponent part} \rangle ::= {}_{10} \langle \text{integer} \rangle$$

$\langle$decimal number$\rangle$ ::= $\langle$unsigned integer$\rangle\,|\,\langle$decimal fraction$\rangle\,|$
$\qquad\qquad\qquad\quad\langle$unsigned integer$\rangle\,\langle$decimal fraction$\rangle$

$\langle$unsigned number$\rangle$ ::= $\langle$decimal number$\rangle\,|\,\langle$exponent part$\rangle\,|$
$\qquad\qquad\qquad\qquad\langle$decimal number$\rangle\,\langle$exponent part$\rangle$

$\langle$number$\rangle$::= $\langle$unsigned number$\rangle\,|\,+\langle$unsigned number$\rangle\,|$
$\qquad\qquad\quad\,-\langle$unsigned number$\rangle$

In this convention a digit, for example, may be any one of the symbols $0, 1, \ldots, 9$. An unsigned integer is either a digit, or an unsigned integer followed by a digit. This provides a recursive technique for constructing all integers. Given any unsigned integer, we construct another one by appending a digit on the right. Similarly, a decimal fraction is defined to be a decimal point followed by any unsigned integer. Using the definitions given here, we may build up a wide class of "numbers," including the following examples:

$$-147.90, \qquad 34.1_{10}-3, \qquad 0.0341, \qquad .0341, \qquad -24.6_{10}17, \qquad 10^3$$

The second, third, and fourth of these examples are equivalent notations for the same number.

With this notation there is associated a precedence graph. We note for example that we cannot determine what a number is unless we know what an unsigned number is. The definition of "unsigned number" in turn depends on the definitions of "decimal number" and "exponent part." These precedence relations we may indicate in the graph shown in Fig. 4.1.

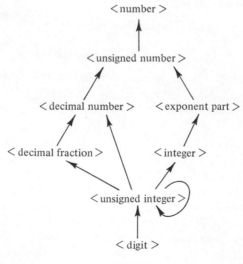

Figure 4.1

The process of definition by BNF is really a process of defining sets by a sequence of set functions. For example, if D represents the set of digits, then $D = \{0, 1, \ldots, 9\}$. Letting U denote the set of unsigned integers, we find that U is defined by the set equation

$$U = D \cup UD,$$

where $UD = \{ud \mid u \in U, d \in D\}$. Continuing in the sequence which we have used as our example, we have six more set equations:

$$I = U \cup \{+\} U \cup \{-\} U \qquad M = U \cup F \cup UF$$

$$F = \{.\} U \qquad\qquad\qquad V = M \cup E \cup ME$$

$$E = \{{}_{10}\} I \qquad\qquad\qquad N = V \cup \{+\} V \cup \{-\} V.$$

This set of definitions involves five distinct set functions:

$$U = f_1(U, D) \qquad M = f_5(U, F)$$

$$I = f_2(U) \qquad\quad V = f_5(M, E)$$

$$F = f_3(U) \qquad\quad N = f_2(V).$$

$$E = f_4(I)$$

We may now substitute into the function defining N the definitions of V and the other sets:

$$
\begin{aligned}
N &= f_2(V) \\
  &= f_2(f_5(M, E)) \\
  &= f_2(f_5(f_5(U, F), f_4(I))) \\
  &= f_2\big(f_5(f_5(U, f_3(U)), f_4(f_2(U)))\big).
\end{aligned}
$$

This equation describes the generation of the set N from the set U. Unfortunately, U is generated from itself and $D, U = f_1(U, D)$, so that the definition appears to end in a circle. However, we may construct U, and hence N, recursively. Let $U_0 = \varnothing$, and $U_{i+1} = f_1(U_i, D), i = 0, 1, 2, \ldots$. Thus U_1 is defined solely in terms of D. That is, the elements of U_1 are simply the digits. Then $U_2 = f_1(U_1, D)$ consists of all single-digit and two-digit unsigned integers, and so forth. Note that $U_i \subset U_{i+1}$. Finally $U = \bigcup_{i=0}^{\infty} U_i$. Thus we have described the process of generating U from the initial set D, and the special symbols $+$, $-$, ., and $_{10}$.

This recursive mode of language definition using set functions has been formalized. The languages resulting are called the *Algol-like* languages since Algol-60 first popularized the use of BNF. It is ironic that because of irregularities in its definition, Algol-60 is not an Algol-like language.

EXERCISES

Consider the following language fragment, in modified BNF.

$$\langle \text{digit} \rangle ::= 0 \,|\, 1 \,|\, \cdots \,|\, 9$$

$$\langle \text{natural} \rangle ::= \{\langle \text{digit} \rangle\}\, k\,(k \geqslant 1)$$

(This means that a $\langle \text{natural} \rangle$ number consists of a sequence of k $\langle \text{digit} \rangle$s, where $k = 1, 2, 3, \ldots$.)

$$\langle \text{natural format part} \rangle ::= \{N\}\, k\,(k \geqslant 1) \,|\, \langle \text{natural} \rangle \{N\}\, k\,(k \geqslant 1)$$

$$\langle \text{natural format} \rangle ::= \text{III } \langle \text{natural format part} \rangle \text{ II} \,|\,$$
$$\text{III } \langle \text{natural format part} \rangle \text{BASE } \langle \text{natural} \rangle \text{II}$$

$$\langle \text{integer format part} \rangle ::= \{I\}\, k\,(k \geqslant 1) \,|\, \langle \text{natural} \rangle \{I\}\, k\,(k \geqslant 1)$$

$$\langle \text{integer format} \rangle ::= \text{III} \pm \langle \text{integer format part} \rangle \text{ II} \,|\,$$
$$\text{III} \pm \langle \text{integer format part} \rangle \text{BASE} \langle \text{natural} \rangle \text{ II}$$

$$\langle \text{real format part} \rangle ::= \{X\}\, k\,(k \geqslant 1) \,|\, \langle \text{natural} \rangle \{X\}\, k\,(k \geqslant 1)$$

$$\langle \text{real format body} \rangle ::= \langle \text{real format part} \rangle . \,|\, . \langle \text{real format part} \rangle \,|\,$$
$$\langle \text{real format part} \rangle . \langle \text{real format part} \rangle$$

$$\langle \text{real format center} \rangle ::= \langle \text{real format body} \rangle \,|\, + \langle \text{real format body} \rangle$$

$$\langle \text{real format} \rangle ::= \text{III } \langle \text{real format center} \rangle \text{ II} \,|\,$$
$$\text{III } \langle \text{real format center} \rangle \text{BASE} \langle \text{natural} \rangle \text{ II}$$

$$\langle \text{string format part} \rangle ::= \{C\}\, k\,(k \geqslant 1) \,|\, \langle \text{natural} \rangle \{C\}\, k\,(k \geqslant 1)$$

$$\langle \text{string format} \rangle ::= \text{III } \langle \text{string format part} \rangle \text{ II}$$

$$\langle \text{type value} \rangle ::= \text{REAL} \,|\, \text{COMPLEX} \,|\, \text{STRING} \,|\, \text{BOOLEAN} \,|\,$$
$$\langle \text{natural format} \rangle \,|\, \langle \text{integer format} \rangle \,|\,$$
$$\langle \text{real format} \rangle \,|\, \langle \text{string format} \rangle$$

(This fragment defines the types of variables which may be defined in the Madcap language. The numerals III and II are specific symbols used by the compiler in format statement decoding.)

1. Draw the precedence graph for this fragment.

2. Write a functional description of $\langle \text{type value} \rangle$ in terms of $\langle \text{digit} \rangle$. (Note that the notation $\{\cdots\}\,(k \geqslant 1)$ implies a recursion.)

Consider the following language fragment, in modified BNF.

$$\langle \text{expsep} \rangle ::= ; \,|\, ,$$

$$\langle \text{such expression} \rangle ::= \ni \langle \text{expression} \rangle$$

$$\langle \text{while expression} \rangle ::= \text{while } \langle \text{expression} \rangle$$

$$\langle \text{compound range part} \rangle ::= \langle \text{expsep} \rangle \langle \text{range expression} \rangle \,|\, \langle \text{range expression} \rangle$$

$$\langle \text{compound range expression} \rangle ::= (\langle \text{range expression} \rangle \{\langle \text{compound range part} \rangle\}\, k\,(k \geqslant 0)) \,|\,$$
$$\{\langle \text{range expression} \rangle \{\langle \text{compound range part} \rangle\}\, k\,(k \geqslant 0)\}$$

$$\langle \text{range primary} \rangle ::= \langle \text{compound range expression} \rangle \,|\,$$
$$\langle \text{compound range expression} \rangle \langle \text{expsep} \rangle$$

⟨range term⟩ ::= ⟨range primary⟩ | ⟨range primary⟩ ⟨such expression⟩

⟨range factor⟩ ::= ⟨range primary⟩ ⟨while expression⟩

⟨range expression⟩ ::= ⟨range term⟩ | ⟨range factor⟩ |
⟨range factor⟩ ⟨such expression⟩ |
⟨while expression⟩ ⟨range term⟩ |
⟨while expression⟩

⟨iterative expression⟩ ::= ⟨range expression⟩ : ⟨expression⟩ |
⟨expression⟩ : ⟨range expression⟩

⟨expression⟩ ::= ⟨iterative expression⟩ | EXP

(This is a small fragment of the Madcap definition of ⟨expression⟩, which actually involves five types of expressions other than ⟨iterative expression⟩s. EXP is a dummy to indicate these other types of expressions.)

3. Draw the precedence graph for this fragment.

4. Write a functional description of ⟨expression⟩ in terms of ⟨expsep⟩.

Using the language fragments defined in this section, write a program to examine a string of at most 30 symbol occurrences:

5. To determine if it is a ⟨digit⟩.

6. To determine if it is a ⟨type value⟩. (REAL is a single symbol, and so on.)

7. To determine if it is an ⟨expression⟩.

5. SEMANTICS OF FORMAL LANGUAGES

The grammars that we have discussed involve only the *syntax* of a formal language—the rules by which grammatically correct sentences are constructed. Nothing has been said about the *semantics* or meaning of a sentence. For example, using definitions from the exercises for Section 4, suppose that R is a defined ⟨range expression⟩. Then "R" is a ⟨compound range expression⟩. Hence "R," is a ⟨range primary⟩, since it is of the form ⟨compound range expression⟩ ⟨expsep⟩. Thus "R," is also a ⟨range term⟩, hence a ⟨range expression⟩. This can be continued indefinitely: "R,;," and "R,;,;" and "R,;,;," and "R,;,;,;" are all range expressions. The meaning of each of these expressions is determined by other considerations, and we may well wish to specify that "R,;,;" is meaningless, even though syntactically (grammatically) correct. A corresponding English example is the sentence, "Colorless green ideas sleep furiously." By ordinary English grammar this is syntactically correct, although meaningless (except possibly poetically). In any compiler or interpreter the syntax checking is quite formal, involving only the internal construction rules (grammar) for the language. The semantic constructs of the language are used in translating the "lines" of program into executable machine code. Some expressions such as "$R,;,;,;$" may be ruled out as meaningless; other expressions will have different meanings for different computer systems. For example, a single-precision arithmetic calculations (say $A = B + C$) will differ on a PDP-10, IBM 370/155, Univac 1108, and CDC Cyber 72, simply because of the different word lengths for the various machines. Even more basically, a syntactically correct program may be semantically incorrect (not perform the intended computation) because of errors in analyzing the problem and writing the program.

6. NATURAL LANGUAGES

Since natural languages are richer and more complex than either formal or programming languages, we would expect their definition to be correspondingly more complex. This is illustrated in the following definition by Kalmár. While the natural language is still basically a semigroup, note that in order to describe its richness 11 different classes of objects relating to the semigroup must be defined.

Kalmár defines a natural language as an 11-tuple $\langle P, R, F, W, C, A, S, M_w, M_s, A_w, A_s \rangle$. (See Marcus [2], pp. 4–6.) The components of this vector specify the structural development of a language from its primitive roots. These roots are the *protosemata P*. For a written language, P is the set of geometrical figures representing the letters or other basic constructs (ideographs) of the language. For a spoken language the protosemata are the basic physical sounds used in pronunciation.

The protosemata are grouped into *graphemes* (written) or *phonemes* (spoken) by the equivalence relation R on the free semigroup P^*. For example, R is the relation which identifies A, a, a, a, and $\mathcal{A}$ as equivalent. The equivalence classes are called *semata*.

The set F of *word forms* is a subset of the free semigroup on the set of semata. The set W of *words* has as elements the sets of related word forms. For example, $\{to\,be, is, are, was, were, ...\}$ and $\{dog, dogs, doggy, ...\}$ are two elements ("words") in W. Note that the elements of W correspond to the primitive symbols in the usual development of a formal language.

The grammar of a natural language is begun with a partition C of W into *word classes* or *parts of discourse*. These classes may be quite general (noun, verb, adjective, ...), or quite specific (nouns referring to biped mammals), depending on the depth of analysis required. It is important to note that C is a partition: a word belongs to exactly one word class.

For each word class c, the *morphological application A* defines a set of grammatical functions applicable to words in c. The purpose of each of these functions is to pick out one of the word forms defining a word. For example, if c is the word class "noun," then functions in $A(c)$ would include "the nominative of ..." and "the accusative of ...." They would not include such functions as "the past tense of ..." or "the superlative of ...." Formally, if $A(c) = G$ and $f \in G$, then f is a set function with the property that for any $w \in c$, $f(w) \in w$.

The set S of *syntactically correct sentences* is a subset of F^*. While this may be defined by an arbitrary generative grammar, in practice the morphological application A plays an important role here. The syntax will specify, for example, that the subject noun of a sentence must be in nominative form; and if it is singular, then the verb form must also be singular.

The sets M_w and M_s consist of *word meanings* and *sentence meanings* respectively. These are roughly the "concepts" and "propositions" of traditional logic. The close similarity of these sets for various languages makes translation from one language to another possible.

Finally, A_w and A_s are functions that define, for each word and sentence respectively, the meanings which are possible for that word or sentence.

The richness of natural languages is mirrored in the complexity of the formal system required to model them. But even this model is inaccurate: well-defined partitions and functions do not exist for natural languages, although they may closely approximate reality. In fairness to comparison with formal languages, it should be noted that Kalmár's system covers a much broader spectrum of structures, from the very primitive sounds or drawings that make up the basic structure of a natural language, through the morphology and syntax of the language, into the complexities of natural semantics. Studies of formal languages generally cover only a small portion of this spectrum.

REFERENCES

1. P. Naur, *et al.* Report on the algorithmic language ALGOL 60. *Commun. ACM*, **3**, (1960) 299–314.
2. S. Marcus, "Algebraic Linguistics; Analytical Models." Academic Press, New York, 1967.

CHAPTER 6

Finite Groups
and Computing

1. DEFINITIONS OF GROUPS AND SUBGROUPS

A group, like a semigroup, has a single binary operation. However, this operation has additional properties beyond those required in a semigroup. These additional properties both limit the class of objects represented by groups, and enable us to say more about the structure of these objects. A *group* is a set G with a binary operator $\cdot$ such that if $a, b, c \in G$, then

(*Closure*)	$a \cdot b \in G$.
(*Associativity*)	$a \cdot (b \cdot c) = (a \cdot b) \cdot c$.
(*Identity*)	There is an element $1 \in G$ such that for any $a \in G$, $1 \cdot a = a \cdot 1 = a$.
(*Inverse*)	For each $a \in G$, there is an element $a' \in G$ such that $a \cdot a' = a' \cdot a = 1$.

A group is *commutative* or *Abelian* if in addition $a \cdot b = b \cdot a$. The multiplication–product terminology is generally used in group theory, although addition–sum terminology is used for commutative groups.

From the definition of a group we can easily establish several basic properties.

P1. The identity element 1 is unique. Suppose for some $a, b \in G$ $a \cdot b = a$. Then

$$b = 1 \cdot b = (a^{-1} \cdot a) \cdot b = a^{-1} \cdot (a \cdot b) = a^{-1} \cdot a = 1.$$

Similarly if $b \cdot a = a$, then $b = 1$.

P2. *Cancellation* holds. That is, for $a, b, c \in G$, if $a \cdot b = a \cdot c$, then $b = c$; and if $b \cdot a = c \cdot a$, then $b = c$. For suppose $a \cdot b = a \cdot c$. Then

$$b = (a^{-1} \cdot a) \cdot b = a^{-1} \cdot (a \cdot b) = a^{-1} \cdot (a \cdot c) = (a^{-1} \cdot a) \cdot c = c.$$

P3. The inverse of an element is unique to that element. For suppose $a \cdot b = 1$. Then, since $a \cdot a^{-1} = 1$, $a \cdot b = a \cdot a^{-1}$.

Hence by P2, $b = a^{-1}$. Similarly, if $b \cdot a = 1$ then $b = a^{-1}$.

Classical examples of groups are the integers under addition, and the nonzero real numbers under multiplication. Note that because of the requirement of the existence of an inverse element, the integers under multiplication do not form a group: there is no integer multiplicative inverse to 2. Observe also that the identity element of the group depends upon the operation defined on the group. For example, 1 is the identity element for the nonzero real numbers under multiplication, but 0 is the identity element for the integers under addition.

These groups are both infinite and commutative, but neither property is required of groups in general. For example, the integers modulo 5 under addition form a finite commutative group (Table 1.1), and the set $\{a, \ldots, f\}$ under the operation defined in Table 1.2 forms a noncommutative group.

TABLE 1.1

	0	1	2	3	4
0	0	1	2	3	4
1	1	2	3	4	0
2	2	3	4	0	1
3	3	4	0	1	2
4	4	0	1	2	3

TABLE 1.2

	a	b	c	d	e	f
a	a	b	c	d	e	f
b	b	a	e	f	c	d
c	c	f	a	e	d	b
d	d	e	f	a	b	c
e	e	d	b	c	f	a
f	f	c	d	b	a	e

The *order* of an element a of a group is the least positive integer n such that $a^n = 1$, where a^n has the usual meaning of $a \cdot a \cdot \cdots \cdot a$, n times. If no such integer exists, then a is said to be of *infinite order*. A group G

is *cyclic* if there is some element $a \in G$ such that every element of G is of the form a^k for some integer k. In this case we say that G is *generated* by a.

Example 1.1

Show that the group of integers modulo 5 under addition is cyclic.

Solution. The group is displayed in Table 1.1. Observe that each element is of the form 1^k:

$$1^1 = 1$$
$$1^2 = 1 + 1 = 2$$
$$1^3 = 1 + 1 + 1 = 3$$
$$1^4 = 1 + 1 + 1 + 1 = 4$$
$$1^5 = 1 + 1 + 1 + 1 + 1 = 0.$$

ANSWER. G is a cyclic group of order 5, generated by 1.

If G is a finite group, then the *order* of G is the number of elements it contains. If G has infinitely many elements, it is of *infinite order*. Observe that if G is cyclic, its order is the same as the order of its generating element.

A *subgroup* of G is a subset of G which is a group in its own right.

Lagrange's Theorem. If H is a subgroup of G, then the order of H is a factor of the order of G.

EXERCISES

Determine if the set with the given multiplication table is a group:

1. Table 1.3.   2. Table 1.4.

3. Table 1.5.   4. Table 1.6.

TABLE 1.3

	a	b	c	d
a	a	c	d	b
b	b	a	c	d
c	d	c	a	b
d	c	d	b	a

TABLE 1.4

	1	2	3	4	5
1	1	2	3	4	5
2	2	3	4	5	1
3	3	4	5	1	2
4	4	5	1	2	3
5	5	1	2	3	4

TABLE 1.6

	a	b	c	d	e	f
a	a	b	c	d	e	f
b	b	d	f	c	a	e
c	c	a	b	e	f	d
d	d	f	e	a	c	b
e	e	c	d	f	b	a
f	f	e	a	b	d	c

TABLE 1.5

	0	1	2	3
0	0	1	2	3
1	1	2	3	4
2	2	3	4	5
3	3	4	5	6

5. Write a program to check a multiplication table of maximum size 25×25 to determine if it represents a group. Test your program on the tables of Exercises 1–4.

6. Show that the group in Table 1.1 is generated by any of its elements other than 0.

7. Show that in general a cyclic group of order n may have elements that are not of order n.

8. Show that the group in Table 1.2 is not cyclic.

9. Let H be an n-element subgroup of G, and suppose that a is a fixed element of G. Determine the number of distinct elements in $\{ah \mid h \in H\}$.

10. (cont.) For a fixed $a \in G$, let $aH = \{ah \mid h \in H\}$. Show that for any $a, b \in G$, either $aH = bH$ or $aH \cap bH = \varnothing$.

11. (cont.) Hence prove LaGrange's theorem by showing that n is a factor of the order of G.

12. Let H and K be two subgroups of a group G. Show that the set intersection $H \cap K$ is a subgroup of G.

13. (cont.) Show that the set union $H \cup K$ of subgroups of G is not generally a subgroup G, but that the set L of all finite products $a_1 a_2 \cdots a_n$, where $a_i \in H \cup K, i = 1, \ldots, n$, is a subgroup of G. L is called the *subgroup generated by H and K*.

2. GROUPS OF GRAPHS

Much of the difficulty in determining graph isomorphism lies in the large number of mappings that may be tried in an attempt to match up two graphs. If G and G' are indeed isomorphic, then they may be regarded as two different labelings of one graph. The problem is to find the mapping that transforms the G labeling into the G' labeling. Mappings that transform a graph into itself (generally with changed labels) are known as *automorphisms* of graph. The set of all automorphisms of a given graph forms the *auto-morphism group* of the graph. This group may vary in size from one element, for the graph of Fig. 2.1, to $n!$ elements for the complete graph K_n. However,

Figure 2.1

the size of the automorphism group alone does not determine the complexity of isomorphism detection. Despite the large number of automorphisms of K_n it is quite simple to determine if a graph is isomorphic to K_n.

Example 2.1

Determine the automorphism group of the graph shown in Fig. 2.2.

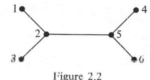

Figure 2.2

Solution.   The key relations here are

1. vertices 2 and 5 are adjacent, each of degree 3;
2. vertices 1 and 3 are adjacent to vertex 2, and each of degree 1;
3. vertices 4 and 6 are adjacent to vertex 5, and each of degree 1.

Represent the graph by the array

$$
\begin{array}{cc}
1 & 4 \\
2 & 5 \\
3 & 6 \\
\end{array}
$$
$$(A) \, .$$

Several other arrays satisfy the three given relations:

1 6	3 4	3 6	4 1	4 3	6 1	6 3
2 5	2 5	2 5	5 2	5 2	5 2	5 2
3 4	1 6	1 4	6 3	6 1	4 3	4 1
(B)	(C)	(D)	(E)	(F)	(G)	(H)

Each of these arrays may be generated from A by a transformation. For example, array B is obtained from A by flipping the right end. Call this transformation r. The transformation l, flipping the left end, produces array C, and the transformation e, exchanging the two ends, produces array E.

188   6   FINITE GROUPS AND COMPUTING

The remaining arrays are all produced by combining these transformations. Array *D* is produced by *rl*, that is, *r* followed by *l*. Arrays *F*, *G*, and *H* are produced by the transformations *er*, *el*, and *erl*, respectively.

Denote these combined tranformations by new letters, $a = rl$, $b = er$, $c = el$, and $d = erl$. Letting *i* denote the identity transformation, which changes nothing, the "multiplication" table for these transformations is given in Table 2.1. It is easy to verify that this is a group, the *dihedral group of order 8, D4*.

TABLE 2.1

	i	a	b	c	d	e	r	l
i	i	a	b	c	d	e	r	l
a	a	i	c	b	e	d	l	r
b	b	c	a	i	r	l	e	d
c	c	b	i	a	l	r	d	e
d	d	e	l	r	i	a	c	b
e	e	d	r	l	a	i	b	c
r	r	l	d	e	b	c	i	a
l	l	r	e	d	c	b	a	i

EXERCISES

Determine the automorphism group for the graph:

1. Fig. 2.3.   2. Fig. 2.4.

3. Fig. 2.5.   4. Fig. 2.6.

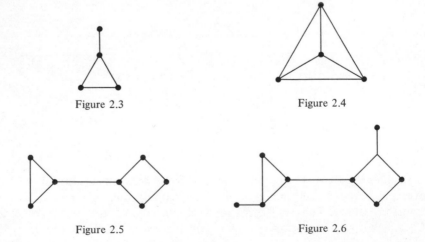

Figure 2.3   Figure 2.4

Figure 2.5   Figure 2.6

*5. Suppose G is a graph of at most 10 vertices, specified by an adjacency matrix. Write a program to determine the automorphism group of G. Test the program on Exercises 1–4.

3. GRAPHS OF GROUPS

Just as we may associate a group with every graph, so also may we associate a directed graph with every group. To do so we need to define two new concepts. Suppose that we are given a set of elements upon which to base a group. That is, the group is defined by multiplying these elements and their inverses together in all possible ways. The elements upon which the group is based are called the *generators* of the group, and the products so defined are called *words*. We may, of course, assign new names to various words in the group. Thus in Example 2.1 the generators are the elements r, l, and e, and the words rl, er, el, and erl have all been given specific names. Any other word in this group, such as $lrerreler$, is a well-defined group element, whose name can be determined by the multiplication table for the group. (This particular element has the name d.) In associating a directed graph with a group, we associate with each group element a vertex, with each generator a directed edge labeled by the name of that generator, and with each word a walk in the digraph. Thus multiplication in the group corresponds to a succession of walks, and a cycle in the graph is a representation for the identity element.

Example 3.1

Determine the graph associated with the group of Example 2.1.

Solution.   Begin with the vertex representing i, and the generators r, l, and e. From i, each generator is an arc directed toward a new vertex, labeled r, l, and e respectively (Fig. 3.1).

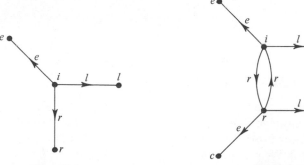

Figure 3.1   Figure 3.2

From another vertex, say r, arcs labeled r, l, and e lead respectively to vertices i since $rr = i$, a since $rl = a$, and c since $re = c$. See Fig. 3.2. Continue in this way to develop the entire graph (Fig. 3.3).

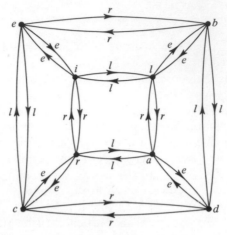

Figure 3.3

Note.   Each word in the group, such as *lrerreler*, can be traced as a walk in the digraph from vertex i to the vertex naming the word (in this case vertex d).

EXERCISES

1. Draw the digraph of the group in Table 1.1.

2. Draw the digraph of the group in Table 1.2, using b and e as the generators.

For the graph, draw the digraph corresponding to its automorphism group (found in Section 2). (Do not expect your first try to be as nice looking as Fig. 3.3. Find the digraph, then rearrange it to look nice.)

3. Fig. 2.2.   4. Fig. 2.3.

5. Fig. 2.4.   6. Fig. 2.5.

4. GENERATORS AND RELATIONS

One of the more common problems encountered is that of equivalence of expressions. Does Fortran statement A mean the same as Fortran statement B? Does program K produce the same results as program L? Can sentences X and Y be translated into the same sentence? For groups, this is known as the *word problem*: given any two words or strings in a group, do they denote the same element?

The general word problem for groups is known to be unsolvable. That is, it is impossible to devise an algorithm that will solve the word problem for all groups. However, much can be done to solve individual cases of this problem.

In this section we study the problem of defining groups by means of generators and relations. The procedures that we use generate words in the group. We then have the problem of determining whether two generated words are equivalent—the word problem. We shall suggest heuristics that are often of use in this determination.

One way of defining a group is by means of a set of generating elements and a set of relations on these elements and the identity element. For example, we might use $\{s, t\}$ as the set of generators, and i to denote the identity element. With no relations (other than $s^0 = t^0 = i$ and $ss^{-1} = s^{-1}s = tt^{-1} = t^{-1}t = i$), we generate the *free group on two generators*, which consists of all strings that can be formed from i and positive and negative integral powers of s and t:

$$\{i, s, s^{-1}, s^2, s^{-2}, s^3, \ldots; t, t^{-1}, t^2, t^{-2}, t^3, \ldots;$$
$$st, st^{-1}, st^2, st^{-2}, \ldots; s^{-1}t, \ldots; ts, \ldots;$$
$$ts^2t, \ldots; s^3t^4s^{-1}t^{-2}s, \ldots\}$$

We may choose to impose equivalence relations such as $s^3 = i$ on this group. The equivalence classes so defined preserve the algebraic structure of the group. That is, suppose a and b are group elements, and $E(a)$ and $E(b)$ are the equivalence classes of a and b respectively. If we define a multiplication by $E(a) \cdot E(b) = \{hk \mid h \in E(a), k \in E(b)\}$, then $E(a) \cdot E(b) = E(ab)$. Thus the equivalence relations define a new group, which may be finite or infinite.

Each relation explicitly defined on a group induces many other relations. For example, the relation $s^3 = i$ induces the relations $s^5 = s^2, t^2s^4t = t^2st$, $t^7s^3ts^5 = t^8s^2$, and so forth. A basic problem in developing a group from generators and relations is to determine the size of the group, so that we know (if the group is finite) when we have generated all of the distinct group elements. This problem involves the word problem since we must decide which words are equivalent under the given relations.

The principal tool in determining the order of a group is a corollary to Lagrange's theorem.

The order of any finite group is a multiple of the order of each of its elements.

The corollary is used in this way. If a is of order 3 and b is of order 7, then any group containing a and b is of infinite order, or else of order $21, 42, 63,$ $84, \ldots$. Thus there must be at least 21 distinct elements. If we find 22 distinct elements, then there must be at least 42 such. And so forth.

Example 4.1

Determine the group defined by $S^3 = T^2 = (ST)^2 = I$.
(I is the group identity.)

Solution.   Since $S^3 = I$ and $T^2 = I$, the order of the group is either a multiple of six, or infinite. Five distinct elements are immediately apparent: I, S, S^2, T, and ST. The mixed relation $(ST)^2 = I$ is a good place to begin looking for other relations. First, multiply on the right by T:

$$STST = I$$

$$STST^2 = IT$$

$$STS = T \qquad \text{(since } T^2 = I).$$

Now multiply this new relation on the right by S^2:

$$ST = TS^2. \qquad (\cdot S^2)$$

Try multiplying it on the left by S^2:

$$TS = S^2 T. \qquad (S^2 \cdot)$$

From the last two relations, observe that whenever S^2 appears with T, it may be replaced by an S (on the other side of the T). Note also that since $STS = T$, the combination STS may always be replaced by T. What about TST? Go back to $STST = I$, and multiply on the left by S^2:

$$TST = S^2. \qquad (S^2 \cdot)$$

Thus any string of more than two characters may be reduced to a one- or two-character string. The sixth element of the group is TS.

ANSWER.   The group is of order 6. Its table is Table 4.1.

TABLE 4.1

	I	S	T	S^2	ST	TS
I	I	S	T	S^2	ST	TS
S	S	S^2	ST	I	TS	T
T	T	TS	I	ST	S^2	S
S^2	S^2	I	TS	S	T	ST
ST	ST	T	S	TS	I	S^2
TS	TS	ST	S^2	T	S	I

A typical word reduction for this group is

$$\underline{S^2 T} S T S T S^2 T S^2 = T\underline{S^2 T} S T S^2 T S^2$$
$$= S T^2 \underline{S T} S^2 T S^2$$
$$= \underline{S^2 T} S^2 T S^2$$
$$= T\underline{S^3} T S^2$$
$$= \underline{T^2} S^2$$
$$= S^2.$$

Two techniques are of general use in determining the group specified by generators and relations. One of these, the determination of the order of each generator, is illustrated in Example 4.2. Knowing that, we have some information about the order of the group. The other technique is to systematically work with strings of increasing length, until we find that all strings of a given length can be reduced to shorter strings. This is illustrated in Example 4.3.

Example 4.2

Determine the group specified by $T^3 = I$, $T^2 S T = S^2$.

Solution. The first question here is that of the order of S. Compute, using the second relation:

$$S^2 = T^2 S T$$
$$S^4 = S^2 \cdot S^2 = T^2 S T \cdot T^2 S T = T^2 S^2 T = T^2 \cdot T^2 S T \cdot T = T S T^2$$
$$S^8 = S^4 \cdot S^4 = T S T^2 \cdot T S T^2 = T S^2 T^2 = T \cdot T^2 S T \cdot T^2 = S.$$

(The technique here is to reduce S^2 to S repeatedly by using the second relation.) Hence $S^7 = I$. Thus S is of order 7, T is of order 3, and hence the group is of order $21, 42, 63, \ldots$, or infinite.

Multiply the second relation on the left by T:

$$S T = T S^2.$$

Thus any S on the left of a T may be moved to the right. The distinct elements are

$$I, \quad S, \quad S^2, \quad S^3, \quad S^4, \quad S^5, \quad S^6, \quad T, \quad TS, \quad TS^2, \quad TS^3, \quad TS^4,$$
$$TS^5, \quad TS^6, \quad T^2, \quad T^2 S, \quad T^2 S^2, \quad T^2 S^3, \quad T^2 S^4, \quad T^2 S^5, \quad T^2 S^6.$$

ANSWER.   The group is of order 21, having the listed elements.

Note. Since in moving S to the right of T, it becomes S^2, the S's begin to pile up. The relation $S^7 = I$ which we discovered saves the day:

$$S^5TS^4TS^3T = S^4TS^2S^4TS^3T = S^3TS^2S^6TS^3T$$

$$= S^2TS^2S^8TS^3T = STS^2S^{10}TS^3T$$

$$= TS^2S^{12}TS^3T \quad = TS^{14}TS^3T$$

$$= TS^7TS^3T \qquad = T^2S^3T$$

$$= T^2S^2TS^2 \qquad = T^2STS^4$$

$$= T^2TS^6 \qquad\quad = T^3S^6$$

$$= S^6.$$

Example 4.3

Find the group generated by $R^4 = S^6 = (RS)^2 = (R^{-1}S)^2 = I$.

Solution. From the given relations, observe that the order of the group, if finite, must be a multiple of 4 and 6, that is, a multiple of 12. Observe that all words of length 1 or 2, namely I, R, S, R^2, RS, SR, and S^2, are apparently distinct. Also note that $R^{-1} = R^3$, so that $(R^{-1}S)^2 = I$ may be rewritten $(R^3S)^2 = I$. Examining the eight strings of length 3, observe that $SRS = R^3$, so there are at most seven distinct strings of this length. That is, there are now potentially 14 distinct strings. Either there must be a way to reduce this number to 12, or there are at least 24 distinct elements in the group. (Why?)

There are 16 words of length 4, but several reductions:

$$R^4 = I \qquad \text{(given)}$$

$$R^3S$$

$$R^2SR$$

$$RSR^2$$

$$SR^3$$

$$R^2S^2$$

$$RSRS = I \qquad \text{(given)}$$

$$RS^2R$$

$$SR^2S$$

$$SRSR = I \qquad (SRS = R^3)$$

$$S^2R^2$$

$$RS^3$$

$$SRS^2 = R^3S \qquad (SRS = R^3)$$
$$S^2RS = SR^3 \qquad (SRS = R^3)$$
$$S^3R$$
$$S^4$$

This produces 11 more unreduced words. Does this mean at least 36 group elements, or are there further reductions? The appearance of such words as R^2SR and RSR^2 suggests a closer examination of the relations $(RS)^2 = (R^3S)^2 = I$. Note that $S^{-1} = S^5$. First,

$$RSRS = I$$
$$SRS = R^3 \qquad (R^3 \cdot)$$
$$SR = R^3S^5 \qquad (\cdot S^5).$$

Then,

$$RSRS = R^3SR^3S$$
$$SRS = R^2SR^3S \qquad (R^{-1} \cdot)$$
$$SR = R^2SR^3 \qquad (\cdot S^{-1})$$
$$SR^2 = R^2S \qquad (\cdot R).$$

From these two relations, $SR = R^3S^5$ and $SR^2 = R^2S$, observe that any R to the right of an S can be moved to its left. Thus there are further reductions, all to a form $R^iS^j, i = 0, 1, \ldots, 3; j = 0, 1, \ldots, 5$.

$$SR = R^3S^5$$
$$RSR = S^5$$
$$SR^2 = R^2S$$
$$S^2R = SR^3S^5 = R^2SRS^5 = R^2R^3S^5S^5 = RS^4$$
$$R^2SR = R^2R^3S^5 = RS^5$$
$$RSR^2 = RR^2S = R^3S$$
$$SR^3 = SR^2R = R^2SR = RS^5$$
$$RS^2R = RSSR = RSR^3S^5 = RSR^2RS^5 = R^3SRS^5 = R^3R^3S^5S^5 = R^2S^4$$
$$SR^2S = R^2S^2$$
$$S^2R^2 = R^2S^2$$
$$S^3R = S^2SR = S^2R^3S^5 = R^2S^2RS^5 = R^2RS^4S^5 = R^3S^3.$$

At this point there are 19 distinct words:

$$I, \quad R, \quad S, \quad R^2, \quad RS, \quad S^2, \quad R^3, \quad R^2S, \quad RS^2, \quad S^3,$$

$$R^3S, \quad R^2S^2, \quad RS^3, \quad S^4, \quad RS^4, \quad S^5, \quad R^3S^3, \quad R^2S^4, \quad RS^5$$

It appears that the group is of order 24, with the remaining elements $R^3S^2, R^2S^3, R^3S^4, R^2S^5, R^3S^5$. (Note that no words of length nine or longer need be considered. Why?)

How do we know these are all distinct? This is left as an exercise.

ANSWER. The group has 24 elements.

Post Mortem. Why did we need the relation $SR^2 = R^2S$? Would not the relation $SR = R^3S^5$ suffice to move all R's to the left of any S's? Unfortunately not, since too many new symbols are generated in the word. Consider SR^2:

$$SR^2 = SRR$$

$$= R^3S^5R \qquad (1)$$

$$= R^3S^4SR$$

$$= R^3S^4R^3S^5 \qquad (2)$$

$$= R^3S^3(SR^2)RS^5$$

In two steps we have generated from SR^2 a longer expression containing SR^2. If it takes k steps to reduce SR^2 to a form R^iS^j, it will take at least k steps to reduce the longer expression. That is, $k > 2+k$, or $0 > 2$. From this contradiction we conclude that reducing SR^2 to the form R^iS^j by use of the relation $SR = R^3S^5$ cannot be done in a finite number of steps. Thus the relation $SR^2 = R^2S$ is definitely necessary.

The problem of determining a group from its generators and relations is essentially a pattern-recognition problem. The techniques illustrated in these examples often, but not always, are useful. As indicated in the post mortem to Example 4.3, it is all too easy to fall into a trap. Nevertheless, a well-written computer program can be of great assistance in determining a group that has been specified by its generators and relations. In addition, if we can determine a standard form for the group elements and specify the transformations necessary to convert any word to this form, as was done in Example 4.3, then we have a very compact means of storing very large and complex groups.

EXERCISES

Determine the group specified by the generators and relations:

1. $R^2 = S^3 = I, (RS)^2 = (SR)^2$.

2. $S^2 = T^2 = (ST)^2$. *Hint:* What is the order of each generator?

3. $S^3 = I, STS = TST$.

4. $A^4 = BAB, B^2 = I$.

5. A single computer program cannot always determine any group from its generators and relations. Nevertheless, write a program that will accept at most three generators and at most six relations as input and materially assist in the discovery of the group so specified. Test your program on Exercises 1–4.

6. If G is a group with identity I, and S an element of G such that $S^p = I$, p a prime, show that the order of S is p.

7. (a) Referring to Example 4.3, show that if not all 24 words given are distinct, then some relation $R^i S^j = I$ holds, with $0 \leqslant i \leqslant 3$ and $0 \leqslant j \leqslant 5$.
   (b) Hence show that either $R^k = I$ for some $k \leqslant 3$, or $S^n = I$ for some $n \leqslant 5$, contrary to the stipulated order of these elements.

5. PERMUTATIONS AND PERMUTATION GROUPS

A *permutation* may be regarded as one of two distinct but related concepts.

1. A permutation is a particular ordered arrangement of the elements in a set, usually finite. For example, there are six permutations of the elements of the set $\{1, 2, 3\}$: 123, 132, 213, 231, 312, 321. In general an n-element set has $n!$ permutations in this sense.

2. A permutation is an operation permuting, or changing the order of the elements in an ordered set. Using again a set of three elements, we may define operations: (i) leave everything where it is; (ii) interchange the first two elements; (iii) interchange the first and third elements; (iv) interchange the second and third elements; (v) put first element in second position, second in third, third in first; (vi) put first element in third position, second in first, third in second. These six operations may be symbolized by

$$\begin{pmatrix} 123 \\ 123 \end{pmatrix}, \quad \begin{pmatrix} 123 \\ 213 \end{pmatrix}, \quad \begin{pmatrix} 123 \\ 321 \end{pmatrix},$$

$$\begin{pmatrix} 123 \\ 132 \end{pmatrix}, \quad \begin{pmatrix} 123 \\ 231 \end{pmatrix}, \quad \begin{pmatrix} 123 \\ 312 \end{pmatrix},$$

where the top row in each case denotes the original positions, the bottom row the positions after permutation. Notice that the six permutations (sense

1) of $\{1, 2, 3\}$ occupy the bottom rows. In this sense also, there are $n!$ permutations on a set of n elements.

We may of course apply several permutations sequentially to a set, with the same net effect as if we had applied one permutation, differently chosen. This leads us to define multiplication of permutations. Let P_1, P_2 and P_3 be permutations. We define the product $P_1 \cdot P_2$ to be P_3 if and only if the effect of applying P_3 to a set is the same as the effect of first applying P_1 and then applying P_2.

Example 5.1

Compute the product $\left(\begin{smallmatrix} 1\,2\,3 \\ 2\,1\,3 \end{smallmatrix}\right) \cdot \left(\begin{smallmatrix} 1\,2\,3 \\ 2\,3\,1 \end{smallmatrix}\right)$.

Solution.   The permutation $\left(\begin{smallmatrix} 1\,2\,3 \\ 2\,1\,3 \end{smallmatrix}\right)$ applied to the ordered set 123 produces the ordered set 213. The permutation $\left(\begin{smallmatrix} 1\,2\,3 \\ 2\,3\,1 \end{smallmatrix}\right)$ is now applied to the ordered set 213. It puts the first element (that is, 2) in the second position, the second element (1) in the third position, and the third element (3) in the first position. The result is the ordered set 321. Thus the effect is the same as that of the permutation $\left(\begin{smallmatrix} 1\,2\,3 \\ 3\,2\,1 \end{smallmatrix}\right)$. Hence $\left(\begin{smallmatrix} 1\,2\,3 \\ 2\,1\,3 \end{smallmatrix}\right) \cdot \left(\begin{smallmatrix} 1\,2\,3 \\ 2\,3\,1 \end{smallmatrix}\right) = \left(\begin{smallmatrix} 1\,2\,3 \\ 3\,2\,1 \end{smallmatrix}\right)$.

ANSWER. $\left(\begin{smallmatrix} 1\,2\,3 \\ 3\,2\,1 \end{smallmatrix}\right)$.

Note.   The product is easily developed by following each individual element. Under the first permutation $1 \to 2$, and under the second $2 \to 3$. Hence under the product, $1 \to 2 \to 3$, or $1 \to 3$. Similarly, $2 \to 1 \to 2$ and $3 \to 3 \to 1$.

Example 5.2

Compute the product $\left(\begin{smallmatrix} 1\,2\,3\,4\,5\,6\,7 \\ 2\,4\,5\,7\,3\,6\,1 \end{smallmatrix}\right)\left(\begin{smallmatrix} 1\,2\,3\,4\,5\,6\,7 \\ 3\,7\,5\,6\,1\,4\,2 \end{smallmatrix}\right)$.

Solution.   Follow each element through the product.

$$1 \to 2 \to 7, \quad 2 \to 4 \to 6, \quad 3 \to 5 \to 1, \quad 4 \to 7 \to 2,$$
$$5 \to 3 \to 5, \quad 6 \to 6 \to 4, \quad 7 \to 1 \to 3.$$

ANSWER. $\left(\begin{smallmatrix} 1\,2\,3\,4\,5\,6\,7 \\ 7\,6\,1\,2\,5\,4\,3 \end{smallmatrix}\right)$.

One may also develop a geometric picture of a permutation by representing the operation as directed arcs in a set of n points (Figs. 5.1 and 5.2). The cycles in these directed graphs suggest the *cycle representation* of a permutation. This representation is obtained by listing the disjoint cycles of the permutation.

Example 5.3

Give cycle representations for the six permutations on $\{1, 2, 3\}$.

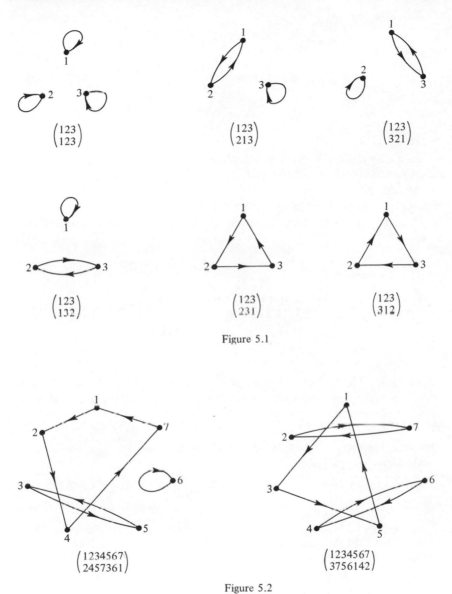

$$\begin{pmatrix}123\\123\end{pmatrix} \qquad \begin{pmatrix}123\\213\end{pmatrix} \qquad \begin{pmatrix}123\\321\end{pmatrix}$$

$$\begin{pmatrix}123\\132\end{pmatrix} \qquad \begin{pmatrix}123\\231\end{pmatrix} \qquad \begin{pmatrix}123\\312\end{pmatrix}$$

Figure 5.1

$$\begin{pmatrix}1234567\\2457361\end{pmatrix} \qquad \begin{pmatrix}1234567\\3756142\end{pmatrix}$$

Figure 5.2

Solution. Under the identity permutation $\begin{pmatrix}123\\123\end{pmatrix}$, each element is mapped into itself in a cycle of length 1. Thus $\begin{pmatrix}123\\123\end{pmatrix}$ is represented $(1)(2)(3)$. The permutation $\begin{pmatrix}123\\132\end{pmatrix}$ has a 1-cycle $(1 \to 1)$ and a 2-cycle $(2 \to 3 \to 2)$. Hence $\begin{pmatrix}123\\132\end{pmatrix}$ is represented $(1)(23)$. Similarly

$$\begin{pmatrix} 123 \\ 213 \end{pmatrix} \quad \text{is} \quad (12)(3), \quad \begin{pmatrix} 123 \\ 321 \end{pmatrix} \quad \text{is} \quad (13)(2),$$

$$\begin{pmatrix} 123 \\ 231 \end{pmatrix} \quad \text{is} \quad (123), \quad \text{and} \quad \begin{pmatrix} 123 \\ 312 \end{pmatrix} \quad \text{is} \quad (132).$$

ANSWER. $(1)(2)(3), (12)(3), (13)(2),$
$(23)(1), (123), (132).$

Since only the cyclic order of the elements is important, the first element listed for any cycle is immaterial. Thus (1427), (4271), (2714), and (7142) all denote the same cycle, but (1472), with a different cyclic order, is a different cycle. By convention, the identity permutation is written simply is (1), and 1-cycles in other permutations are omitted when no confusion arises.

Example 5.4

Give cycle representations for the permutations

$$\begin{pmatrix} 1234567 \\ 2457361 \end{pmatrix} \quad \text{and} \quad \begin{pmatrix} 1234567 \\ 3756142 \end{pmatrix}.$$

Solution.   The cycles can be read from the graphs (Fig. 5.2), or computed by following successive transformations of elements, much as in the product computation. First, take $\begin{pmatrix} 1234567 \\ 2457361 \end{pmatrix}$. Begin with $1: 1 \to 2 \to 4 \to 7 \to 1$. Hence (1247) is a cycle. Choose an element not in the cycle, say $3: 3 \to 5 \to 3$. Hence (35) is a cycle. Then (6) is the remaining cycle.
Similarly, for $\begin{pmatrix} 1234567 \\ 3756142 \end{pmatrix}$,

$$1 \to 3 \to 5 \to 1, \quad 2 \to 7 \to 2, \quad \text{and} \quad 4 \to 6 \to 4.$$

ANSWER. $(1247)(35), (135)(27)(46).$

Note that we do not list the 1-cycle (6) in the first answer. Note also that the answer (1247)(35) is a product of cycles, as is the answer (135)(27)(46).

The definition of the product is retained, and computations may be carried out directly in the cycle notation. The only caution is to remember that the notation is cyclic, so that in (1247), for example, 7 is transformed into 1.

Example 5.5

Compute the products $(12) \cdot (123)$ and $(1247)(35) \cdot (135)(27)(46)$.

Solution. For the first product,

$$
\begin{array}{ll}
\text{(left) (right)} & \\
1 \to 2 \to 3 & \\
3 \to 3 \to 1 & (13) \\
\hline
2 \to 1 \to 2 & (2)
\end{array}
$$

Hence the product is $(13)(2)$, or (13). For the second product,

$$
\begin{array}{ll}
\text{left) (right)} & \\
1 \to 2 \to 7 & \\
7 \to 1 \to 3 & \\
3 \to 5 \to 1 & (173) \\
\hline
2 \to 4 \to 6 & \\
6 \to 6 \to 4 & \\
4 \to 7 \to 2 & (264) \\
\hline
5 \to 3 \to 5 & (5)
\end{array}
$$

Hence the product is $(173)(264)$.

ANSWER. $(13), (173)(264).$

While the cyclic order of elements within any one cycle is important, the order in which disjoint cycles are listed is immaterial, just as the order of elements listed in a set is unimportant. Thus $(173)(264)$ and $(264)(173)$ represent the same permutation. However, if the cycles are not disjoint, the order of listing is quite important. For example, $(123)(124) = (14)(23)$, but $(124)(123) = (13)(24)$.

Transpositions or 2-cycles are particularly important in permutation theory since any cycle of length n can be represented as a product of $n-1$ non-disjoint transpositions,

$$(a_1 a_2 a_3 \cdots a_n) = (a_1 a_2)(a_1 a_3) \cdots (a_1 a_n).$$

Moreover, since the square of any transposition is the identity $[(ab)(ab) = (a)(b)]$, the number of transportations used in representing a given permutation is not fixed, but may be extended by any even number. For example,

$$(1247) = (12)(14)(17)$$

$$= (12)(14)(17)(23)(23)$$

$$= (12)(14)(17)(23)(23)(34)(34),$$

and so forth.

A permutation is *odd* or *even* if the number of transpositions used to represent it is odd or even, respectively.

It will be shown in the exercises that the set of all permutations on n elements forms a group, the *symmetric group of degree n*, $\mathscr{S}_n$, and that the set of all even permutations on n elements forms a subgroup of this, the *alternating group of degree n*, $\mathscr{A}_n$.

A *permutation matrix* is a square matrix having a single 1 in each row and column, 0's everywhere else. There are clearly $n!$ such matrices, each representing one of the permutations of $\mathscr{S}_n$. Matrix multiplication corresponds exactly to the permutation product which we have defined.

Example 5.6

Give the permutation matrices for the six permutations on three elements, and for the permutations (1247)(35) and (135)(27)(46).

Solution.   In the matrix, $a_{ij} = 1$ if and only if $i \to j$ under the permutation.

ANSWER.

$$(1): \begin{bmatrix} 100 \\ 010 \\ 001 \end{bmatrix}, \quad (12): \begin{bmatrix} 010 \\ 100 \\ 001 \end{bmatrix},$$

$$(13): \begin{bmatrix} 001 \\ 010 \\ 100 \end{bmatrix}, \quad (23): \begin{bmatrix} 100 \\ 001 \\ 010 \end{bmatrix},$$

$$(123): \begin{bmatrix} 010 \\ 001 \\ 100 \end{bmatrix}, \quad (132): \begin{bmatrix} 001 \\ 100 \\ 010 \end{bmatrix},$$

$$(1247)(35): \begin{bmatrix} 0100000 \\ 0001000 \\ 0000100 \\ 0000001 \\ 0010000 \\ 0000010 \\ 1000000 \end{bmatrix},$$

$$(135)(27)(46): \begin{bmatrix} 0010000 \\ 0000001 \\ 0000100 \\ 0000010 \\ 1000000 \\ 0001000 \\ 0100000 \end{bmatrix}.$$

EXERCISES

Compute the product of the permutations:

1. $\binom{12345}{24513} \cdot \binom{12345}{25314}$.

2. $\binom{1234567}{4123567} \cdot \binom{1234567}{5432176}$.

3. $(135)(24) \cdot (15423)$.

4. $(12)(345)(67) \cdot (1234)(567)$.

5. $(152)(376) \cdot (1425)(36) \cdot (174)(2653)$.

6. $(17)(26)(35) \cdot (147)(2653) \cdot (1234567)$.

Compute all distinct powers of the permutation:

7. (12345).

8. $(12)(435)$.

9. (136254).

10. (123456).

11. Write a computer program to multiply two permutations on at most 10 symbols, in cycle form as input. Test your program on Exercises 3–6.

12. Let π_1, π_2, and π_3 be permutations on n elements. Show that $(\pi_1 \pi_2)\pi_3 = \pi_1(\pi_2 \pi_3)$.

13. (cont.) Show that the set of all permutations on n elements forms a group. What is the order of the group?

14. (cont.) Show that the set of all even permutations on n elements forms a group. What is its order?

15. Show that no permutation can be both odd and even.

16–21. Represent the permutations of Exercises 1–6 as matrices and compute the products.

6. PERMUTATION GENERATORS

Permutations are sufficiently important in computer applications that a large number of algorithms have been written for generating them. The prime criterion for a permutation generator is that it generate all permutations on n objects, without repetition. Beyond that, a specific order of generation, or speed of generation are the most important criteria.

The most commonly specified order for permutations is *lexicographic order*. For four objects, this order is given in Table 6.1. Unfortunately, the fastest generators do not produce a lexicographic ordering of the permutations.

TABLE 6.1

abcd	bacd	cabd	dabc
abdc	badc	cadb	dacb
acbd	bcad	cbad	dbac
acdb	bcda	cbda	dbca
adbc	bdac	cdab	dcab
adcb	bdca	cdba	dcba

Four of the best permutation generators from the *Communications of the Association for Computing Machinery* (*Comm. ACM*) are presented here.[†] These are written in Algol-60, which is fully described in *Comm. ACM*, May 1960.

Algorithm 1[†]

```
procedure PERM(x, n);   value n;   integer n;   array x;
begin own integer array p, d[2:n];   integer k, q;   real t;
    if first then initialize:
    begin for k := 2 step 1 until n do
        begin p[k] := 0;   d[k] := 1 end;
        first := false
    end initialize;
    k := 0;
    INDEX: p[n] := q := p[n] + d[n];
        if q = n then
        begin d[n] := −1;   go to LOOP end;
        if q ≠ 0 then go to TRANSPOSE;
        d[n] := 1; k := k + 1;
        LOOP: if n > 2 then begin
            n := n − 1; go to INDEX end LOOP;
        Final exit: q := 1;   first := true;
        TRANSPOSE: q := q + k; t := x[q]; x[q] := x[q + 1]; x[q + 1] := t
end PERM;
```

[†] Algorithm 1 from H. F. Trotter, Algorithm 115, *Comm. ACM* **5** (1962) 434–435; copyright 1962, Association for Computing Machinery, Inc. Reprinted by permission of the Association for Computing Machinery.

Algorithm 2[†]

```
procedure ECONOPERM(x, n);   value n;   integer n;   array x;
begin own integer array q[2; n];
    integer k, l, m;   real t;
```

```
  l := 1;   k := 2;
  if first then begin first := false; go to label end initialization;
loop: if q[k] = k then
    begin if k < n then
        begin k := k + 1; go to loop end
        else begin first := true; go to finish end
    end;
  n := k - 1;
label: for m := 2 step 1 until n do q[m] := 1; q[k] := q[k] + 1;
transpose: t := x[l]; x[l] := x[k]; x[k] := t; l := l + 1; k := k - 1;
    if l < k then go to transpose;
finish:
end of procedure ECONOPERM
```

† Algorithm 2 from R. J. Ord-Smith, Algorithm 308, *Comm. ACM* **10** (1967) 452; copyright 1967, Association for Computing Machinery, Inc. Reprinted by permission at Association for Computing Machinery.

Algorithm 3 †

```
procedure BESTLEX(x, n);   value n;   integer n;   array x;
begin own integer array q[2:n];   integer k, m;   real t;
    if first then
    begin first := false;
        for m := 2 step 1 until n do q[m] := 1
    end of initialization process;
    if q[2] = 1 then
    begin q[2] := 2; t := x[1]; x[1] := x[2]; x[2] := t;
        go to finish
    end;
    for k := 2 step 1 until n do
        if q[k] = k then q[k] := 1 else go to trstart;
first := true; k := n; go to trinit;
trstart: m := q [k]; t := x[m]; x[m] := x[k]; x[k] := t; q[k] := m + 1;
    k = k - 1;
trinit: m := 1;
transpose: t := x[m]; x[m] := x[k]; x[k] := t; m := m + 1; k := k - 1;
    if m < k then go to transpose:
finish:
end of procedure BESTLEX
```

† Algorithm 3 from R. J. Ord-Smith, Algorithm 323, *Comm. ACM* **11** (1968) 117; copyright 1968, Association for Computing Machinery, Inc. Reprinted by permission of Association for Computing Machinery.

Algorithm 4[†]

new *ECONOPERM*:

modify *BESTLEX* by replacing

entire *trstart* sequence by *trstart*: $q[k] := q[k] + 1$.

Relative timings for these algorithms are given in Table 6.2 (from *Comm. ACM*).

TABLE 6.2

	t_7	t_8	Number of transpositions
Old *ECONOPERM*	6.2	50.6	$1.175n!$
BESTLEX	6	47	$1.53n!$
New *ECONOPERM*	5.9	45	$1.175n!$
PERM	5.6	43	$n!$

t_7	time for permutations on seven objects
t_8	time for permutations on eight objects
Number of transpositions	limiting number of required transpositions

Note that all but *PERM* require more than $n!$ transpositions to produce $n!$ permutations.

[†] Algorithm 4 from R. J. Ord-Smith, Remark on Algorithm 308, *Comm. ACM* **12** (1969) 638; copyright 1969, Association for Computing Machinery, Inc. Reprinted by permission of Association for Computing Machinery.

Example 6.1

Analyze the operation of the algorithm *PERM*.

Solution. *PERM* is a procedure that will produce a new permutation of an ordered set x each time it is called. The first two lines of the procedure consist of needed declarations. The declaration "**value** n" is a *call by value*, indicating that each time the procedure is called, the variable n will have the value defined in the calling program, regardless of any changes in its value effected by prior use of the procedure. The arrays p and d are vectors with components $p_2, p_3, \ldots, p_n$, and $d_2, d_3, \ldots, d_n$. They control the choice of transposition.

Lines 3–7 of the procedure constitute the initialization phase. If *first* has the value "true," then initialization takes place. The entries of p are all set to 0; those of d are all set to 1. Then *first* is set to "false," so that the initialization procedure is bypassed on subsequent procedure calls.

To illustrate use of *PERM*, assume $n = 4$, and let $x = \langle a, b, c, d \rangle$.

First Call.   Initialize: $p = \langle 0,0,0 \rangle$, $d = \langle 1,1,1 \rangle$. The statement *INDEX* sets p_4 and q to 1. Control passes to *TRANSPOSE*, which interchanges x_1 and x_2, producing the permutation $x = \langle b,a,c,d \rangle$.

Second Call.   Initialization is bypassed. From the first call, $p = \langle 0,0,1 \rangle$ and $d = \langle 1,1,1 \rangle$. *INDEX* sets p_4 and q to 2. Then *TRANSPOSE* interchanges x_2 and x_3, producing $x = \langle b,c,a,d \rangle$.

Third Call. $p = \langle 0,0,2 \rangle$, $d = \langle 1,1,1 \rangle$, $x = \langle b,c,d,a \rangle$.

Fourth Call. $p = \langle 0,0,3 \rangle$, $d = \langle 1,1,1 \rangle$; $p_4 = q = 4$. Since $q = n$, d_4 is set to -1 and control passes to *LOOP*, which changes n to 3 and passes control back to *INDEX*. Then $p_3 = q = 1$, and control passes to *TRANS-POSE*, producing an interchange of x_1 and x_2: $x = \langle c,b,d,a \rangle$.

Fifth Call.   Because of the call by value, n has the value 4, despite the change to 3 in the previous procedure call.

The full sequence of operation for 24 procedure calls is given in Table 6.3. Note that a shuttles back and forth in these permutations. After the twenty-fourth call, *first* is reset to "true," so that the sequence is reinitialized.

TABLE 6.3

p	d	x	p	d	x
—	—	abcd	1,3,4	1,−1,−1	dcab
0,0,0	1,1,1	bacd	1,3,3	1,−1,−1	dacb
0,0,1	1,1,1	bcad	1,3,2	1,−1,−1	adcb
0,0,2	1,1,1	bcda	1,3,1	1,−1,−1	adbc
0,0,3	1,1,1	cbda	1,2,0	1,−1,1	dabc
0,1,4	1,1,−1	cbad	1,2,1	1,−1,1	dbac
0,1,3	1,1,−1	cabd	1,2,2	1,−1,1	dbca
0,1,2	1,1,−1	acbd	1,2,3	1,−1,1	bdca
0,1,1	1,1,−1	acdb	1,1,4	1,−1,−1	bdac
0,2,0	1,1,1	cadb	1,1,3	1,−1,−1	badc
0,2,1	1,1,1	cdab	1,1,2	1,−1,−1	abdc
0,2,2	1,1,1	cdba	1,1,1	1,−1,−1	abcd[a]
0,2,3	1,1,1	dcba			

[a] Termination; *first* set to "true".

EXERCISES

1. Analyze the action of *PERM* for $n = 5$.

2. Analyze the action of old *ECONOPERM*.

3. Analyze the action of *BESTLEX*.

4. Analyze the action of new *ECONOPERM*.

CHAPTER 7

Partial Orders and Lattices

1. INTRODUCTION

In Chapter 1 we defined a partial order to be a relation ρ that is reflexive ($x\rho x$), antisymmetric (if $x\rho y$ and $y\rho x$, then $x = y$), and transitive (if $x\rho y$ and $y\rho z$, then $x\rho z$). In this chapter we study partial orders, particularly the special type of partial order called a lattice. Lattices have application both to explicit definition (Chapter 3), and to the development of Boolean algebra (Chapter 8). More general types of partial orders apply to any classification scheme. In particular, recall that $<$ and $<_{\mathscr{L}}$ are both partial orders (the latter being linear) on Gorn's universal address set (Chapter 3).

2. PARTIAL ORDERS

Whenever there is a noncyclic relationship on the elements of a set, this relationship provides a natural partial order for the set. For example, any cycle-free directed graph displays explicitly a partial order on the vertices of the graph. In particular, the flowchart for a loop-free program displays a partial order on the computation steps, the order in which the computations are performed. Indeed, small loops can even be accommodated into the order if each entire loop is considered as a single entity.

Other partial orders arise whenever analysis and classification of any set are undertaken. For example, the subset relation $A \subseteq B$ is a natural partial order. Note that the word *partial* is significant: the order relation need not be everywhere defined. If $A = \{1, 2\}$, $B = \{2, 3\}$, and $C = \{1, 2, 3\}$, then $A \subseteq C$ and $B \subseteq C$, but neither A nor B is a subset of the other, despite the fact that they have an element in common.

Often the useful partial orders are related to the linear order of the integers, or to an alphabetical order.

Example 2.1

Let S be the set of integers greater than 1. Show that the relation $\prec$ defined by $x \prec y$ if x is a divisor of y is a partial order on S, and diagram the partial order.

Solution. Beginning with any integer, $\prec$ leads to chains of multiples of that integer:

$$2 \prec 4 \prec 8 \prec 16 \prec 32 \prec \cdots$$

$$2 \prec 6 \prec 12 \prec 24 \prec 48 \prec \cdots$$

$$2 \prec 10 \prec 20 \prec 40 \prec \cdots$$

$$\vdots$$

$$3 \prec 6 \prec 12 \prec 24 \prec \cdots$$

$$3 \prec 9 \prec 18 \prec \cdots$$

$$3 \prec 15 \prec 30 \prec \cdots$$

$$\vdots$$

$$4 \prec 12 \prec 24 \prec \cdots$$

$$\cdots$$

Observe that the relation $\prec$ is reflexive, antisymmetric, and transitive. Hence it is a partial order on S. Observe also that the relation does not hold between each pair of numbers (for example, neither $2 \prec 3$ nor $3 \prec 2$ holds), and that the chains being developed link together at various points (for example, $2 \prec 6 \prec 12 \prec \cdots$, $3 \prec 6 \prec 12 \prec \cdots$, and $4 \prec 12 \prec 24 \prec \cdots$). The key to diagramming this order is the number of prime factors of each integer. If the ith row from the bottom of the diagram contains all integers having exactly i prime factors, not necessarily distinct, then the diagram comes easily.

ANSWER.   See Fig. 2.1.

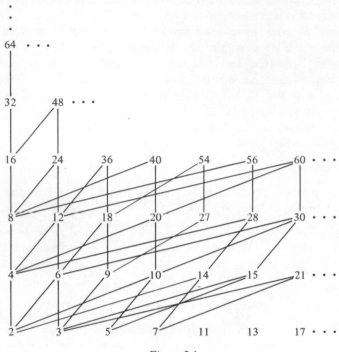

Figure 2.1

Let S be a partially ordered set (under $\leqslant$), $R \subseteq S$, $R \neq \emptyset$. Then $a \in S$ is an *upper (lower) bound for* R if for each $x \in R$, $x \leqslant a$ $(a \leqslant x)$. The *supremum of* R (sup R) is the least upper bound of R, if such exists. That is, $b = \sup R$ if b is an upper bound of R and for each upper bound a of R, $b \leqslant a$. Similarly, the *infimum of* R (inf R) is the greatest lower bound of R, if it exists.

EXERCISES

1. Show that there is a natural partial order on the set of trees (words) obtained by repeated applications of d-elimination to a given tree (word). (See Chapter 3.)

2. (cont.) Show that there is a natural partial order on the set of trees (words) obtained by repeated applications of d-introduction to a given tree (word). How is this order related to that induced by d-elimination?

3. Show that sup R and inf R, if they exist, are each unique.

3. LATTICES

Lattices are defined in terms of two operations analogous to set intersection and union. These induce a partial order analogous to subset ordering. The triple $\langle L, \wedge, \vee \rangle$ is a *lattice* if L is a set, and $\wedge$ (*meet*) and $\vee$ (*join*) are binary operations on L satisfying

(*Associativity*)	$L_1.$	$(a \wedge b) \wedge c = a \wedge (b \wedge c)$
	$L_2.$	$(a \vee b) \vee c = a \vee (b \vee c)$
(*Commutativity*)	$L_3.$	$a \wedge b = b \wedge a$
	$L_4.$	$a \vee b = b \vee a$
(*Absorption*)	$L_5.$	$a \wedge (a \vee b) = a$
	$L_6.$	$a \vee (a \wedge b) = a.$

Note that meet and join are defined for all pairs of elements of L, and yield *unique* results. That is, if $a \wedge b = x$ and $a \wedge b = y$, then $x = y$. Note also the *duality* in the defining conditions. Each condition on meet holds also for join, and vice versa. In effect, there are two operations on L, either one of which may be taken as "meet," and the other as "join."

Example 3.1

Show that the set operations union and intersection define a lattice on the set of subsets of a given set. (Recall the "subset lattice" of Chapter 1.)

Solution. Associativity, commutativity, and absorption hold for both union and intersection. Hence we may define a lattice with union as meet, and intersection as join, or with intersection as meet, and union as join.

Example 3.2

Let N be the set of positive integers, and define $\wedge$ and $\vee$ as the greatest common divisor and the least common multiple, respectively. Show that $\langle N, \wedge, \vee \rangle$ is a lattice.

Solution. The greatest common divisor (gcd) of two integers is the largest integer which is a factor (divisor) of both; the least common multiple (lcm) is the smallest integer which is a multiple of both. Clearly gcd and lcm are commutative.

Factor integers a, b, and c into prime powers:

$$a = p_1^{a_1} p_2^{a_2} \cdots p_n^{a_n}, \qquad b = p_1^{b_1} p_2^{b_2} \cdots p_n^{b_n}, \qquad c = p_1^{c_1} p_2^{c_2} \cdots p_n^{c_n},$$

where the a_i, b_i, and c_i are nonnegative integers. Then

$$a \wedge b = \gcd(a, b) = p_1^{\min(a_1, b_1)} p_2^{\min(a_2, b_2)} \cdots p_n^{\min(a_n, b_n)},$$

where $\min(a_i, b_i)$ is the smaller of a_i and b_i; and $a \vee b$ is similarly defined using $\max(a_i, b_i)$.

Let $\prod_{i=1}^{n} p_i^{x_i} = p_1^{x_1} p_2^{x_2} \cdots p_n^{x_n}$. Then

$$(a \wedge b) \wedge c = \prod_{i=1}^{n} p_i^{\min(a_i, b_i)} \wedge \prod_{i=1}^{n} p_i^{c_i}$$

$$= \prod_{i=1}^{n} p_i^{\min(\min(a_i, b_i), c_i)}$$

$$= \prod_{i=1}^{n} p_i^{\min(a_i, b_i, c_i)}$$

$$= \prod_{i=1}^{n} p_i^{\min(a_i, \min(b_i, c_i))}$$

$$= \prod_{i=1}^{n} p_i^{a_i} \wedge \prod_{i=1}^{n} p_i^{\min(b_i, c_i)}$$

$$= a \wedge (b \wedge c).$$

Next,

$$a \wedge (a \vee b) = \prod_{i=1}^{n} p_i^{a_i} \wedge \prod_{i=1}^{n} p_i^{\max(a_i, b_i)}$$

$$= \prod_{i=1}^{n} p_i^{\min(a_i, \max(a_i, b_i))}.$$

If $a_i < b_i$, then

$$\min(a_i, \max(a_i, b_i)) = \min(a_i, b_i) = a_i.$$

If $a_i \geqslant b_i$, then

$$\min(a_i, \max(a_i, b_i)) = \min(a_i, a_i) = a_i.$$

Thus $a \wedge (a \vee b) = \prod_{i=1}^{n} p_i^{a_i} = a$. The properties L_2, L_4, and L_6 are shown similarly.

Example 3.3

Let $\mathscr{F}$ be the set of real functions defined on $[0, 1]$. Define $\wedge$, $\vee$ by, for each $x \in [0, 1]$:

$$(f \wedge g)(x) = \min(f(x), g(x)),$$

$$(f \vee g)(x) = \max(f(x), g(x)).$$

Show that $\langle \mathscr{F}, \wedge, \vee \rangle$ is a lattice.

Solution. The solution to Example 3.2 depended on the fact that gcd and lcm can be defined in terms of minimum and maximum functions. Here meet and join are defined directly in terms of minimum and maximum. The solution is exactly parallel to that of Example 3.2.

Use of the two absorption properties shows that meet is idempotent, that is, $a \wedge a = a$. In particular

$$a \wedge a = a \wedge (a \vee (a \wedge b)) \qquad \text{(by } L_6\text{)}$$

$$= a \qquad \text{(by } L_5\text{)}.$$

Similarly, join is idempotent, $a \vee a = a$.

From idempotency it follows that $a \wedge b = a \vee b$ if and only if $a = b$. On the one hand, if $a = b$ then $a \wedge b = a \vee b = a$. On the other hand, if $a \wedge b = a \vee b$, then

$$a = a \vee (a \wedge b) = a \vee (a \vee b) = (a \vee a) \vee b = a \vee b$$

$$= b \vee a = (b \vee b) \vee a = b \vee (b \vee a) = b \vee (a \vee b)$$

$$= b \vee (a \wedge b) = b \vee (b \wedge a) = b.$$

The lattice operations provide a natural partial order, defined by $a \leqslant b$ if and only if $a \wedge b = a$. Alternatively, this order can be defined by $a \leqslant b$ if and only if $a \vee b = b$. The order provides a useful way to diagram a lattice, following the basic scheme of Fig. 3.1. The subset lattice diagram for $\{a, b, c\}$ is an example of this (Fig. 3.2).

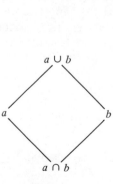

Figure 3.1

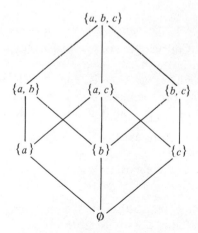

Figure 3.2

Example 3.4

Diagram the lattice of factors of 210, under gcd and lcm.

Solution.   The factors of 210 are 1, 2, 3, 5, 6, 7, 10, 14, 15, 21, 30, 35, 42, 70, 105, and 210. The same prime factorization idea as used in Example 2.1 provides the basis of this diagram.

ANSWER.   See Fig. 3.3.

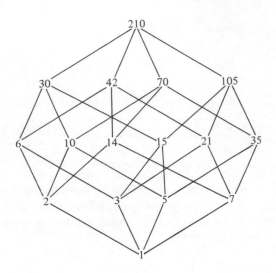

Figure 3.3

Conversely, a partial order on a set S can often be used to define a lattice. In order that this be possible, it is *not* required that any two elements be comparable, that is, that $a \leqslant b$ or $b \leqslant a$. However, it *is* required that any two elements of S have a unique greatest lower bound and a unique least upper bound in common. That is, for any $a, b \in S$,

1. there exists a $c \in S$ such that $c \leqslant a$, $c \leqslant b$, and if $d \leqslant a$ and $d \leqslant b$, then $d \leqslant c$;
2. there exists a $c' \in S$ such that $a \leqslant c'$, $b \leqslant c'$, and if $a \leqslant d'$ and $b \leqslant d'$, then $c' \leqslant d'$.

Note that although the uniqueness of c and c' is not explicitly stated, it follows easily from the requirements that are stated. The lattice is then defined by $a \wedge b = c$, $a \vee b = c'$.

Example 3.5

Let $S = \{a, b, c, d, e, f, g, h, i\}$, and suppose that the order relation $\leqslant$ is specified by

$$a \leqslant b \leqslant e \leqslant g \leqslant i, \qquad a \leqslant c \leqslant f \leqslant g,$$
$$a \leqslant d \leqslant f \leqslant h \leqslant i, \qquad d \leqslant e.$$

Show that meet ($\wedge$) and join ($\vee$) can be defined so that $\langle S, \wedge, \vee \rangle$ is a lattice.

Solution. It is necessary to determine unique least upper bounds and greatest lower bounds for all pairs of elements. These are given by Table 3.1. (Verify the uniqueness of the table entries.)

TABLE 3.1

x	y	$\mathrm{glb}(x, y)$	$\mathrm{lub}(x, y)$	x	y	$\mathrm{glb}(x, y)$	$\mathrm{lub}(x, y)$
a	a	a	a	c	i	c	i
a	b	a	b	d	d	d	d
a	c	a	c	d	e	d	e
a	d	a	d	d	f	d	f
a	e	a	e	d	g	d	g
a	f	a	f	d	h	d	h
a	g	a	g	d	i	d	i
a	h	a	h	e	e	e	e
a	i	a	i	e	f	d	g
b	b	b	b	e	g	e	g
b	c	0	e	e	h	d	i
b	d	0	e	e	i	e	i
b	e	b	e	f	f	f	f
b	f	0	g	f	g	f	g
b	g	b	g	f	h	f	h
b	h	0	i	f	i	f	i
b	i	b	i	g	g	g	g
c	c	c	c	g	h	f	i
c	d	0	f	g	i	g	i
c	e	0	g	h	h	h	h
c	f	c	f	h	i	h	i
c	g	c	g	i	i	i	i
c	h	c	h				

ANSWER.   The lattice diagram is Fig. 3.4.

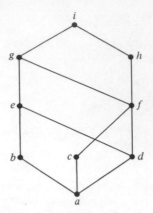

Figure 3.4

In summary, the important lattice properties are

Associativity of meet and join,
Commutativity of meet and join,
Uniqueness of meet and join,
Duality of meet and join,
Absorption laws,
Idempotency: $a \wedge a = a \vee a = a$,
$a \wedge b = a \vee b$ if and only if $a = b$,
$a \leqslant b$ if and only if $a \wedge b = a$,
$a \leqslant b$ if and only if $a \vee b = b$.

EXERCISES

1. Complete Example 3.2 by establishing L_2, L_4, and L_6.

2. Show that $a \vee a = a$.

3. Show that $a \wedge b = a$ if and only if $a \vee b = b$.

4. Show that if $a \leqslant x$ and $a \leqslant y$, then $a \leqslant x \wedge y$.

5. Diagram the subset lattice of $\{1, 2, 3, 4\}$.

6. Diagram the lattice of the factors of 150 under gcd and lcm.

7. Consider the functions $f(x) = x^2$, $g(x) = x/2$, and $h(x) = 1 - x$ on $[0, 1]$. For these functions define a lattice by minimization and maximization, as in Example 3.3.

8. Show that Fig. 3.5 is not a lattice diagram.

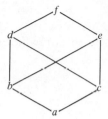

Figure 3.5

9. Let H and K denote subgroups of a group G. Show that the subgroups of G form a lattice if $H \wedge K$ is the set intersection of H and K, and $H \vee K$ is the subgroup generated by H and K. (See Exercises 6 and 7, Chapter 6, Section 1.)

**10. (See Exercise 1, Section 2.) Show that the partially ordered set of words obtained by d-elimination applied to a given word is a lattice. *Note:* While this seems to be true, it is in fact an open research problem.

4. SPECIALIZED LATTICES

Typically in mathematics, as we impose more conditions on a structured concept, we eliminate many objects from study. Thus, although Fig. 3.5 represents a partial order, it does not represent a lattice. Hence it is not an object that can be studied in lattice theory. In compensation for this wholesale elimination of objects, those objects that remain have a much richer structure and can be described in more detail. Everything that is true about semigroups is true about groups, but the latter have additional properties; everything that is true about partial orders is true about lattices, but the latter have a richer structure.

In this section we will enrich this structure still further by restricting lattices in various ways. For each restriction, we will give examples of lattices that do and do not satisfy the restriction. Our purpose in this section and the next is to define concepts leading to Boolean algebras. For a more detailed account of lattice theory see Szasz [1].

Most of the lattices we shall encounter have both a maximal and a minimal element. We say that a lattice L is *bounded* if there exist elements $0, 1 \in L$ such that for any $x \in L$, $0 \leqslant x \leqslant 1$. The *lower bound* is 0 and the *upper bound* is 1. A subset lattice is obviously bounded, with $\varnothing$ as 0 and the set itself is 1. At the other extreme, the lattice of integers under the usual order

$$\cdots < -3 < -2 < -1 < 0 < 1 < 2 < 3 < \cdots$$

(the "chain of integers") is clearly unbounded.

Refer back to Fig. 2.1. While this is not a lattice (for example, $2 \wedge 3$ is missing), it can be made into one by including 1 as a divisor of all prime numbers (Fig. 4.1). The resulting lattice has a lower bound, but no upper bound.

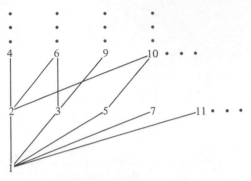

Figure 4.1

Although every finite lattice is bounded (see Exercise 1), not every bounded lattice is finite. For example, the lattice of real numbers between (and including) 0 and 1 under the usual order is bounded and infinite.

A *sublattice* of a lattice L is a subset of L which is itself a lattice under the meet and join of L. Thus if R is a sublattice of L, and $a, b \in R$, then $a \wedge b \in R$ and $a \vee b \in R$. If $a \leqslant b$ in L, then the *interval* $[a, b]$ in L is the set of all $x \in L$ such that $a \leqslant x \leqslant b$. Thus the interval $[a, b]$ is a sublattice of L with lower bound a and upper bound b.

Bounds are important in lattice theory since they permit us to define complements. In particular, if L is a bounded lattice, then $x \in L$ is a *complement* for $u \in L$ if $u \wedge x = 0$ and $u \vee x = 1$. It is easy to show that 0 and 1 always have complements. (See Exercise 3.) However, it is not necessary that any other element in a bounded lattice have a complement. For example, in the chain lattice of Fig. 4.2, $a \wedge 0 = 0$ but $a \vee 0 = a$; and for any $x \neq 0$, $a \wedge x = a$. Hence a has no complement. Similarly, b and c do not have complements in this lattice.

Even if a complement to an element exists, it need not be unique. Thus in the lattice of Fig. 4.3 both a and c are complements of b. Note, however, that a and c each have a unique complement, namely b. If an element u has a unique complement, this complement is denoted by u'. A lattice L is *complemented* if each element of L has a complement in L, and *uniquely complemented* if each element has a unique complement in L. The lattice of Fig. 4.3 is thus complemented, but not uniquely complemented. The complete subset lattice for any set (containing all subsets of the set) is an example of a uniquely complemented lattice.

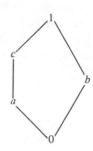

Figure 4.2 Figure 4.3

A lattice L is *distributive* if for any a, b, c in L

$$a \wedge (b \vee c) = (a \wedge b) \vee (a \wedge c) \qquad \text{and} \qquad a \vee (b \wedge c) = (a \vee b) \wedge (a \vee c).$$

Note that meet distributes over join *and* join distributes over meet, as we should expect because of the duality principle. The complete subset lattices once again provide examples of distributive lattices, while the lattices of Figs. 4.3 and 4.4 are nondistributive. For the lattice in Fig. 4.3, $c \wedge (a \vee b) = c \wedge 1 = c$, while $(c \wedge a) \vee (c \wedge b) = a \vee 0 = a$. For the lattice in Fig. 4.4, $a \wedge (b \vee c) = a \wedge 1 = a$, while $(a \wedge b) \vee (a \wedge c) = 0 \vee 0 = 0$. In fact, any nondistributive lattice contains a sublattice isomorphic to one of the lattices of Figs. 4.3 and 4.4.

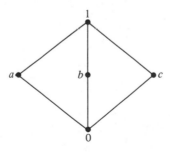

Figure 4.4

A lattice L is *complete* if every subset of L has both a supremum and an infimum. Any finite lattice is complete (see Exercise 11); the simplest incomplete lattice is the chain of integers, $\cdots < -3 < -2 < -1 < 0 < 1 < 2 < 3 \cdots$.

Finally, a lattice L is *modular* if any $a, b, c \in L$ with $a \leqslant c$ satisfy $a \vee (b \wedge c) = (a \vee b) \wedge c$. The subgroup lattice of any Abelian group is modular (see Exercise 13), but the subgroup lattice of a non-Abelian group is generally not modular. For example, in the subgroup lattice of the alternating group A_4, shown in Fig. 4.5, $c \vee (a \wedge h) \neq (c \vee a) \wedge h$. In another example, even when d-elimination in Gorn trees produces a lattice, this lattice may well be nonmodular. (See Exercise 21.)

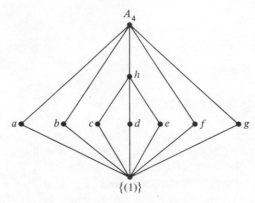

$$a = \{(1), (123), (132)\} \qquad e = \{(1), (14)(23)\}$$

$$b = \{(1), (124), (142)\} \qquad f = \{(1), (134), (143)\}$$

$$c = \{(1), (12)(34)\} \qquad g = \{(1), (234), (243)\}$$

$$d = \{(1), (13)(24)\} \qquad h = \{(1), (12)(34), (13)(24), (14)(23)\}$$

Figure 4.5

It can be shown that every nonmodular lattice contains a sublattice isomorphic to the lattice of Fig. 4.3. Thus there is a relationship between distributive and modular lattices. (See Exercise 12.) Looking again at Figs. 4.3 and 4.4, we note that in both lattices a and c are complements of b. In addition, we observe that in Fig. 4.3 a and c are comparable, that is, $a \leqslant c$. The conditions for distributivity and modularity may thus be stated in this manner.

A lattice L is *distributive* if no interval $[a,b]$ of L contains an element having two distinct complements in $[a,b]$.

A lattice is *modular* if no interval $[a,b]$ of L contains an element having two distinct comparable complements in $[a,b]$.

EXERCISES

1. Show that any finite lattice is bounded.

2. Show that if a bounded lattice has more than one element, then $0 \neq 1$.

3. Show that $0' = 1$, and $1' = 0$.

4. (a) Show that the statement "If u has a unique complement u', then $u'' = (u')' = u$." is not generally true.
   (b) Show that this statement is true in a uniquely complemented lattice.

5. Let n be a positive integer. Under what conditions is the lattice of divisors of n complemented?

6. Show that either one of the conditions for a distributive lattice implies the other one.

7. Show that in any lattice the distributive laws hold under each of the following conditions:

   (a) any of the elements a, b, c is 0;   (b) any of the elements a, b, c is 1;
   (c) any two of the elements a, b, c are equal; (d) $a \leqslant b \wedge c$;
   (e) $a \geqslant b \vee c$;   (f) $b \leqslant a \wedge c$ (or $c \leqslant a \wedge b$);
   (g) $b \geqslant a \vee c$ (or $c \geqslant a \vee b$).

8. Let n be a positive integer. Under what conditions is the lattice of divisors of n distributive?

9. Show that a complemented distributive lattice is uniquely complemented. *Hint:* Show that in a distributive lattice, if b and c are complements of a, then so are $b \vee c$ and $b \wedge c$. Then show that this implies that $b = c$.

10. Prove that every finite lattice is complete.

11. In any lattice, show that the modularity law holds under each of the following conditions:

    (a) any of the elements a, b, c is 0;   (b) any of the elements a, b, c is 1;
    (c) any two of the elements a, b, c are equal; (d) $b \leqslant a$;
    (e) $a \leqslant b \leqslant c$;   (f) $c \leqslant b$.

12. Show that every distributive lattice is modular.

13. Show that the subgroup lattice of any Abelian group is modular.

    Determine whether the lattice has each of the properties complemented, uniquely complemented, distributive, modular:

14. Fig. 4.6.   15. Fig. 4.7.

16. Fig. 4.8.   17. Fig. 4.9.

18. Fig. 4.10.   19. Fig. 4.11.

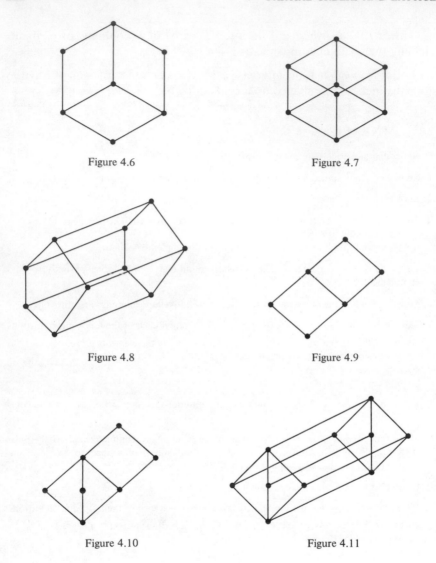

Figure 4.6   Figure 4.7

Figure 4.8   Figure 4.9

Figure 4.10   Figure 4.11

20. Using the results of Exercises 7, 11, and 12 write a program to check a lattice of at most 10 points for the properties complemented, uniquely complemented, distributive, and modular. Assume that the lattice is given by ordered pairs $\langle a, b \rangle$, where $a \leqslant b$, and there is no other x such that $a \leqslant x \leqslant b$. Test your program on the lattices of Figs. 4.6—4.11.

21. Let $T = ddxyz$ be a (prefix form) Gorn tree, where $d\alpha_1 \alpha_2 =_{df} +\alpha_1 + \alpha_2 \alpha_1$. Show that the diagram of d-eliminations from T is a lattice under the order $T_1 \leqslant T_2$ if and only if T_1 is obtainable from T_2 by d-eliminations. Determine whether the lattice is modular.

5. ATOMIC LATTICES

An *atom* in a lattice with 0 is a nonzero element a such that for any x, $x \leqslant a$ implies $x = a$ or $x = 0$. For any elements a and b in a lattice, define $a < b$ (and $b > a$) to mean that $a \leqslant b$ and $a \neq b$. Given a non-zero element x in a lattice with 0, either x is an atom, or there is some element a_1 such that $x > a_1 > 0$. If a_1 is not an atom, then we may similarly find an element a_2 such that $x > a_1 > a_2 > 0$. This procedure may be repeated: $x > a_1 > a_2 > \cdots > a_n > 0$. Now if the lattice is finite, we must eventually find an element a_N that is an atom. (Why?) Thus in any finite lattice with 0, for any nonzero element x there is an atom a such that $a \leqslant x$. We express this by saying that the lattice is *atomic*. Not every infinite lattice is atomic.

If a is an atom, then for any element x in the lattice, either $a \leqslant x$ or $a \wedge x = 0$. This arises from the fact that $a \wedge x \leqslant a$, and hence (since a is an atom) $a \wedge x = a$ or $a \wedge x = 0$.

If L is a bounded atomic lattice, and $x \subset L$, we let $A(x)$ be the set of all atoms a such that $a \leqslant x$. In particular, $A(1)$ is the set of all atoms of L.

Example 5.1

Let L be the lattice of Fig. 5.1. Determine $A(x)$ for each $x \in L$.

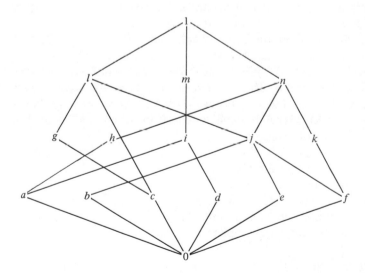

Figure 5.1

Solution. First determine the set of atoms. These are the elements immediately above ($>$) the 0 of the lattice. Hence $A(1) = \{a, b, c, d, e, f\}$. At the other extreme, $A(0) = \varnothing$. For each $x \in L$, $A(x) = \{y \mid y \leqslant x$ and $y \in A(1)\}$. For instance, $A(l) = \{b, c, e, f\}$.

ANSWER.

$$
\begin{array}{ll}
A(0) = \varnothing, & A(a) = \{a\}, \\
A(b) = \{b\}, & A(c) = \{c\}, \\
A(d) = \{d\}, & A(e) = \{e\}, \\
A(f) = \{f\}, & A(g) = \{c\}, \\
A(h) = \{a\}, & A(i) = \{a, d\}, \\
A(j) = \{b, e, f\}, & A(k) = \{f\}, \\
A(l) = \{b, c, e, f\}, & A(m) = \{a, d\}, \\
A(n) = \{a, b, e, f\}, & A(1) = \{a, b, c, d, e, f\}.
\end{array}
$$

These sets have a number of interesting properties. First, $A(x \wedge y) = A(x) \cap A(y)$. Suppose $a \in A(x \wedge y)$. Then $a \leqslant x \wedge y$. But $x \wedge y \leqslant x$ and $x \wedge y \leqslant y$. Hence $a \leqslant x$ and $a \leqslant y$, or $a \in A(x) \cap A(y)$. Thus $A(x \wedge y) \subseteq A(x) \cap A(y)$. Now if $a \in A(x) \cap A(y)$, then $a \leqslant x$ and $a \leqslant y$. Hence $a \leqslant x \wedge y$ (see Exercise 4, Section 3), so that $a \in A(x \wedge y)$. Thus $A(x) \cap A(y) \subseteq A(x \wedge y)$, so equality is established.

If L is a complemented distributive lattice and y is a complement of $x \in L$, then $x \wedge y = 0$. Hence $A(x) \cap A(y) = A(x \wedge y) = A(0) = \varnothing$. Thus $A(y)$ is contained in the (set) complement of $A(x)$, $A(y) \subseteq A(1) - A(x)$. We wish to show that $A(y) = A(1) - A(x)$. Thus take $a \in A(1) - A(x)$. Then $a \wedge x = 0$. If $a \wedge y = 0$, then $a \wedge (x \vee y) = (a \wedge x) \vee (a \wedge y) = 0$. But $a \wedge (x \vee y) = a \wedge 1 = a$. That is, if $a \wedge y = 0$ then $a = a \wedge (x \vee y) = 0$. But since a is an atom, $a \neq 0$. Hence we must have $a \wedge y = a$, so that $a \in A(y)$, and $A(1) - A(x) \subseteq A(y)$. We conclude that in a complemented distributive lattice, if y is a complement of x, then $A(y) = A(1) - A(x)$.

In any lattice, clearly if $x = y$ then $A(x) = A(y)$. However, if the lattice is complemented and distributive, the converse holds: if $A(x) = A(y)$, then $x = y$. To show this, let z be a complement of x. Then $A(z) = A(1) - A(x) = A(1) - A(y)$. Thus z is a complement of y. Hence x and y are both complements of z. But the lattice, being complemented and distributive, is uniquely complemented (see Exercise 9, Section 4); hence $x = y$.

Similarly, in any distributive lattice, if $a_1, \ldots, a_k$ are atoms then $A(a_1 \vee \cdots \vee a_k) = \{a_1, \ldots, a_k\}$. We leave the proof of this as an exercise.

Properties of Bounded Atomic Lattices

Lattice	*Property*
any	$A(x \wedge y) = A(x) \cap A(y)$
distributive	$A(a_1 \vee \cdots \vee a_k) = \{a_1, \ldots, a_k\}$
complemented distributive	$A(x') = A(1) - A(x)$
complemented distributive	$A(x) = A(y)$ if and only if $x = y$
complemented distributive	$A(x \vee y) = A(x) \cup A(y)$

EXERCISES

1. Prove that if $a_1, \ldots, a_k$ are atoms in a distributive lattice, then $A(a_1 \vee \cdots \vee a_k) = \{a_1, \ldots, a_k\}$.

2. Show that if $a, b \in L$, a complemented distributive lattice, then $A(a) \cup A(b) = A(a \vee b)$. *Hint:* $A(a') = A(1) - A(a) = \overline{A(a)}$.

REFERENCE

1. G. Szasz, "Introduction to Lattice Theory." Academic Press, New York, 1963.

CHAPTER 8

Boolean Algebras

1. INTRODUCTION

We define a *Boolean algebra* to be a complemented distributive lattice. In discussing Boolean algebras we change our terminology and notation from that of meet and join to that of multiplication and addition respectively. Since a Boolean algebra is a lattice, both multiplication and addition are associative and commutative, and the absorption and idempotency properties hold as they do for any lattice. Since the lattice is complemented, there exist elements 0 and 1 that act respectively as the additive and multiplicative identities. Since 0 and 1 are the lower and upper bounds respectively for the lattice, we also have the properties

$$0 + a = a, \qquad 1 + a = 1,$$

$$0 \cdot a = 0, \qquad 1 \cdot a = a,$$

for any element a in the Boolean algebra. Finally, since the lattice is distributive we have the distributive laws

$$a(b+c) = ab + ac, \qquad a + bc = (a+b)(a+c).$$

226

Use of the words *addition* and *multiplication* draws attention to the analogy between Boolean algebras and ordinary algebras. As we proceed differences between these will become apparent. Some of these differences are worth noting now. For example, the absorption and idempotency properties do not hold for ordinary algebra. In addition, while Boolean addition distributes over Boolean multiplication, the corresponding statement is not true for ordinary addition and multiplication. Finally, we note that in ordinary arithmetic $1 + a$ is not generally equal to 1, whereas equality between these holds for Boolean algebras.

We call attention also to the fact that each element a in a Boolean algebra has a complementary element, denoted by a'. This complementary element is simply the complement whose existence is guaranteed by the definition of a Boolean algebra as a complemented lattice. Since the lattice is also distributive, we know that the complement to any given element is unique. (See Exercise 9, Chapter 7, Section 4.)

2. PROPERTIES OF BOOLEAN ALGEBRAS

In the previous chapter, the duality properties of lattices were noted. The operations meet and join have exactly corresponding properties. Since a Boolean algebra is a lattice, this same duality property holds for Boolean algebras. It is extremely useful in establishing properties of the algebras.

Principle of Duality   If a property can be established for a Boolean algebra, then the property derived from the given one by interchanging the operations of addition and multiplication, and the elements 0 and 1, holds and can be established by corresponding changes in the proof of the given property.

Among the most useful properties of a Boolean algebra are *DeMorgan's laws*, discovered by the British mathematician Augustus De Morgan. These laws provide the means of converting multiplication into addition, and vice versa. They are

$$(ab)' = a' + b', \qquad (a+b)' = a'b'.$$

We know that $(ab)'$ is the unique complement to the element ab. Thus if we can establish that $a' + b'$ has the properties that we expect of the complement to ab, it follows that $a'+b'$ is another expression for $(ab)'$. Let us establish these properties:

$$(ab)(a'+b') = aba' + abb' \qquad (ab)+(a'+b') = (a+a'+b')(b+a'+b')$$
$$= 0b + a0 \qquad\qquad\qquad\qquad = (1+b')(1+a')$$
$$= 0 + 0 \qquad\qquad\qquad\qquad\quad = 1\cdot 1$$
$$= 0. \qquad\qquad\qquad\qquad\qquad = 1.$$

Thus we have established the first of DeMorgan's laws. To illustrate the application of the principle of duality, we give a parallel derivation of the second law:

$$(a+b) + a'b' = (a+b+a')(a+b+b') \qquad (a+b)a'b' = aa'b' + ba'b'$$
$$= (1+b)(a+1) \qquad\qquad\qquad\qquad = 0b' + 0a'$$
$$= 1\cdot 1 \qquad\qquad\qquad\qquad\qquad = 0 + 0$$
$$= 1. \qquad\qquad\qquad\qquad\qquad\quad = 0.$$

Each line in these second derivations is the dual of the corresponding line in the first derivations. Thus having the first derivations, the second are mechanically obtained.

We observe in passing that since 0 and 1 are complements in any complemented lattice, and since they are distinct in any lattice with two or more elements, these properties hold for Boolean algebras. Thus a 1-element Boolean algebra consists of a single element which may be written either 0 or 1; a 2-element Boolean algebra consists of the two elements 0 and 1.

There do not exist Boolean algebras for an arbitrary number of elements. For example, if there is a third element a in a Boolean algebra, then there must also be a fourth element a'. We note that a' cannot be either 0 or 1 since the latter two elements are each other's unique complement. However, if $a' = a$, then the equations $aa' = aa = a$, and $aa' = 0$ force us to conclude that $a = 0$. Hence a' cannot be the element a, and must in fact be a fourth distinct element. Thus there does not exist a three-element Boolean algebra.

EXERCISES

1. Show that with but one exception any finite Boolean algebra has an even number of elements.

2. Write the addition and multiplication tables for a two-element and a four-element Boolean algebra.

3. Show that there is no six-element Boolean algebra.

3. BOOLEAN ALGEBRAS AND SET ALGEBRAS

It is easy to verify that a set algebra for any finite set is a complemented distributive lattice. Thus each such set algebra is an example of a Boolean algebra. The converse is also true: each finite Boolean algebra is isomorphic to the complete set algebra for some given finite set. In particular, this implies that each finite Boolean algebra contains 2^n elements, for some nonnegative integer n. In this section we establish this fact.

To establish the isomorphism with set algebras, we wish to exhibit for any particular finite Boolean algebra a set algebra that is isomorphic to it. This is done using the concept of an "atom," introduced in Chapter 7. Recall that for any lattice element a we use $A(a)$ to denote the set of all atoms contained in a. Recall also that in an atomic lattice, $A(a) = A(b)$ if and only if $a = b$. That is, the mapping $a \to A(a)$ is a one-to-one mapping from the elements of an atomic lattice onto the collection of atomic sets for that lattice.

Now, since a Boolean algebra is a complemented distributive lattice, if it is finite, it is an atomic lattice. Thus we have a one-to-one mapping from the elements of the Boolean algebra to the atomic sets. In Section 5, Chapter 7, we showed that this mapping preserves algebraic operations in the sense that multiplication and complementation within the Boolean algebra translate directly into set intersection and set complementation on the sets $A(a)$. In addition, through the use of DeMorgan's laws we can show that addition in the Boolean algebra corresponds to set union among these sets of atoms. (See Exercise 2 in Section 5, Chapter 7.) Thus each finite Boolean algebra is isomorphic to the set algebra on the set whose elements are the atomic sets for the given Boolean algebra.

Example 3.1

Determine a set algebra isomorphic to the eight-element Boolean algebra.

Solution. The addition and multiplication tables for the 8-element Boolean algebra are given in Table 3.1. From these tables it follows that the atoms of the algebra are a, b, and c. Thus $A(a) = \{a\}$, while $A(a') = \{b, c\}$, for example, since $a'a = 0$, $a'b = b$, and $a'c = c$.

ANSWER. See Table 3.2.

Since the set algebras depend only on the number of elements in a set, and not the particular elements, it can be rigorously established that any two finite Boolean algebras with the same number of elements are isomorphic. We note in passing that there are infinite Boolean algebras, and

TABLE 3.1

+	0	a	a'	b	b'	c	c'	1
0	0	a	a'	b	b'	c	c'	1
a	a	a	1	c'	b'	b'	c'	1
a'	a'	1	a'	a'	1	a'	1	1
b	b	c'	a'	b	1	a'	c'	1
b'	b'	b'	1	1	b'	b'	1	1
c	c	b'	a'	a'	b'	c	1	1
c'	c'	c'	1	c'	1	1	c'	1
1	1	1	1	1	1	1	1	1

	0	a	a'	b	b'	c	c'	1
0	0	0	0	0	0	0	0	0
a	0	a	0	0	a	0	a	a
a'	0	0	a'	b	c	c	b	a'
b	0	0	b	b	0	0	b	b
b'	0	a	c	0	b'	c	a	b'
c	0	0	c	0	c	c	0	c
c'	0	a	b	b	a	0	c'	c'
1	0	a	a'	b	b'	c	c'	1

TABLE 3.2

Element	Set of atoms
0	$\varnothing$
a	$\{a\}$
a'	$\{b, c\}$
b	$\{b\}$
b'	$\{a, c\}$
c	$\{c\}$
c'	$\{a, b\}$
1	$\{a, b, c\}$

that the situation for these is somewhat more complicated. The above discussion depends on the fact that any finite Boolean algebra is atomic. It is not true that every infinite Boolean algebra is atomic. In particular, set algebras based on an infinite set but not containing all subsets of the given set provide examples of nonatomic Boolean algebras.

EXERCISES

1. Give an example of an infinite atomic Boolean algebra.

2. Give an example of an infinite nonatomic Boolean algebra. *Hint:* Does interchanging the interpretations of addition and multiplication in your answer to Exercise 1 help?

4. BOOLEAN FUNCTIONS

In any algebraic system we may define functions mapping the algebra into itself. For Boolean algebras, certain of these functions are termed *Boolean functions*. A Boolean function is any function on a Boolean algebra which can be derived from the constant function and the projection functions by use of the Boolean operations addition, multiplication, and complementation. That is, if $x_1, \ldots, x_n$ are variables over of a Boolean algebra, and a is a fixed element of the algebra, then $f(x_1, \ldots, x_n) = a$ is a Boolean function, the *constant function*. In addition, for each $i = 1, \ldots, n$, the function $g_i(x_1, \ldots, x_n) = x_i$ is a Boolean function, the *ith projection function*. In particular, if $n = 1$, the function $g(x) = x$ is the *identity function*. Furthermore, given any two Boolean functions f and g of n variables, new Boolean functions are defined through the use of the three Boolean operations:

$$h(x_1, \ldots, x_n) = (f(x_1, \ldots, x_n))'$$

$$j(x_1, \ldots, x_n) = f(x_1, \ldots, x_n) + g(x_1, \ldots, x_n)$$

$$k(x_1, \ldots, x_n) = f(x_1, \ldots, x_n) \cdot g(x_1, \ldots, x_n).$$

Iteration of this process a finite number of times results in the development of the complete class of Boolean functions on n variables.

Informally, the Boolean functions may be written as polynomials in constants and variables of the algebra, together with their complements, such as the following:

$$a' + xy, \qquad ax + bx'y + xyz', \qquad ax'y + bxy'.$$

(In contrast to ordinary polynomials, no exponents larger than 1 are needed. Why?)

Because of DeMorgan's laws, the absorption laws, and other properties of Boolean algebras, there is no unique expression for a given Boolean function. However, we can define a standard or *canonical form* to which all Boolean functions may be transformed. This form is

$$f(x_1, \ldots, x_n) = \sum f_{e_1 e_2 \cdots e_n} x_1^{e_1} x_2^{e_2} \cdots x_n^{e_n},$$

where $f_{e_1 e_2 \cdots e_n} = f(e_1, e_2, \ldots, e_n)$, the e_i take the values 0 and 1, and $x_i^{e_i}$ is interpreted as x_i or x_i' accordingly as e_i has the value 1 or 0. The sum is over all 2^n combinations of values of the e_i. Thus

$$f(x) = f_1 x + f_0 x', \qquad f(x, y) = f_{11} xy + f_{10} xy' + f_{01} x'y + f_{00} x'y',$$

and so forth. We establish that this is the case for the single variable Boolean function, leaving the case for bivariate Boolean functions as an exercise.

We consider all possible forms of f.

Case 1. f is a constant function, $f(x) = a$. Then $f_1 = f_0 = a$.

$$f_1 x + f_0 x' = ax + ax' = a(x + x') = a \cdot 1 = a = f(x).$$

Case 2. f is the identity function, $f(x) = x$. Then $f_1 = 1$, $f_0 = 0$.

$$f_1 x + f_0 x' = 1 \cdot x + 0 \cdot x' = x = f(x).$$

Case 3. Suppose f is expressible in canonical form, and let $h(x) = (f(x))'$.

$$
\begin{aligned}
h(x) = (f(x))' &= (f_1 x + f_0 x')' \\
&= (f_1 x)'(f_0 x')' \\
&= (f_1' + x')(f_0' + x) \\
&= f_1' f_0' + f_1' x + f_0' x' + x'x \\
&= f_1' f_0' 1 + f_1' x + f_0' x' \\
&= f_1' f_0'(x + x') + f_1' x + f_0' x' \\
&= f_1' f_0' x + f_1' f_0' x' + f_1' x + f_0' x' \\
&= f_1' x(f_0' + 1) + f_0' x'(f_1' + 1) \\
&= f_1' x \cdot 1 + f_0' x' \cdot 1 \\
&= f_1' x + f_0' x' \\
&= h_1 x + h_0 x'.
\end{aligned}
$$

Case 4. Suppose f and g are expressible in canonical form, and let $j(x) = f(x) + g(x)$.

$$
\begin{aligned}
j(x) = f(x) + g(x) &= f_1 x + f_0 x' + g_1 x + g_0 x' \\
&= (f_1 + g_1)x + (f_0 + g_0)x' \\
&= j_1 x + j_0 x'.
\end{aligned}
$$

Case 5. Suppose f and g are expressible in canonical form, and let $k(x) = f(x)g(x)$.

$$
\begin{aligned}
k(x) = f(x)g(x) &= (f_1 x + f_0 x')(g_1 x + g_0 x') \\
&= f_1 g_1 x + f_1 g_0 xx' + f_0 g_1 x'x + f_0 g_0 x' \\
&= f_1 g_1 x + f_0 g_0 x' \\
&= k_1 x + k_0 x'.
\end{aligned}
$$

Thus in all cases, the canonical form

$$f(x) = f_1 x + f_0 x'$$

is a valid representation of the function.

Example 4.1

Let 0, a, a', and 1 be the elements of a four-element Boolean algebra. Construct the canonical form for the function $f(x, y)$ with values shown in Table 4.1.

TABLE 4.1

x	y	$f(x, y)$
0	0	a
0	1	0
1	0	a'
1	1	1

Solution. The canonical form for $f(x, y)$ is $f(x, y) = f_{11}xy + f_{10}xy' + f_{01}x'y + f_{00}x'y'$, regardless of the number of elements in the Boolean algebra. Thus

$$f(x, y) = 1 \cdot xy + a' \cdot xy' + 0 \cdot x'y + a \cdot x'y'.$$

ANSWER. $f(x, y) = xy + a'xy' + ax'y'$.

It will be noted that regardless of the number of elements in a Boolean algebra the coefficients in the canonical form for a function are the function values at the "corners," that is, at points whose coordinates are all 0 and 1. If these function values are given, then the function is completely determined; if, however, other function values are given, then there may be no Boolean functions, or several Boolean functions that satisfy the given values, as shown in the following examples.

Example 4.2

Construct all Boolean functions over a four-element Boolean algebra with the values given in Table 4.2.

TABLE 4.2

x	y	$f(x, y)$
0	a	a
1	1	1
a'	a	a'
a	1	a

Solution.   Substitute the table values into the canonical form

$$f(x, y) = f_{11}xy + f_{10}xy' + f_{01}x'y + f_{00}x'y'.$$

$\langle x, y \rangle = \langle 0, a \rangle$: $a = f_{01}a + f_{00}a'$

$\langle x, y \rangle = \langle 1, 1 \rangle$: $1 = f_{11}$

$\langle x, y \rangle = \langle a', a \rangle$: $a' = f_{10}a' + f_{01}a$

$\langle x, y \rangle = \langle a, 1 \rangle$: $a = f_{11}a + f_{01}a'.$

Multiply the first equation by a: $a = f_{01}a$. Thus $f_{01} = 1$ or $f_{01} = a$.
Multiply the third equation by a: $0 = f_{01}a$. Thus $f_{01} = 0$ or $f_{01} = a'$. This
contradicts the previously determined values.

ANSWER.   There is no such Boolean
function.

Example 4.3

Construct all Boolean functions over a four-element Boolean
algebra with the values given in Table 4.3.

TABLE 4.3

x	y	$f(x, y)$
0	a	a
1	1	a'
a'	a	a
a	1	0

Solution.   Substitute the values into the canonical form.

$\langle 0, a \rangle$: $a = f_{01}a + f_{00}a'$ (1)

$\langle 1, 1 \rangle$: $a' = f_{11}$ (2)

$\langle a', a \rangle$: $a = f_{10}a' + f_{01}a$ (3)

$\langle a, 1 \rangle$: $0 = f_{11}a + f_{01}a'.$ (4)

Multiply (1) by a: $f_{01}a = a$. Hence $f_{01} = a$ or $f_{01} = 1$.
Multiply (1) by a': $f_{00}a' = 0$. Hence $f_{00} = 0$ or $f_{00} = a$.
Multiply (3) by a: $f_{01}a = a$. Hence $f_{01} = a$ or $f_{01} = 1$.
Multiply (3) by a': $f_{10}a' = 0$. Hence $f_{10} = 0$ or $f_{10} = a$.
From (2) and (4), $f_{01}a' = 0$. Hence $f_{01} = 0$ or $f_{01} = a$.

From the first and last of these, $f_{01} = a$. Hence the allowable coefficient values are:

$$f_{11} = a', \qquad f_{10} = 0 \quad \text{or} \quad a, \qquad f_{01} = a, \qquad f_{00} = 0 \quad \text{or} \quad a.$$

ANSWER.
$$f(x, y) = a'xy + ax'y,$$
$$f(x, y) = a'xy + ax'y + ax'y',$$
$$f(x, y) = a'xy + axy' + ax'y,$$
$$f(x, y) = a'xy + axy' + ax'y + ax'y'.$$

It is important to note that the elements a and a' that appear in these examples are distinct elements of the Boolean algebras, and not variables which take on the value 0 and 1.

EXERCISES

Determine all Boolean functions (on a four-element Boolean algebra) having the given values:

1. Table 4.4.
2. Table 4.5.
3. Table 4.6.
4. Table 4.7.
5. Table 4.8.

TABLE 4.4

x	y	$f(x, y)$
0	a	0
1	a	1
a'	0	a'
a	a'	a

TABLE 4.5

x	y	$f(x, y)$
a	a'	a
a	1	a
1	a'	a
0	a	a

TABLE 4.6

x	y	z	$f(x, y, z)$
0	0	0	a'
0	0	1	0
0	1	0	a'
0	1	1	1
1	0	0	a
1	0	1	0
1	1	0	a
1	1	1	1

TABLE 4.7

x	y	z	$f(x, y, z)$
0	1	a	1
a'	a	0	0
a	1	1	a'
1	a	0	0
a'	1	1	1
a	1	a	0
0	a'	1	a'
0	1	a'	a

TABLE 4.8

x	y	z	$f(x, y, z)$
a'	0	1	a
0	a'	a	a
a'	1	a	a
a	0	a'	a'
a	a	a	a'
1	0	1	1
a'	0	a	1
0	a	a'	0

5. SWITCHING CIRCUITS

Because bistable electronic elements are inexpensive and simple to construct, virtually all circuitry for digital computers is designed on the basis of the two-element Boolean algebra. Actual design of computer elements involves many considerations other than the Boolean algebra. Since propagation of a signal through an electronic element or along a wire takes time, timing considerations are of great importance. Characteristics of various types of electronic elements give rise to certain wave forms for the signals; these must be recognized and considered in the design of a computer circuit. In addition, one must consider the fact that no electronic element is perfectly reliable, and design the circuitry to avoid or minimize problems arising from failure of specific elements. In this section, however, we shall ignore these important points, and concentrate solely on the underlying Boolean algebra.

While circuit elements may be designed with a variety of characteristics, we shall concentrate on two types of elements, the *and-gate* and the *or-gate*. These correspond respectively to the Boolean operations of multiplication and addition, and take their names from the fact that the Boolean product ab is 1 if and only if both a *and* b are 1, while the Boolean sum $a+b$ is 1 if and only if either a *or* b is 1. Complementation is typically achieved through an electronic inverter, which interchanges high and low voltages, where a high voltage represents 1 and a low voltage represents 0. The conventional symbols for the and-gate and the or-gate are shown in Fig. 5.1. Inversion is denoted by a circle on the input or output line. We illustrate the use of these symbols with two examples.

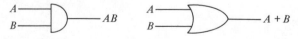

Figure 5.1

Example 5.1

Design a circuit to realize the Boolean function

$$f(x, y, z) = (xy)' + x'z.$$

Solution. The expressions xy and $x'z$ are realized by and-gates (Fig. 5.2). Note the circle on x line in Fig. 5.2b, denoting inversion or complementation. After inversion of the output from xy, the signals are combined via an or-gate.

ANSWER.   See Fig. 5.3.

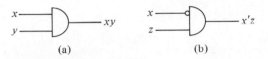

(a)   (b)

Figure 5.2

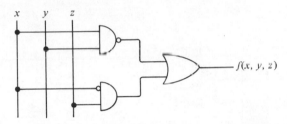

Figure 5.3

Example 5.2

Determine the Boolean function realized by the circuit of Fig. 5.4.

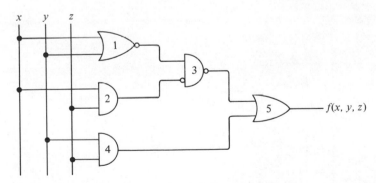

Figure 5.4

Solution. The output from gate 1 with the inverter is $(x + y)'$; from gate 2, xz. The latter is inverted and combined with the former signal at gate

3, an and-gate with inverter. The output is $((x+y)'(xz)')'$. The output from gate 4, yz, is combined with this by gate 5, an or-gate.

ANSWER. $((x+y)'(xz)')'+yz.$

Note: This simplifies to the expression $x+y$. Thus the same function could be realized by a single or-gate with inputs x and y.

EXERCISES

1.   Show that $((x+y)'(xz)')'+yz = x+y.$

Using and- and or-gates and inverters, design a circuit to realize the Boolean function:

2. $x(y+x'(z+y')).$ 3. $xyz+x'z+y'z'.$

4. $x(y'z+y(xz+y'))+x'(z+y(z+x')).$ 5. $w(x+y(z+x')+y')+w'x'y'z'.$

Determine the Boolean function realized by the given switching circuit:

6.   Fig. 5.5.   7.   Fig. 5.6.

8.   Fig. 5.7.   9.   Fig. 5.8.

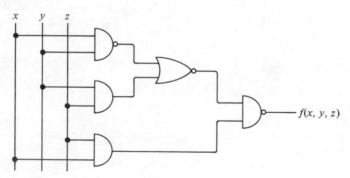

Figure 5.5

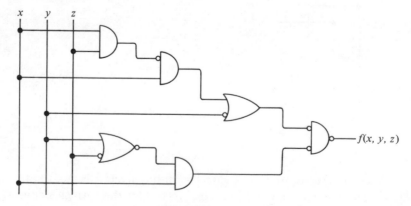

Figure 5.6

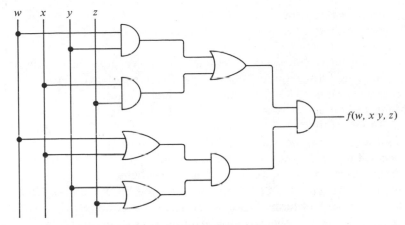

Figure 5.7

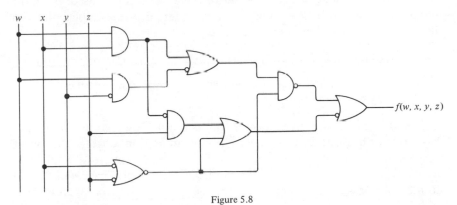

Figure 5.8

6. BOOLEAN FUNCTION MINIMIZATION

Since several different expressions may represent the same Boolean function, it is desirable in switching circuits to use an expression that is in some sense minimal, particularly if the function represented is to be used many times by a logic designer in the construction of a computer. For this reason, it is useful to have techniques available to determine a minimal expression for representing a given function. The term *minimal* is understood to mean minimal only with respect to a particular form of expression. If we change the allowable form of the expression, or if we allow other operators, as we shall do in Chapter 9, it may be possible to obtain an expression that is simpler than the best obtainable within the originally specified form. We shall restrict our attention to the *sum of products* form.

Expressions in this form consist of a Boolean sum of terms, each term being a Boolean product of variables and their complements. Thus for example, $ab + ac'$ is an expression in sum of products form. Note that through the use of the distributive law we can obtain a simpler equivalent expression $a(b + c')$. (This form requires one and-gate instead of two.) However, the latter expression is no longer in sum of products form.

Many techniques exist for minimizing Boolean functions. Common to most of these is the concept of a *prime implicant*. If $A = \{a_1, \ldots, a_n\}$ is a set of Boolean expressions, we say that the Boolean expression b *covers* or *subsumes* A precisely when $b + x = b$ if and only if $x \in A$. Note that if $n > 1$, covering terms contain fewer variables than the terms they cover. For example, the term ab and ab' together are covered by the term a. Note also that not every set of terms has a cover term. For example, $\{ab, a'b'\}$ has no cover term. A *prime implicant* is a minimal covering term, that is, one which covers a maximal set of terms. The technique used in most minimization algorithms is to determine the set of prime implicants, and then eliminate those that are not necessary.

Example 6.1

Find a minimal sum of products expression equivalent to the expression $pq + pr + q'r$.

Solution. Since the given expression contains three terms, there are $2^3 - 1$ sets of terms to be examined. Upon examination, observe that most of these sets have no prime implicants. In fact, the only prime implicants are the three given terms pq, pr, and $q'r$. That is, each term covers itself, and no other coverage exists. This does not mean that the given expression is minimal. Expand each term of the given expression to contain all three variables, using the relation $a = ab + ab'$:

$$pq + pr + q'r = (pqr + pqr') + (pqr + pq'r) + (pq'r + p'q'r)$$

$$= pqr + pqr' + pq'r + p'q'r.$$

In this expanded form there are $2^4 - 1$ sets of terms to be examined. Two of these have cover terms, namely

$$\{pqr, pqr'\}, \quad \text{covered by} \quad pq,$$

and $\{pq'r, p'q'r\}, \quad \text{covered by} \quad q'r.$

Together these two sets contain all terms of the expanded expression. Hence the middle term pr of the original expression is not needed.

ANSWER. $pq + q'r$.

Note: This may be verified by evaluating the two expressions as p, q, and r take the values 0 and 1.

The *Karnaugh map* is a simple visual device used for minimizing Boolean expressions of only a few variables. It consists of a rectangle divided into squares, each square corresponding to a particular product of all the variables either complemented or uncomplemented. These squares are arranged in such a way that contiguous squares differ in exactly one variable being complemented or uncomplemented. The Karnaugh map must be thought of as covering the surface of a sphere, so that squares on opposite edges of the map are contiguous, and all four corner squares of the map have a point in common. Karnaugh maps for one, two, three, and four variables are given in Fig. 6.1. In these maps the label of a particular

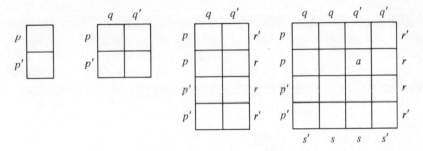

Figure 6.1

square is the product of the variables indicated on the edges of the rectangle. For example, the square marked a in the four-variable map represents the term $pq'rs$. Karnaugh maps may be used with the expressions involving five, six, or more variables, but visualization of the results becomes increasingly difficult. For example, in using a five-variable map we must remember that each square has five variables associated with it, one of which labels each half of the map (Fig. 6.2). Conventionally, to represent

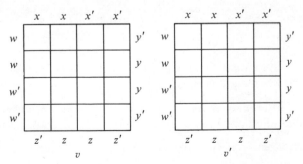

Figure 6.2

a term in a Karnaugh map we expand the term to a sum of terms, each containing all of the variables, and label the corresponding squares in the map with the digit 1. The squares that do not represent the terms may be either labeled 0 or left blank.

Example 6.2

Use a Karnaugh map to find a minimal sum of products expression equivalent to $pq+pr+q'r$.

Solution. Expand the terms to produce the Karnaugh map for this expression (Fig. 6.3a). In this map circle the squares representing the original terms of the expression. Note that because of the property of contiguous squares in a Karnaugh map, prime implicants appear as maximal sized rectangles containing 2^n squares for some n, completely filled with ones. Thus in fact, the three circled rectangles are the prime implicants for the given expression. Observe that two of the squares within the map are each covered by two of the rectangles indicated by the circles. Hence, it is possible to eliminate the horizontal rectangle in the map, and maintain a complete cover (Fig. 6.3b). Thus only the terms pq (left rectangle) and $q'r$ (right rectangle) are necessary to completely cover the expression.

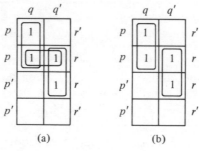

(a)   (b)

Figure 6.3

ANSWER. $pq+q'r$.

Example 6.3

Find the minimal sum of products expression represented by the Karnaugh map of Fig. 6.4.

Solution.   Find the maximal covering rectangles. There are four of them (Fig. 6.5). Note that four corner squares form a larger square. (Algebraically,

$$pqr's' + pq'r's' + p'q'r's' + p'qr's' = pr's' + p'r's'$$
$$= r's'.)$$

Note also that the upper horizontal rectangle $(pqr's' + pqr's = pqr')$ is not needed since its two squares are covered by other rectangles. However the other three rectangles are needed.

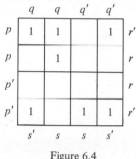

Figure 6.4

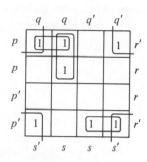

Figure 6.5

ANSWER. $r's' + pqs + p'q'r'$.

Note: The square $p'q'r's'$ is doubly covered. We could have avoided this by using $p'q'r's$ in place of $p'q'r'$, but the latter term is simpler.

There are times when the value of a particular term does not matter, perhaps because it is known that that particular term will not occur in practice. Such situations are called *don't care conditions*, and are represented by d on a Karnaugh map. Squares marked in this way may be assigned either value, 0 or 1, as the logic designer wishes. For example, the square representing $p'qr$ in Fig. 6.6 may be assigned either the value 0 or, preferably

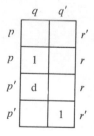

Figure 6.6

in this case, the value 1. The latter assignment allows the simpler term qr to be used, rather than pqr.

Karnaugh maps may also be used with Boolean algebras having more than two elements.

Example 6.4

Find the simplest Boolean function with the values given in Table 6.1.

Solution. Using the techniques of Section 5, find the possible coefficient values (Table 6.2). Hence there are $2 \cdot 1 \cdot 2 \cdot 2 \cdot 2 \cdot 1 \cdot 1 \cdot 4 = 64$ functions having

TABLE 6.1

x	y	z	$f(x, y, z)$
0	a	1	a'
a'	1	1	a'
a	0	a'	1
1	a'	0	a
a'	a	a'	0
1	0	a'	a
a	1	0	a'
0	a'	a	1

TABLE 6.2

f_{111}	$1, a'$
f_{110}	0
f_{101}	$0, a$
f_{100}	$1, a$
f_{011}	$0, a'$
f_{010}	a'
f_{001}	1
f_{000}	$0, a, a', 1$

the given values. Do not list and examine all these functions! Since the function in question has three variables, each coefficient represents a square on a three-variable Karnaugh map (Fig. 6.7). Enter the possible values of the coefficients in each square (Fig. 6.8). This suggests a choice of coefficients

	y	y'	
x	f_{110}	f_{100}	z'
x	f_{111}	f_{101}	z
x'	f_{011}	f_{001}	z
x'	f_{010}	f_{000}	z'

Figure 6.7

	y	y'	
x	0	$1, a$	z'
x	$1, a'$	$0, a$	z
x'	$0, a'$	1	z
x'	a'	$0, a, a', 1$	z'

Figure 6.8

to yield a relatively simple expression (Fig. 6.9). However, observe that $1 = a' + a$. Thus the term $x'y'$ derived from the squares $x'y'z$ and $x'y'z'$ can be expanded:

$$1 \cdot x'y' = (a' + a)x'y'$$
$$= a'x'y' + ax'y'$$
$$= (a'x'y'z + a'x'y'z') + (ax'y'z + ax'y'z').$$

Hence if the coefficients of the squares are represented as $a' + a$, the a' portion can be combined with the squares $x'yz$ and $x'yz'$, while the a portion can be combined with squares $xy'z$ and $xy'z'$. See Fig. 6.10.

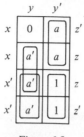

Figure 6.9   Figure 6.10

ANSWER. $ay' + a'(x' + yz)$.

Note that the simplest sum of products expression is $ay' + a'x' + a'yz$.

Because of the visual difficulties with Karnaugh maps for six or more variables, a variety of other techniques have been devised for minimizing Boolean functions. One of the best known of these is the Quine–McCluskey technique. This technique is algebraic, and depends upon a systematic expansion of the given terms to include all of the variables, in order to locate the prime implicants.[†] While the technique can be utilized on a computer, the expansion in the middle stages involves a considerable amount of computer storage. When done by hand, this same expansion makes the technique quite laborious.

As we have discussed them, both the Karnaugh map and the Quine–McCluskey technique are used to implement a direct representation of a given Boolean function. However, at times it may be more economical to implement the complement of the complement of a function, either in a product of sums form, or in a complemented sum of products form. For example, if a four-variable function involves thirteen of the sixteen possible terms, it may be simpler to implement its three-term complement, and then complement that. We now present a technique that involves complete expansion of the given function, as does the Quine–McCluskey technique, but has the advantage that it yields the prime implicants for both the function and its complement. We use the function $pq + pr + q'r$ to illustrate this technique, called *symmetric tabular expansion*.

[†] For further details on the Quine–McCluskey technique see [1].

Step 1.   Expand each term of the function to include all variables, eliminating any duplicate terms produced. (Illustration: $pq + pr + q'r$ expands to $pqr + pqr' + pq'r + p'q'r$.)

Step 2.   Encode each expanded term, using 1 to denote an unprimed variable and 0 to denote a primed variable. Note that this assumes a standard order for the variables. (Illustration: $111 + 110 + 101 + 001$.)

Step 3.   Create a table whose rows are labeled with the expanded encoded terms, and whose columns are labeled with the codes for all terms not included in the expression. The (i, j)-entry, t_{ij}, in the table is the symmetric difference (exclusive or) of the row i and column j labels, given by

$$1 \triangle 1 = 0 \triangle 0 = 0, \qquad 1 \triangle 0 = 0 \triangle 1 = 1.$$

(Illustration: Table 6.3.)

TABLE 6.3

	000	010	011	100
110	110	100	101	010
111	111	101	100	011
101	101	111	110	001
001	001	011	010	101

Step 4.   Steps 4 and 5 produce the prime implicant *prototypes* for each row. The same procedure applied to columns yields the corresponding prototypes for each column. For each row i, and each j and k, $j \neq k$, if $t_{ij} \cdot t_{ik} = t_{ij}$, eliminate t_{ik}. Note that if column prime implicants are to be produced also, then t_{ik} cannot be destroyed, but only marked so that it is no longer used in row calculations. (Illustration: In the first row of Table 6.3 the second entry causes elimination of the first and third entries.) This step corresponds to absorption, where each entry is interpreted as the sum of the variables corresponding to ones in the entry. For example, in the first row of Table 6.3 the expressions $p + q$ (110) and $p + r$ (101) are absorbed by p (100) since $p(p + q) = p(p + r) = p$. (In this step and Step 5 the primes are ignored since they are irrelevant.)

Step 5.   Continuing the interpretation given in Step 4, in each row (or column) the terms which were not eliminated are multiplied together and expanded into a sum of products form whose terms are the prime implicant prototypes. For example, 100 and 010 produce 110 ($p \cdot q = pq$); 010 and 101 produce 110 and 011 ($q \cdot (p + r) = pq + qr$). (Illustration: Table 6.4.)

TABLE 6.4

	000	010	011	100	
110	110	100	101	010	11
111	111	101	100	011	110, 101
101	101	111	110	001	101, 011
001	001	011	010	101	011
	101	110	110	011	
	011	101			

Step 6. The prime implicants are now derived by interpreting each prototype as follows. A zero corresponds to a variable which is ignored; a one corresponds to a variable which occurs in the prime implicant as it appears, primed or unprimed, in the corresponding row or column label. (Illustration: the prime implicants for the given expression are pq, pr, and $q'r$; those for its complement are $p'r'$, $q'r'$, and $p'q$. See Table 6.5.)

TABLE 6.5

Row label	Prime implicants	Column label	Prime implicants
pqr' (110)	pq	$p'q'r'$ (000)	$p'r', q'r'$
pqr (111)	pq, pr	$p'qr'$ (010)	$p'q, p'r'$
$pq'r$ (101)	$pr, q'r$	$p'qr$ (011)	$p'q$
$p'q'r$ (001)	$q'r$	$pq'r'$ (100)	$q'r'$

Example 6.5

Find the simplest sum of products expression equivalent to

$$pqr' + pq'r + pr's' + p'qr's + p'q's'.$$

Solution. Expansion of this expression yields eight terms, leaving eight terms for the complement. The development of the prototypes is given in Table 6.6. The prime implicants for the expression are given in Table 6.7. The prime implicants $pq'r$, $qr's$, and $q's'$ are essential, since they are the only ones covering the terms $pq'rs$, $p'qr's$, $p'q'r's'$, and $p'q'rs$. However, they also cover all remaining terms except $pqr's'$. For this either pqr' or $pr's'$ is necessary.

TABLE 6.6

	0001	0011	0100	0110	0111	1001	1110	1111	
1100	1101	1111	1000	1010	1011	0101	0010	0011	1110, 1011
1101	1100	1110	1001	1011	1010	0100	0011	0010	1110, 0111
1010	1011	1001	1110	1100	1101	0011	0100	0101	1110, 0101
1011	1010	1000	1111	1101	1100	0010	0101	0100	1110
1000	1001	1011	1100	1110	1111	0001	0110	0111	1011, 0101
0101	0100	0110	0001	0011	0010	1100	1011	1010	0111
0000	0001	0011	0100	0110	0111	1001	1110	1111	0101
0010	0011	0001	0110	0100	0101	1011	1100	1101	0101
	1101	1101	1101	1101	1011	0111	0110	0110	
	0111	1011		0110	0110				

TABLE 6.7

Row label	Prime implicants
$pqr's'$ (1100)	pqr', $pr's'$
$pqr's$ (1101)	pqr', $qr's$
$pq'rs'$ (1010)	$pq'r$, $q's'$
$pq'rs$ (1011)	$pq'r$
$pq'r's'$ (1000)	$pr's'$, $q's'$
$p'qr's$ (0101)	$qr's$
$p'q'r's'$ (0000)	$q's'$
$p'q'rs'$ (0010)	$q's'$

ANSWER. $pqr' + pq'r + qr's + q's'$,

$$pq'r + pr's' + qr's + q's'.$$

Note.   The first expression is preferable since it involves one less primed variable.

Development of the prime implicants for the columns in Example 6.5 produces two expressions for the complement of the given expression. One of these is $p'qs' + p'rs + qr + q'r's$. Thus the given expression may also be represented by $(p'qs' + p'rs + qr + q'r's)'$ or

$$(p + q' + s)(p + r' + s')(q' + r')(q + r + s'),$$

neither of which is in the requested sum of products form.

EXERCISES

Use a Karnaugh map to determine a minimal sum of products expression equivalent to the given one:

1. $xyz + x'z + y'z'$.

2. $x(y'z + y(xz + y')) + x'(z + y(z + x'))$.

3. $w(x + y(z + x') + y') + w'x'y'z'$.

4. $wx'y + wx'z + w(y' + z) + w'x(y' + z')$.

5. $v'wxy + vw'xy + vwx'y + vwxy' + vxz + x'y'z + w'xy$

6. Extend Example 6.5 by computing the prime implicants and the coverings for the complement of the given expression.

Use symmetric tabular expansion to determine a minimal sum of products expression equivalent to the given one.

7. $x(y'z + y(xz + y')) + x'(z + y(z + x'))$.

8. $wx'y + wx'z + w(y' + z) + w'x(y' + z')$.

9. $v'wxy + vw'xy + vwx'y + vwxy' + vxz + x'y'z + w'xy$.

10. $u(v + w(x + y(z + x) + v) + w') + u'vx'z + u'wx'y' + v'wz'$.

11. $u'xy' + v'wz' + u'xz' + v'yw + x'y'z + uvw$.

12. Compare the prime implicants obtained in Exercises 7, 8, and 9 with the Karnaugh maps of the functions (Exercises 2, 4, and 5). Where do the prime implicants appear?

13–17. For the expressions in Exercises 7–11, develop alternative forms by finding the prime implicants for the complement of each function.

7. COMPUTER ARITHMETIC

Although the circuitry of a computer is constructed from Boolean elements, it is designed to perform a wide variety of functions. In this section we examine the problem of developing arithmetic capabilities from the Boolean elements. Because of the bistable nature of an electronic element, it is natural to use binary arithmetic as the internal arithmetic of a computer. This has the effect of greatly simplifying the Boolean circuitry necessary to perform the arithmetic.

Multiplication of bits, or single binary digits, is exactly the same as Boolean multiplication. That is, the product ab has the the value of 1 if and only if both a and b have the value 1. For addition, things are somewhat more complicated. The problem arises from the fact that in binary arithmetic $1 + 1 = 10$. The result is not a single bit, but a pair of bits. If we think of the sum of any two bits as being a pair of bits, then we may call the first bit the "carry bit" and the second bit the "sum bit." We observe that the carry bit has the value 1 if and only if both bits being added are 1. Thus the carry bit is the Boolean product of the two bits being added. The sum bit, however, is not represented by the Boolean sum since it has a value of 0 when both bits being added have the value 1. Thus the sum is represented by a Boolean function that has the value 1 if and only if precisely one of the bits being added has the value 1. One Boolean expression that has this property is $ab' + a'b$. This is a minimal sum of products expression for the sum bit. Another expression, which is not a sum of products expression, for the sum bit is $(a+b)(ab)'$. Note that using and-gates and or-gates, implementation of the former expression requires three gates and two inverters, while implementation of the latter expression requires three gates and one inverter.

The construction of circuitry to add binary numerals rather than single bits involves the construction and proper interlinking of circuitry to compute both the sum and carry bit for each bit position in the numeral.

Example 7.1

Construct a circuit that given bits a and b as input will produce the sum (s) and carry (c) bits as output.

Solution. Since the sum bit may be computed by $(a+b)(ab)'$, and the carry bit by ab, we need a circuit combining and- and or-gates to produce these functions. Note that included in the sum expression is the term ab, so we can use a single and-gate for that and the carry bit.

ANSWER.   See Fig. 7.1.

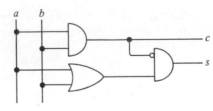

Figure 7.1

The circuit constructed in Example 7.1 is called a *half-adder*. A *full adder* sums both the two sum bits and the carry bit from the addition of bits immediately to the right. It may be constructed from two half-adders (Fig. 7.2).

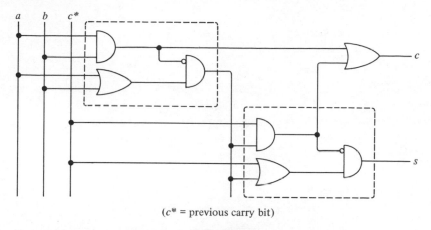

(c^* = previous carry bit)

Figure 7.2

EXERCISE

1. Verify that the full adder works correctly.

REFERENCE

1. R. Korfhage, "Logic and Algorithms." Wiley, New York, 1967.

CHAPTER 9

The Propositional Calculus

1. INTRODUCTION

Boolean algebra is very closely related to the system of elementary logic known as the propositional calculus. This logical system is concerned with *propositions*, or statements that can either be declared to be true, or declared to be false. Note that this system does not allow for such ideas as "possibly true," or "indeterminate." The propositions are generally represented by letters $p, q, r, \ldots$, and the *truth values* by T (true) and F (false). The values T and F correspond directly to the Boolean constants 1 and 0. In this chapter we develop the concepts of the propositional calculus, and show its relationship to Boolean algebra and set algebra.

It is entirely possible to define other propositional calculi, which admit such concepts as "possibly true." These generally do not correspond to Boolean algebras, but rather to systems called Post algebras, or to other algebraic systems. However, since these do not presently have great relevance to computing, we shall not discuss them.

2. FUNDAMENTAL DEFINITIONS

Basically, we wish to be able to make statements about the truth or falsity of certain combinations of propositions, given statements about the truth or falsity of the individual propositions. Thus we must develop a set of logical operators and define their properties. Since a single proposition can only be either true or false, the only thing that we can do to it is to change its truth value. The *negation* or *denial* of a proposition p is that proposition, denoted by $\sim p$ or $\neg p$, which has the value "false" if and only if p has the value "true," and the value "true" if and only if p has the value "false." This is symbolized in Table 2.1. Note the correspondence to Boolean complementation.

TABLE 2.1

p	$\sim p$
T	F
F	T

Since each of two propositions p and q may take the values T and F independently of the other, there are four possible combinations of truth values: TT, TF, FT, and FF. For each these combinations in turn, a combined statement about p and q may take on either the value T or the value F. Hence there are $2^4 = 16$ possible *truth functions* to be defined for two propositions. These 16 are shown in Table 2.2. The interpretation

TABLE 2.2

p	q	1	2	3	4	5	6	7	8	9	10	11	12	13	14	15	16
T	T	T	T	T	T	T	T	T	T	F	F	F	F	F	F	F	F
T	F	T	T	T	T	F	F	F	F	T	T	T	T	F	F	F	F
F	T	T	T	F	F	T	T	F	F	T	T	F	F	T	T	F	F
F	F	T	F	T	F	T	F	T	F	T	F	T	F	T	F	T	F

of some of the columns of this table is quite obvious. Column 1, for example, is a proposition that is always true. This is termed a *tautology*. Similarly, column 16 is *contradiction*, a statement that is always false. Scanning further, we find that column 4 is the statement p, and column 13 is the statement $\sim p$; column 6 represents q, and column 11 represents $\sim q$.

In fact, we note that throughout the table the negation of column k is column $17 - k$. Thus we need only discuss the interpretation of columns 2, 3, 5, 7, and 8. The interpretation of the remaining columns follows from the fact that they are negations of these.

In column 2 we find the value T when either one or both of the propositions p and q had the value T. This corresponds to the English word *or*, used in its inclusive sense. The logical operator for this is called *disjunction* and written $p \vee q$. Skipping to column 8, we find there the value T if and only if both p and q have the value T. This "and" function is called *conjunction* and written $p \wedge q$. Note that disjunction and conjunction are precisely Boolean addition and multiplication.

The function represented in column 7 has the value T if and only if both p and q have the same truth value. Hence this function, written $p \equiv q$, is called *equivalence*. In column 10 we find the negation of this function, *inequivalence*. It has the value T if and only if precisely one of the variables p and q has the value T. Thus it corresponds to the exclusive use of the English word *or*. Note that this function can be interpreted as yielding the sum bit for binary addition.

The English statement "if p then q," or "p implies q," is generally considered to be true if both p and q are true, and false if p is true and q is false. Thus of the remaining columns in the table, this statement corresponds to column 5. The logical operator is called the *conditional*, or *material implication*, and is written $p \supset q$. Similarly, column 3 corresponds to the statement "if q then p," or "p is implied by q." Table 2.3 reproduces Table 2.2 with the appropriate logical operator written above each column.

TABLE 2.3

p	q	"true"	$p \vee q$	$p \subset q$	p	$p \supset q$	q	$p \equiv q$	$p \wedge q$	$p \vert q$	$p \not\equiv q$	$\sim q$	$p \not\supset q$	$\sim p$	$p \not\subset q$	$p \downarrow q$	"false"
T	T	T	T	T	T	T	T	T	T	F	F	F	F	F	F	F	F
T	F	T	T	T	T	F	F	F	F	T	T	T	T	F	F	F	F
F	T	T	T	F	F	T	T	F	F	T	T	F	F	T	T	F	F
F	F	T	F	T	F	T	F	T	F	T	F	T	F	T	F	T	F

The interpretation of column 5 as the conditional or "if ... then ..." operator requires further discussion. Why is it appropriate to consider "if p then q" to be true whenever p is false? This interpretation is appropriate since we are more concerned with the form of a logical argument than with its content. Often we do not know when such a statement as $p \supset q$ is made whether p is true or false. Consider the statement, "If it snows tomorrow, then ...." We do not know whether "it snows tomorrow" is true or false; yet we may wish to use this conditional statement in a more extensive discussion, knowing that its form is correct independently of the truth or falsity of "it snows tomorrow." A mathematical example: "If f is differentiable at a point x_0, then f is continuous at x_0." What if f is not differentiable at x_0? The conditional statement then makes no assertion about the continuity of f. (It may or may not be continuous at x_0.) However, we may wish to make use of the conditional statement in a general mathematical argument, without knowing the differentiability of the function under discussion. The form of the statement is correct; the argumentation should not halt just because the function is not known to be differentiable. And from computing, "IF ALPHA .GT. $X + 7$, GO TO 13" is a valid statement in a Fortran program, even though the truth of "ALPHA .GT. $X + 7$" is not known when the program is written, and will possibly change during execution of the program. So that the statement $p \supset q$ does not disrupt an argument or a procedure whenever p is false, we assign it the value T.

EXERCISE

1. Negation, disjunction, and conjunction may be interpreted as Boolean complementation, addition, and multiplication, respectively. Interpret the other logical operators as Boolean functions.

3. TRUTH TABLES

In determining the truth value of an expression such as $p \vee q$, we need only know the truth values of the expressions p and q, not their particular forms. Thus both p and q can be names for more complex logical expressions. This indicates that we can construct quite complex logical expressions, and determine their truth values by examination of the truth values of their components. The standard format for this procedure is called a *truth table*. Each line of the table carries one particular combination of truth values for the variables, with the computation of the truth value of the logical expression for this particular combination of truth values of the variables.

Example 3.1

Determine the truth values of the expression $p \supset (q \vee \sim p)$.

Solution. The truth table contains four lines, one for each combination of truth values of p and q. The subexpression $\sim p$ is evaluated first, followed by the subexpression $q \vee \sim p$. Then the complete expression is evaluated. In the truth table, the subexpression values are indicated below the appropriate operator; the expression values are in the boxed columns.

ANSWER.   See Table 3.1.

TABLE 3.1

p	q	$p \supset$	$(q \vee$	$\sim p)$
T	T	T	T	F
T	F	F	F	F
F	T	T	T	T
F	F	T	T	T
		(3)	(2)	(1)

Note that the given expression is equivalent to (that is, has the same truth values as) $p \supset q$.

Example 3.2

Determine the truth values of the expression

$$(p \vee (q \supset r)) \equiv ((p \vee \sim r) \supset q).$$

Solution.   Since there are three variables, the truth table has $8 = 2^3$ lines. There are five subexpressions to evaluate before attacking the main expression.

ANSWER.   See Table 3.2.

TABLE 3.2

p	q	r	$(p \vee$	$(q \supset r))$	$\equiv$	$((p \vee$	$\sim r)$	$\supset q)$
T	T	T	T	T	T	T	F	T
T	T	F	T	F	T	T	T	T
T	F	T	T	T	F	T	F	F
T	F	F	T	T	F	T	T	F
F	T	T	T	T	T	F	F	T
F	T	F	F	F	F	T	T	T
F	F	T	T	T	T	F	F	T
F	F	F	T	T	F	T	T	F
			(2)	(1)	(6)	(4)	(3)	(5)

Example 3.3

Determine the truth values of the expression

$$p \supset (q \supset (r \supset (\sim p \supset (\sim q \supset \sim r)))).$$

Solution. In this example, much work is eliminated by noting that $p \supset q$ has the value T whenever p has the value F. The order of evaluation is shown in Table 3.3. Note that the result is independent of the subexpression $\sim q \supset \sim r$.

TABLE 3.3

p	q	r	p	$\supset$	$(q$	$\supset$	$(r$	$\supset$	$(\sim p$	$\supset$ $(\sim q$ $\sim r))))$
T	T	T	(11)	T	(10) T	(9) T	(7) F	(8) T		
T	T	F	(6)	T	(5) T	(4) T				
T	F	T	(3) {	T	(2) { T					
T	F	F		T	{ T					
F	T	T		T						
F	T	F	(1) {	T						
F	F	T		T						
F	F	F		T						

ANSWER. The expression is a tautology.

The truth table can also be used in the opposite mode. That is, given a table with function values, we can determine from it an expression for the function.

Example 3.4

Determine a logical expression having the truth values shown in Table 3.4.

TABLE 3.4

p	q	r	$f(p,q,r)$
T	T	T	T
T	T	F	F
T	F	T	T
T	F	F	T
F	T	T	F
F	T	F	F
F	F	T	T
F	F	F	F

Solution. The unknown expression first has the value T in the first line of the table. Observe that the expression $p \wedge q \wedge r$ has the value T precisely for this line and no other. The unknown expression next has the value T in the third line of the table, where p and r have the value T, and q has the value F, that is, $\sim q$ has the value T. The expression $p \wedge \sim q \wedge r$ has the value T precisely for this line and no other. Hence the expression $(p \wedge q \wedge r) \vee (p \wedge \sim q \wedge r)$ has the value T for the first and third lines of the truth table, and no other. Similarly, use the expressions $p \wedge \sim q \wedge \sim r$ and $\sim p \wedge \sim q \wedge r$ to represent the T's in the fourth and seventh lines of the table.

ANSWER.

$$(p \wedge q \wedge r) \vee (p \wedge \sim q \wedge r)$$
$$\vee (p \wedge \sim q \wedge \sim r) \vee (\sim p \wedge \sim q \wedge r)$$

is a logical expression satisfying the truth table.

Note that the expression thus determined is not the only, nor usually the simplest, representation of the logical function. The expression $(p \wedge r) \vee (p \wedge \sim q) \vee (\sim q \wedge r)$ also satisfies Table 3.4.

The particular expression determined in Example 3.4 for the function is called a *disjunctive normal form* for the function. It is a disjunction of terms, each of which is a conjunction of logical variables and their negations. Since the procedure used in Example 3.4 is completely general, we conclude that any logical function, in any number of variables, can be expressed in terms of conjunction, disjunction, and negation.

We call attention to the exact correspondence between the disjunctive normal form and the Boolean sum of products form. Because of this correspondence, we have a procedure for representing any logical condition, translating it into a Boolean condition, using any of the various Boolean minimization techniques, retranslating, and arriving at a simpler equivalent logical expression. Conversely, we may at times wish to use the wealth of logical operators to determine an expression that is simpler than the minimal sum of products expression.

Example 3.5

Determine the minimal number of operators needed to express the Boolean function $r + pq + p'q'$.

Solution. A check of the Karnaugh map reveals that this Boolean function is in minimal sum of products form. In this form, six operators are needed: in circuit terms, two and-gates, two or-gates, and two inverters. The equivalent logical expression is $r \vee (p \wedge q) \vee (\sim p \wedge \sim q)$. Examine its truth table (Table 3.5). Now think. The function has the value T for the first two and last two lines of the table; and for these lines $p \equiv q$. Hence

TABLE 3.5

p	q	r	r ∨	(p ∧	q) ∨ (~p ∧ ~q)	
T	T	T	T			
T	T	F	T	T		
T	F	T	T			
T	F	F	F	F	F	F
F	T	T	T			
F	T	F	F	F	F	F
F	F	T	T			
F	F	F	T	F	T	T

the function has the value T whenever $p \equiv q$. For the middle four lines, the function has the same value as r. Thus the function can be represented using only two operators: $(p \equiv q) \vee r$. (Check to be sure this representation is valid.)

ANSWER. Two operators ($=$ and $\vee$) are required.

Note that the development of "equivalence" gates would permit very simple circuitry to realize this function (Fig. 3.1). The corresponding and- and or-gate circuitry is shown in Fig. 3.2.

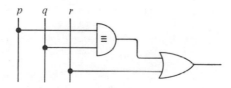

Figure 3.1

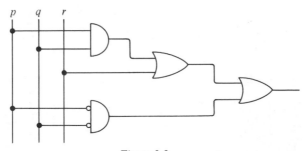

Figure 3.2

EXERCISES

Determine the truth values of the logical expression:

1. $\big((p \supset (q \equiv \sim p)) \wedge (\sim p \vee \sim q)\big) \supset \sim (q \equiv ((q \equiv \sim p) \wedge p))$.

2. $\big(p \equiv ((q \equiv (r \equiv \sim p)) \supset \sim q)\big) \supset (\sim r \vee \sim (p \wedge q))$.

3. $(p \vee q) \supset ((p \vee r) \supset (r \vee q))$.

4. $\big(p \supset ((q \vee r) \wedge s)\big) \supset \big((\sim p \equiv \sim s) \supset ((p \vee r) \equiv (q \vee \sim s))\big)$.

5. Find another two-operator expression for the function in Table 3.5.

Determine a logical expression having the given truth table:

6. Table 3.6.   7. Table 3.7.

8. Table 3.8.   9. Table 3.9.

TABLE 3.6

p	q	r	$f(p,q,r)$
T	T	T	F
T	T	F	T
T	F	T	T
T	F	F	T
F	T	T	T
F	T	F	F
F	F	T	T
F	F	F	F

TABLE 3.7

p	q	r	$f(p,q,r)$
T	T	T	T
T	T	F	F
T	F	T	T
T	F	F	T
F	T	T	T
F	T	F	T
F	F	T	F
F	F	F	T

TABLE 3.8

p	q	r	s	$f(p,q,r,s)$
T	T	T	T	T
T	T	T	F	T
T	T	F	T	F
T	T	F	F	F
T	F	T	T	F
T	F	T	F	T
T	F	F	T	F
T	F	F	F	T
F	T	T	T	T
F	T	T	F	F
F	T	F	T	F
F	T	F	F	T
F	F	T	T	T
F	F	T	F	F
F	F	F	T	F
F	F	F	F	F

TABLE 3.9

p	q	r	s	$f(p,q,r,s)$
T	T	T	T	F
T	T	T	F	T
T	T	F	T	T
T	T	F	F	T
T	F	T	T	F
T	F	T	F	F
T	F	F	T	T
T	F	F	F	F
F	T	T	T	T
F	T	T	F	F
F	T	F	T	F
F	T	F	F	F
F	F	T	T	T
F	F	T	F	T
F	F	F	T	T
F	F	F	F	F

Use logical operators to find the simplest expression equivalent to the Boolean expression, that is, to minimize the number of operator occurrences:

10. $pq + p'r + pq'r'$.

11. $p'q(p' + r') + q'(p + rp' + r')$.

12. $pq'r + p'(q + r(s' + q')) + pq's + q'(r + s)$.

13. $pr' + p'qs + p'rs' + p'q'rs + qr's'$.

14. Write a program accepting as input a truth table (maximum of five variables, 32 lines), and producing a logical expression satisfying that table. Test on Exercises 6–9.

4. WELL-FORMED FORMULAS

We have tacitly assumed that logical expressions "make sense," that is, are grammatically correct. But as with any language, not any assemblage of letters, parentheses, and logical operators constitutes a proposition. In this section we specify syntactic rules for the propositional calculus, which constitute a context-free grammar. It is convenient to use BNF notation to describe these rules. There are three specification statements.

$$\langle \text{letter} \rangle ::= a \,|\, b \,|\, c \,|\, \cdots \,|\, y \,|\, z$$

$$\langle \text{binop} \rangle ::= \supset \,|\, \vee \,|\, \wedge \,|\, \equiv$$

$$\langle \text{wff} \rangle ::= \langle \text{letter} \rangle \,|\, \sim (\langle \text{wff} \rangle) \,|\, (\langle \text{wff} \rangle) \langle \text{binop} \rangle (\langle \text{wff} \rangle)$$

The term "wff" means *well-formed formula*, the traditional name for grammatically correct propositions. Note that nothing in the grammar specifies the truth values of a wff. The list of $\langle \text{binop} \rangle$s could be extended to include all other logical operators ($|$, $\downarrow$, $\not\equiv$, $\not\supset$, $\subset$, $\not\subset$), but this is not really necessary.

While use of these grammatical rules produces wffs, the formulas often seem buried in parentheses. For example, the formula which we have informally written

$$(p \equiv ((q \equiv (r \equiv \sim p)) \supset \sim q)) \supset (\sim r \vee \sim (p \wedge q))$$

becomes the wff

$$((p) \equiv (((q) \equiv ((r) \equiv (\sim(p)))) \supset (\sim(q)))) \supset ((\sim(r)) \vee (\sim((p) \wedge (q)))).$$

To avoid this overparenthesization, we adopt two additional rules.

A. Parentheses around single letters are deleted. (The wff in our example then becomes

$$(p \equiv ((q \equiv (r \equiv (\sim p))) \supset (\sim q))) \supset ((\sim r) \vee (\sim(p \wedge q))).)$$

B. The *hierarchy* or *order of precedence* $\sim$, $\wedge$, $\vee$, $\supset$, $\equiv$ is observed. Further, among occurrences of the same operator the precedence is left-to-right. This means that in restoring parentheses in an expression where they are missing, those associated with $\sim$ are restored first, followed by those associated with $\wedge$, then with $\vee$, then with $\supset$, and finally with $\equiv$. If two occurrences of an operator, say $\wedge$, are encountered, the parentheses associated with the left-most occurrence are restored before those associated with the right-most occurrence. Thus $\sim p \vee q$ becomes $(\sim p) \vee q$, rather than $\sim(p \vee q)$; and $p \supset q \supset r$ becomes $(p \supset q) \supset r$, rather than $p \supset (q \supset r)$. With these hierarchy rules, $\sim(p \vee q)$ and $p \supset (q \supset r)$ must be written with parentheses. In removing parentheses, the sequence is reversed. Using the minimum number of parentheses, the example wff becomes

$$\bigl(p \equiv (q \equiv (r \equiv \sim p))\bigr) \supset \sim q) \supset \sim r \vee \sim(p \wedge q).$$

This is perhaps less readable than the original form we used, but still valid.

EXERCISES

1. Show that left-to-right precedence is important only for the conditional. That is, show that $(p \supset q) \supset r$ and $p \supset (q \supset r)$ are not equivalent expressions, but that $(p \text{ op } q) \text{ op } r$ is equivalent to $p \text{ op } (q \text{ op } r)$ if "op" is $\equiv$, $\vee$, or $\wedge$.

Properly restore parentheses in the logical expression:

2. $p \supset q \equiv \sim r \vee \sim(p \wedge q \vee r)$.

3. $p \equiv q \supset q \equiv (p \vee r) \supset q \wedge r \supset p \vee q$.

4. $p \vee q \wedge r \vee (s \supset r \wedge \sim p \vee q \supset \sim(s \equiv p) \vee r)$.

5. $p \wedge q \wedge r \vee s \vee \sim p \supset \sim q \supset \sim r \equiv \sim s \equiv p$.

Remove as many parentheses as possible from the logical expression:

6. $((p) \vee (q)) \equiv \bigl(\bigl(\sim((p) \vee (\sim((r) \supset (s))))\bigr) \supset \bigl((\sim(s)) \equiv (\sim((p) \wedge (q)))\bigr)\bigr)$.

7. $\bigl(((p) \supset ((q) \supset (r))\bigr) \supset \bigl((((s) \supset (\sim(p))) \supset (\sim(q))) \supset ((\sim(p)) \supset (r))\bigr) \supset \bigl(((p) \equiv (r)) \equiv (\sim((q) \equiv (s)))\bigr)$.

8. $(p) \equiv \bigl(\sim((q) \equiv (\sim((r) \equiv (\sim((p) \equiv (\sim(q)))) \equiv (\sim(r)))) \equiv (p))\bigr)$.

9. $\bigl(((p) \supset (q)) \supset ((p) \supset ((q) \supset (p)))\bigr) \supset (q)) \supset \bigl((((p) \supset (q)) \supset (p)) \supset (q)\bigr)$.

Write the required computer program. If your computer language does not accept logical operator symbols, use the convention

p'	for	$\sim p$
pq	for	$p \wedge q$
$p + q$	for	$p \vee q$
$p * q$	for	$p \supset q$
$p = q$	for	$p \equiv q$

Assume that the longest input expression will fit on one card, or one line if you are using a terminal:

10. To restore parentheses. (Test on Exercises 2–5.)

11. To eliminate parentheses. (Test on Exercises 6–9.)

12. To determine the truth table for an expression. (Test on Exercises 1–4, Section 3.)

5. MINIMAL SETS OF OPERATORS

In Section 3 we established that conjunction, disjunction, and negation are sufficient to represent any logical expression. In switching circuit terms, and-gates, or-gates, and inverters are sufficient to represent any logical or Boolean expression. Actually we can do better. DeMorgan's laws provide us with the means of replacing conjunction by disjunction, or disjunction by conjunction, so that we really need only negation and one of the operators, conjunction and disjunction.

We are trading simplicity in circuit element manufacture for complexity within the circuit itself. If we can manufacture or-gates much more simply and cheaply than and-gates, it may pay to replace all and-gates in a logical circuit with or-gates. However, in doing so we will generally have to add a number of inverters to the circuit, which may negate the economics of or-gate production.

Carrying this to an extreme, we can in fact construct all logical functions from only one operator, either the *nand* operator $(p \mid q)$ or the *nor* operator $(p \downarrow q)$. As a matter of fact, nand- and nor-gates are quite economical to produce, and are widely used in computer circuitry. Note that they are simply the negations of and- and or-gates. The representation of negation, conjunction, and disjunction in terms of nand and nor are given in Table 5.1.

TABLE 5.1

	$\mid$	$\downarrow$
$\sim p$	$p \mid p$	$p \downarrow p$
$p \wedge q$	$(p \mid q) \mid (p \mid q)$	$(p \downarrow p) \downarrow (q \downarrow q)$
$p \vee q$	$(p \mid p) \mid (q \mid q)$	$(p \downarrow q) \downarrow (p \downarrow q)$

Example 5.1

Find an expression for $(p \not\equiv q) \supset r$ using only (a) the nand operator, (b) the nor operator.

Solution. The conversion of $\not\equiv$ and $\supset$ to conjunction, disjunction, and negation, and then the conversion of these using Table 5.1 will solve the problem. However, the result may be two very inelegant expressions. (Try it!) Apply a little thought. The expression $p \supset q$ is equivalent to $\sim(p \wedge \sim q)$, or to $p \,|\, \sim q$. Hence $(p \not\equiv q) \supset r$ can be written $(p \not\equiv q) \,|\, \sim r$, or $(p \not\equiv q) \,|\, (r\,|\,r)$. Conversion of $p \not\equiv q$ is still necessary. Now $p \,|\, \sim q$ has the value F precisely when p has the value T and q had the value F. Similarly, $\sim p \,|\, q$ has the value F when p has the value F and q the value T. Hence $p \equiv q$ is equivalent to $(p\,|\,\sim q) \wedge (\sim p\,|\,q)$, and $p \not\equiv q$ is the negation of this, namely $(p\,|\,\sim q)\,|\,(p\,|\,\sim q)$, or $(p\,|\,(q\,|\,q))\,|\,((p\,|\,p)\,|\,q)$.

For the equivalent expression using only nor, observe that $p \downarrow q$ has the value F for three of the four truth value combinations, and hence is "usually" false. Since the given expression "usually" has the value T, you might expect a rather long nor representation of it. Observe that $p \vee q$ is equivalent to $\sim(p \downarrow q)$. Now by Example 3.5 and Exercise 5, Section 3, the given function is equivalent to $(p \equiv q) \vee r$. Thus the function may be expressed $\sim((p \equiv q) \downarrow r)$, or $((p \equiv q) \downarrow r) \downarrow ((p \equiv q) \downarrow r)$. An expression for $p \equiv q$ is needed. Observe that $p \equiv q$ is equivalent to $(p \downarrow q) \vee (\sim p \downarrow \sim q)$, or to $\sim((p \downarrow q) \downarrow (\sim p \downarrow \sim q))$.

ANSWER.

(a) $((p\,|\,(q\,|\,q))\,|\,((p\,|\,p)\,|\,q))\,|\,(r\,|\,r)$;

(b) $((((p \downarrow q) \downarrow ((p \downarrow p) \downarrow (q \downarrow q))) \downarrow ((p \downarrow q)$
$\downarrow ((p \downarrow p) \downarrow (q \downarrow q)))) \downarrow r)$
$\downarrow ((((p \downarrow q) \downarrow ((p \downarrow p) \downarrow (q \downarrow q)))$
$\downarrow ((p \downarrow q) \downarrow ((p \downarrow p) \downarrow (q \downarrow q)))) \downarrow r)$.

Switching circuit symbols for nand- and nor-gates are shown in Fig. 5.1.

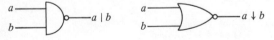

Figure 5.1

Instead of representing a nand-gate as shown, it could be represented as an or-gate with inverted (negated) inputs; and similarly a nor-gate could be represented as an and-gate with inverted inputs (Fig. 5.2).

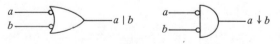

Figure 5.2

Example 5.2

Devise nand and nor circuits for the half-adder.

Solution. In Section 7, Chapter 8 it was observed that the sum bit is the "exclusive or" of the bits, and the carry bit is the "and" of the bits. "Nand" and "nor" expressions for these functions are given in Table 5.2. Note that negation has been left in the tabular expressions since $\sim p$ is simply $p|p$ or $p \downarrow p$. Circuits for the half-adder are given in Figs. 5.3 and 5.4.

TABLE 5.2

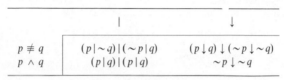

	$\mid$	$\downarrow$
$p \not\equiv q$	$(p\mid\sim q)\mid(\sim p\mid q)$	$(p\downarrow q)\downarrow(\sim p\downarrow\sim q)$
$p \wedge q$	$(p\mid q)\mid(p\mid q)$	$\sim p\downarrow\sim q$

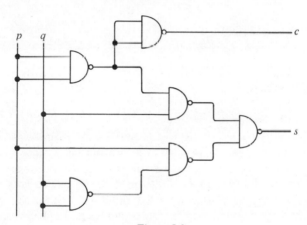

Figure 5.3

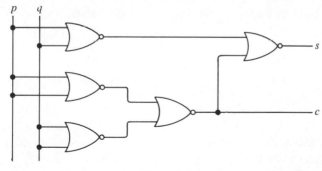

Figure 5.4

EXERCISES

1. Show that $\{\sim, \supset\}$ is a minimal set of logical operators.

2. Show that $\{\sim, \equiv\}$ is not a minimal set of logical operators. *Hint:* Show that $p \wedge q$ cannot be expressed in terms of $\sim$ and $\equiv$.

3. Write a computer program to generate all nand expressions in two variables, having at most five operator occurrences, and determine their truth values.

4. Write a computer program to generate all nor expressions in two variables, having at most five operator occurrences, and determine their truth values.

Use the input conventions and assumptions stated for the Exercises in Section 4. Write the program and test it on the expressions of Exercises 2–5, Section 4:

5. To convert a logical expression entirely to nand operators.

6. To convert a logical expression entirely to nor operators.

6. POLISH NOTATION

One of the difficulties with ordinary logical (or arithmetic) notation is illustrated by the formula

$$(p \wedge (\sim(\sim(q \wedge p) \vee (\sim(r \supset p)) \vee \sim p) \wedge \sim(p \supset r)) \equiv r) \vee p.$$

It is difficult visually to determine which part of the formula belongs with which other part. The processing of such a formula on a computer involves the development and manipulation of various stacks of subexpressions, as the program determines the proper evaluation of the formula. This process is sufficiently unwieldy that most compilers transform arithmetic assignment statements and logical statements into another notation before processing them.

The *Polish notation* was developed by the Polish logician, Jan Łukasiewicz. It may be used in either a prefix form, as Łukasiewicz developed, or in a postfix form, which is more common in compilers. We shall work primarily with the prefix form, pointing out the corresponding developments for the postfix form. In prefix form, all operators are written immediately to the left of their operand list, as Cpq for $p \supset q$ and Epq for $p \equiv q$; in postfix form the operators are written immediately to the right, as pqA for $p \vee q$, and pN for $\sim p$. (In Polish notation the symbols N, C, A, K, and E are used for $\sim$, $\supset$, $\vee$, $\wedge$, and $\equiv$, respectively, to avoid confusion with ordinary notation.) Łukasiewicz observed that if each operator is associated with a fixed number of operands, use of this type of notation eliminates the need for parentheses.

Example 6.1

Convert the expressions
(a) $(p \supset (q \supset r)) \equiv (\sim(p \vee r) \wedge \sim q)$
(b) $(\sim(p \equiv \sim(q \supset r))) \equiv ((p \vee r) \supset \sim(\sim p \vee \sim q))$
to Polish prefix notation.

Solution. Successful conversion depends on properly determining which operand expressions are associated with each operator. This is shown by the nesting of the parentheses, where corresponding parentheses have the same number:

$$\underset{1}{(}p \underset{2}{\supset} (q \supset r\underset{21}{))} \equiv \underset{3}{(}\sim(\underset{4}{}p \vee r\underset{4}{)} \wedge \sim q\underset{3}{)},$$

$$\underset{1}{(}\sim(\underset{2}{}p \equiv \sim(\underset{3}{}q \supset r\underset{321}{)))} \equiv (\underset{45}{(}p \vee r\underset{5}{)} \supset \sim(\underset{6}{}\sim p \vee \sim q\underset{64}{))}.$$

Convert the first formula to Polish notation working from the inside out:

$$(p \underset{\uparrow}{\supset} (q \underset{\uparrow}{\supset} r)) \equiv (\underset{\uparrow}{\sim}(p \vee r) \wedge \underset{\uparrow}{\sim} q)$$

$$(p \underset{\uparrow}{\supset} Cqr) \equiv (\underset{\uparrow}{\sim}Apr \wedge Nq)$$

$$CpCqr \equiv (NApr \underset{\uparrow}{\wedge} Nq)$$

$$CpCqr \underset{\uparrow}{\equiv} KNAprNq$$

$$ECpCqrKNAprNq.$$

For variety, convert the second formula from the outside in:

$$(\sim(p \equiv \sim(q \supset r))) \underset{\uparrow}{\equiv} ((p \vee r) \supset \sim(\sim p \vee \sim q))$$

$$E\underset{\uparrow}{(}\sim(p \equiv \sim(q \supset r)))((p \vee r) \underset{\uparrow}{\supset} \sim(\sim p \vee \sim q))$$

$$EN(\underset{\uparrow}{p} \equiv \sim(q \supset r)) C(\underset{\uparrow}{p} \vee r) \underset{\uparrow}{\sim}(\sim p \vee \sim q)$$

$$ENEp \underset{\uparrow}{\sim}(q \supset r) CAprN(\underset{\uparrow}{\sim} p \vee \sim q)$$

$$ENEpN(\underset{\uparrow}{q} \supset r) CAprNA \underset{\uparrow}{\sim} p \underset{\uparrow}{\sim} q$$

$$ENEpNCqrCAprNANpNq.$$

ANSWER.
(a) $ECpCqrKNAprNq$,
(b) $ENEpNCqrCAprNANpNq$.

Example 6.2

Convert the expressions of Example 6.1 to Polish postfix notation.

Solution.   The same technique works. Just place the operators on the right rather than on the left.

(a)

$$(p \supset (q \supset r)) \equiv (\sim(p \vee r) \wedge \sim q)$$

$$(p \supset qr\text{C}) \equiv (\sim pr\text{A} \wedge q\text{N})$$

$$pqr\text{CC} \equiv (pr\text{AN} \wedge q\text{N})$$

$$pqr\text{CC} \equiv pr\text{AN}q\text{NK}$$

$$pqr\text{CC}pr\text{AN}q\text{NKE}.$$

(b)

$$(\sim(p \equiv \sim(q \supset r))) \equiv ((p \vee r) \supset \sim(\sim p \vee \sim q))$$

$$(\sim(p \equiv \sim(q \supset r)))((p \vee r) \supset \sim(\sim p \vee \sim q))\text{E}$$

$$(p \equiv \sim(q \supset r))\text{N}(p \vee r)\sim(\sim p \vee \sim q)\text{CE}$$

$$p \sim(q \supset r)\text{EN}pr\text{A}(\sim p \vee \sim q)\text{NCE}$$

$$p(q \supset r)\text{NEN}pr\text{A} \sim p \sim q\text{ANCE}$$

$$pqr\text{CNEN}pr\text{A}p\text{N}q\text{NANCE}.$$

ANSWER.
(a) $pqr\text{CC}pr\text{AN}q\text{NKE}$,
(b) $pqr\text{CNEN}pr\text{A}p\text{N}q\text{NANCE}$.

A prime advantage of Polish notation is that in scanning a formula from right to left for prefix notation, or left to right for postfix notation, all operands are encountered before their operators. Thus in using a compiler or interpreter, whenever an operator is encountered code for it can be immediately compiled, or it can be immediately interpreted and executed.

One of the problems with any complex formula is that, due to an error in its construction, it may be meaningless. In ordinary notation the search for well-formed formulas involves a determination of the nesting of parentheses. Using a Polish notation, the determination becomes a simple numerical procedure involving a single scan of the formula. Every symbol in a formula is given a weight: +1 for constants and variables, 0 for unary operators and −1 for binary operators. Intuitively, each new variable or constant constitutes a new piece of formula, while each binary operator combines two pieces of formula into one, thereby reducing the number of pieces of formula. Unary operators do not affect the number of pieces of formula. Partial sums are formed of these weights, starting at the right-hand end of the formula if prefix notation is used, or the left-hand end if postfix notation is used. A formula is well-formed if and only if each partial sum is positive and the total sum is one. Intuitively, at each stage of the summation we have at least one piece of formula, and at the final stage we have exactly one formula.

Example 6.3

Determine if the formula CECpEpNqCprNq is well-formed.

Solution. Since operators occur to the left, this is a prefix formula. Begin the summation at the right:

C	E	C	p	E	p	N	q	C	p	r	N	q	
−1	−1	−1	+1	−1	+1	0	+1	−1	+1	+1	0	+1	values
1	2	3	4	3	4	3	3	2	3	2	1	1	partial sums

Since all partial sums are positive and the total sum is 1, the formula is well-formed.

ANSWER.
The formula is well-formed.

Example 6.4

Determine if the formula ppqNECprCEqNC is well-formed.

Solution. With operators on the right, this is a postfix formula. Begin the summation at the left:

	p	p	q	N	E	C	p	r	C	E	q	N	C
values	+1	+1	+1	0	−1	−1	+1	+1	−1	−1	+1	0	−1
partial sums	1	2	3	3	2	1	2	3	2	1	2	2	1

Since all partial sums are positive and the total sum is 1, the formula is well-formed.

ANSWER.
The formula is well-formed.

Example 6.5

Determine if the formulas
(a) KpACpqNqp
(b) pqNCNAprNA
(c) CKApANpqNCApNqp
are well-formed.

Solution.   (a) (prefix)

$$\text{K } p \text{ A C } p \; q \text{ N } q \; p$$
$$2 \; 3 \; 2 \; 3 \; 4 \; 3 \; 2 \; 2 \; 1 \quad \text{partial sums}$$

Although all partial sums are positive, the total sum is not 1. Hence the formula is not well-formed. It consists of two (that is, total sum) pieces, KpACqpNq and p, each of which is well-formed.

(b) (postfix)

$$p \; q \text{ N C N A } p \; r \text{ N A}$$
$$\text{partial sums} \quad 1 \; 2 \; 2 \; 1 \; 1 \; 0 \; 1 \; 2 \; 2 \; 1$$

Although the total sum is 1, not all partial sums are positive. Hence the formula is not well-formed. The partial sum 0 indicates a deficiency in the formula. Note that the left portion, pqNCN, is well-formed. It constitutes one operand for the first A, but the other operand is missing.

(c) (prefix)

$$\text{C K A } p \text{ A N } p \; q \text{ N C A } p \text{ N } q \; p$$
$$0 \; 1 \; 2 \; 3 \; 2 \; 3 \; 3 \; 2 \; 1 \; 1 \; 2 \; 3 \; 2 \; 2 \; 1 \quad \text{partial sums}$$

Although the partial sums are positive, the total sum is 0. Hence the formula is not well-formed. The situation is similar to that in (b). The formula from K on is well-formed, constituting one of the operands for C. But the other operand is missing.

> ANSWER.   None of the formulas is
> well-formed.

We note at this point that the Polish prefix notation is precisely the prefix notation developed in Chapter 3, Section 4 for representing Gorn trees. A simple algorithm exists for determining the depth function for a tree. We shall give this algorithm for the prefix notation, operating from left to right; for the postfix notation, use the same algorithm, operating from right to left.

Step 1.   Form the partial sum sequence $\Sigma_1, \ldots, \Sigma_n$, numbering from the left-hand end (for prefix formulas). If the formula is well-formed, go to Step 2.

Step 2. Beginning with the left-hand end, form two sequences $d_1, \ldots, d_n$ and $c_1, \ldots, c_n$ as follows.

  (a) $d_1 = 0$, $c_1 = 0$.

  (b) for $i = 1, \ldots, n-1$:

    if $\Sigma_{i+1} > \Sigma_i$, set $d_{i+1} = d_i + 1$, $c_{i+1} = 0$;

    if $\Sigma_{i+1} = \Sigma_i$, set $d_{i+1} = d_i + 1$, $c_{i+1} = 1$;

    if $\Sigma_{i+1} < \Sigma_i$, set $d_{i+1} = d_k$, $c_{i+1} = 1$, $c_k = 1$, where k is the greatest integer $\leq i$ such that $c_k = 0$.

Step 3. The result of this procedure is that $c_1 = 0$, $c_i = 1$ $(i = 2, \ldots, n)$, and d_i is the depth of the ith symbol in the tree.

Example 6.6

Determine the depth function for CECpEpNqCprNq.

Solution. It was determined in Example 6.3 that this formula is well-formed. Begin with that determination and compute d_i and c_i.

$$\text{C} \quad \text{E} \quad \text{C} \, p \, \text{E} \, p \, \text{N} \, q \, \text{C} \, p \, r \, \text{N} \quad q$$

partial sums:	1	2	3	4	3	4	3	3	2	3	2	1	1
	$(\Sigma_1$	Σ_2			$\cdots$							$\Sigma_{13})$	
d_i:	0	1	2	3	3	4	4	5	2	3	3	1	2
c_i:	0	0̸	0̸	0̸	1	0̸	1	1	1	0̸	1	1	1
		1	1	1		1				1			

The changed values of c_2, c_3, c_4, c_6, and c_{10} arise from the computation of d_{12}, d_9, d_5, d_7, and d_{11} respectively.

ANSWER. The tree with depth indicated is given in Fig. 6.1.

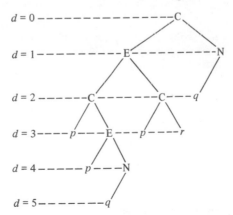

Figure 6.1

We leave the development of an algorithm to determine the address for each symbol from the depth function as an exercise.

EXERCISES

Convert the formula to Polish prefix and postfix notation:

1. $((p \vee \sim q) \supset (p \wedge \sim r)) \equiv (\sim p \supset (q \vee r))$.

2. $\sim(\sim p \equiv \sim(q \supset \sim r)) \wedge (p \supset (q \vee r) \equiv (\sim r \vee \sim p))$.

3. $((p \supset \sim r) \supset ((q \wedge s) \equiv (p \wedge \sim s))) \supset (p \vee \sim q \vee r \wedge s)$.

4. 4. $((p \vee \sim q \vee s) \wedge (q \vee r \vee \sim s)) \supset (((\sim p \vee q) \supset (\sim r \vee s)) \equiv (p \equiv s))$.

Test the formula to determine if it is well-formed:

5. ACpEpNqAKprCNrNEpq.   6. pqrAqNpNsKCNpNCrpNqACE.

7. EACpqrNpsNAKpqNsKACpNCqs.   8. spEpNqAKNprCApNprsAKspCqKApKCE.

9. pqKrNpAqsNrNKApNqAK.   10. CAsEpACqNpAKrNpqs.

Convert the formula to ordinary infix notation:

11. CAKEpqNrAKpNrNqp.   12. KAEpEqrENpApqKANprNq.

13. pqrKANpNqrEEprKApNCE.   14. CpCqKNqNCAprKANpNAqrp.

15. pNqANpqNprpKANACpEqAK.   16. rqNANpCqNEprqNENAC.

17. Develop an algorithm to determine the address of each symbol in a Polish string from the depth function.

18–23. Determine the depth function and address for each symbol in the formulas of Exercises 11–16.

24. Write a program to convert a formula from infix notation to Polish prefix and postfix notation. Use the input assumptions and conventions of Section 4, and test your program on Exercises 1–4.

Write a program to perform the desired operation on either a Polish prefix or postfix expression. Assume that the input string fits on one card or one terminal line, that N, C, A, K, and E are reserved for logical operators, and that any other letter is a propositional variable:

25. To test for a well-formed formula. Test on Exercises 5–10.

26. To convert a formula to infix notation. Test on Exercises 11–16.

27. To determine the depth function and symbol addresses. Test on Exercises 18–23.

7. PROOFS IN LOGIC

The concept of proof from a system of axioms has much in common with the concept of sentence description in formal linguistics. Using a descriptive grammar, word strings in a language are combined (parsed) into various intermediate phrases, and ultimately (if the original string is a sentence) into the string "sentence." Similarly in logic axioms are combined into intermediate strings, ultimately leading to a single entity called a "theorem," using the rules of inference for the logic.

A *set of axioms* is a set of well-formed formulas. A *rule of inference* is a procedure, production, or algorithm for generating a new well-formed formula from given ones. Even though axioms and rules of inference could be chosen quite arbitrarily, they are invariably chosen with one criterion in mind: the theorems generated by the axiom and rule-of-inference system should be precisely the tautologies. This implies that the axioms are tautologies, and that the rules of inference generate only tautologies when used on tautologies.

Various other criteria may also be used in defining the axioms and rules of inference. For example, economy suggests that no formula be included in the axiom set if it can be generated from the axioms already chosen. This is called *independence*. Another aspect of economy might be that there be a minimal number of axioms. That is, given an independent set of five axioms, and a different independent set of four axioms that generate the same theorems, we might prefer to use the smaller set as the basis upon which we build. Esthetics may dictate a choice of axioms involving particular operators.

A *proof* is a sequence $F_1, F_2, ..., F_n$, $n \geq 1$, of well-formed formulas such that each F_i is either an axiom or a previously generated theorem, or is generated from earlier formulas in the sequence by one of the rules of inference. A proof is a proof of the last formula listed (F_n), which is called a *theorem*. (Note that logicians do not admit the use of "a previously generated theorem." Logically such use is unnecessary and hence unesthetic; practically it makes a world of difference. Allowing the use of previous theorems amounts to d-introduction.)

We shall briefly demonstrate two axiom systems, a rather common one, and one with a single axiom and a single rule of inference. In each case the proof trees (see Chapter 3, Section 6) have axioms or previous theorems at the end points. Intermediate points are generated using the rules of inference, and the root is the theorem to be proven.

System 1

Axiom 1. $p \supset (q \supset p)$.

Axiom 2. $(p \supset (q \supset r)) \supset ((p \supset q) \supset (p \supset r))$.

Axiom 3. $(\sim p \supset \sim q) \supset ((\sim p \supset q) \supset p)$.

Inference Rule 1 (modus ponens). From A and $A \supset B$, generate B.

Inference Rule 2 (substitution). From string A, well-formed formula B, and propositional variable p, generate C by substituting B for each occurrence of p in A.

Substitution is a rather tricky rule which must be carefully applied. For example, suppose in the formula $p \supset q$ we wish to substitute $p \supset q$ for p, and p for q. The result should be $(p \supset q) \supset p$. But observe:

$$p \supset q$$

$$(p \supset q) \supset q \qquad \text{(substitute for } p\text{)}$$

$$(p \supset p) \supset p \qquad \text{(substitute for } q\text{)}.$$

Also

$$p \supset q$$

$$p \supset p \qquad \text{(substitute for } q\text{)}$$

$$(p \supset q) \supset (p \supset q) \qquad \text{(substitute for } p\text{)}.$$

To achieve the correct result, we must protect the proper variables (in this case q) by changing them as an early step in substitution:

$$p \supset q$$

$$p \supset r \qquad \text{substitute } r \text{ for } q$$

$$(p \supset q) \supset r \qquad \text{substitute for } p$$

$$(p \supset q) \supset p \qquad \text{substitute } p \text{ for } r.$$

Note that by changing the substitution p for q to the two-step substitution r for q, p for r we have protected the q given in the original formula, and achieved a proper substitution.

Use of these rules of inference is illustrated in the proofs of three theorems of System 1.

Theorem S1.1. $p \supset p$.

1. $(p \supset (q \supset r)) \supset ((p \supset q) \supset (p \supset r))$ (Ax 2)
2. $(p \supset ((p \supset p) \supset r)) \supset ((p \supset (p \supset p)) \supset (p \supset r))$ (Sub, q)
3. $(p \supset ((p \supset p) \supset p)) \supset ((p \supset (p \supset p)) \supset (p \supset p))$ (Sub, r)
4. $p \supset (q \supset p)$ (Ax 1)

5. $p \supset ((p \supset p) \supset p)$ (Sub, q)
6. $(p \supset (p \supset p)) \supset (p \supset p)$ (MP: 3, 5)
7. $p \supset (p \supset p)$ (Sub, q)
8. $p \supset p$ (MP: 6, 7)

The proof tree for Theorem S1.1 is given in Fig. 7.1. A node in the tree has one immediate successor if substitution is used; two if modus ponens is used.

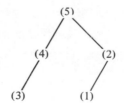

Figure 7.1

Theorem S1.2. $(\sim p \supset p) \supset p$.

1. $(\sim p \supset \sim q) \supset ((\sim p \supset q) \supset p)$ (Ax 3)
2. $(\sim p \supset \sim p) \supset ((\sim p \supset p) \supset p)$ (Sub, q)
3. $p \supset p$ (Th. S1.1)
4. $\sim p \supset \sim p$ (Sub, p)
5. $(\sim p \supset p) \supset p$ (MP: 2, 4)

The proof tree is Fig. 7.2.

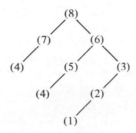

Figure 7.2

Theorem S1.3. $(q \supset r) \supset ((p \supset q) \supset (p \supset r))$.

1. $(p \supset (q \supset r)) \supset ((p \supset q) \supset (p \supset r))$
2. $p \supset (q \supset p)$
3. $p \supset (s \supset p)$

4. $((p \supset (q \supset r)) \supset ((p \supset q) \supset (p \supset r)))$
$\supset (s \supset ((p \supset (q \supset r)) \supset ((p \supset q) \supset (p \supset r))))$
5. $s \supset ((p \supset (q \supset r)) \supset ((p \supset q) \supset (p \supset r)))$
6. $(q \supset r) \supset ((p \supset (q \supset r)) \supset ((p \supset q) \supset (p \supset r)))$
7. $(q \supset r) \supset (s \supset (q \supset r))$
8. $(q \supset r) \supset (p \supset (q \supset r))$
9. $(p \supset (s \supset r)) \supset ((p \supset s) \supset (p \supset r))$
10. $(p \supset (s \supset t)) \supset ((p \supset s) \supset (p \supset t))$
11. $((q \supset r) \supset (s \supset t)) \supset (((q \supset r) \supset s) \supset ((q \supset r) \supset t))$
12. $((q \supset r) \supset ((p \supset (q \supset r)) \supset t))$
$\supset (((q \supset r) \supset (p \supset (q \supset r))) \supset ((q \supset r) \supset t))$
13. $((q \supset r) \supset ((p \supset (q \supset r)) \supset ((p \supset q) \supset (p \supset r))))$
$\supset (((q \supset r) \supset (p \supset (q \supset r))) \supset ((q \supset r) \supset ((p \supset q) \supset (p \supset r))))$
14. $((q \supset r) \supset (p \supset (q \supset r))) \supset ((q \supset r) \supset ((p \supset q) \supset (p \supset r)))$
15. $(q \supset r) \supset ((p \supset q) \supset (p \supset r))$

The proof tree is Fig. 7.3.

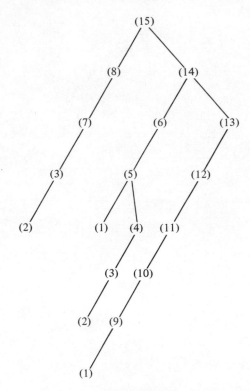

Figure 7.3

System 2

Axiom 1. $(p|(q|r))|((p|(r|p))|((s|q)|((p|s)|(p|s))))$.

Inference Rule 1.   From A and $A|(B|C)$ generate C.

Inference Rule 2.   Substitution, as in System 1.

While this system has the austere beauty of a single operator and a single axiom, this austerity causes some difficulty in generating proofs. Note that the axiom is in the form $A|(B|C)$, where

$$A \text{ is } p|(q|r), \quad B \text{ is } p|(r|p), \quad C \text{ is } (s|q)|((p|s)|(p|s)).$$

The first proof is initiated by substituting for p, q, and r respectively these three formulas.

Theorem S2.1. $(s|(p|(r|p)))|(((p|(q|r))|s)|((p|(q|r))|s))$.

1. $(p|(q|r))|((p|(r|p))|((s|q)|((p|s)|(p|s))))$
2. $(t|(q|r))|((t|(r|t))|((s|q)|((t|s)|(t|s))))$
3. $(t|(u|r))|((t|(r|t))|((s|u)|((t|s)|(t|s))))$
4. $(t|(u|v))|((t|(v|t))|((s|u)|((t|s)|(t|s))))$
5. $((p|(q|r))|(u|v))|(((p|(q|r))|(v|(p|(q|r))))$
     $|((s|u)|(((p|(q|r))|s)|((p|(q|r))|s))))$
6. $((p|(q|r))|((p|(r|p))|v))|(((p|(q|r))|(v|(p|(q|r))))$
     $|((s|(p|(r|p)))|(((p|(q|r))|s)|((p|(q|r))|s))))$
7. $((p|(q|r))|((p|(r|p))|((s|q)|((p|s)|(p|s)))))|$
     $(((p|(q|r))|(((s|q)|((p|s)|(p|s)))|(p|(q|r))))$
     $|((s|(p|(r|p)))|(((p|(q|r))|s)|((p|(q|r))|s))))$
8. $(s|(p|(r|p)))|(((p|(q|r))|s)|((p|(q|r))|s))$

The proof tree for Theorem S2.1 is Fig. 7.4 (overleaf).

Because of the relative simplicity of the propositional calculus, some of the work in artificial intelligence has concentrated on developing theorem proving programs for elementary logic. Such programs rely heavily on heuristics—conceptual aids which may or may not help solve a problem, and on the setting of intermediate goals. Suppose, for example, that a theorem proving program is asked to prove $p \supset p$ from our first axiom system. The "reasoning" in the program might go as follows.

1. The expression $p \supset p$ is shorter than any of the axioms. Therefore modus ponens is needed. Hence think of $p \supset p$ as the B in $A \supset B$, and search for a suitable A.

2. Since $p \supset p$ does not involve negation, Axiom 3 is probably of no help. Concentrate on the first two axioms. (Note that while modus ponens is definitely needed, we cannot say at this stage that Axiom 3 is definitely not needed.)

3. Look at Axiom 1. Substituting p for q provides a suitable B; the A is then p. Then either immediately or after some trials, depending on the program, the program decides that p is probably a poor choice for A and abandons the effort. However, this possibility is saved in case we get desperate.

4. Look at Axiom 2. The B here is $(p \supset q) \supset (p \supset r)$, far too complicated. However, substituting p for r places the desired expression at the end of this modified axiom:

$$(p \supset (q \supset p)) \supset ((p \supset q) \supset (p \supset p)).$$

5. Establish subgoals: Find some q so that both $p \supset (q \supset p)$ and $p \supset q$ can be proven. The plan is

(a) establish $(p \supset (q \supset p)) \supset ((p \supset q) \supset (p \supset p))$;
(b) establish $p \supset (q \supset p)$;
(c) use modus ponens to deduce $(p \supset q) \supset (p \supset p)$;
(d) establish $p \supset q$;
(e) use modus ponens to deduce $p \supset p$.

6. The first of these expressions, $p \supset (q \supset p)$, is Axiom 1: any q will work. Therefore concentrate on the second expression, $p \supset q$.

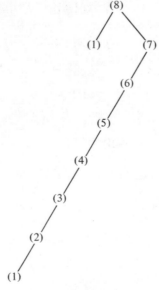

Figure 7.4

7. Axiom 1 is the simplest. Will $p \supset q$ fit this mold? Yes, by choosing $p \supset p$ (or $q \supset p$ or $r \supset p$, and so on) as the q of $p \supset q$. Thus the subgoals become $p \supset ((p \supset p) \supset p)$ and $p \supset (p \supset p)$, both of which can be proven from Axiom 1. Hence the plan of Step 5 can be executed, and the theorem proven.

A theorem prover of this type uses various heuristics which are not guaranteed to work: if negation is involved, try starting with Axiom 3 (otherwise, ignore Axiom 3); try the simplest axiom first; try the given expression as B in $A \supset B$, and search for a suitable A; set up a choice of one or more subgoals, and work at these. Since the logic involved is quite simple, theorem provers based on these and similar heuristics have been quite successful.

In effect, a theorem prover of this type begins with the root, and attempts to construct a tree from this. Thus at step 5, for example, we have tentatively established the subtree of Fig. 7.1 which is shown in Fig. 7.5. If this can successfully be extended to a full tree, a proof has been found.

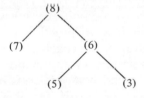

Figure 7.5

EXERCISE

**1. A readable account of theorem proving for the propositional calculus is given in [1]. Read it, and utilize the ideas in the construction of a theorem proving program for System 1. Test your program on the following theorems:

$$p \supset p$$

$$(\sim p \supset p) \supset p$$

$$(q \supset r) \supset ((p \supset q) \supset (p \supset r))$$

$$(p \supset q) \supset ((q \supset r) \supset (p \supset r))$$

$$(p \supset (q \supset r)) \supset (q \supset (p \supset r))$$

$$(\sim p \supset \sim \sim p) \supset p$$

$$\sim \sim p \supset p$$

$$(\sim \sim \sim p \supset p) \supset \sim \sim p$$

$$p \supset \sim \sim p$$

8. SETS AND WORDSETS

We have shown the correspondence between set algebra and Boolean algebra, and between Boolean algebra and propositional calculus. We round out the picture by discussing the relationship between set algebra and propositional calculus. This relationship is in fact simply that which is reflected in our ordinary language describing set operations. For example, we define $A \cup B$ as $\{x \mid x \in A$ or $x \in B$ or both$\}$. More formally put,

$$a \in A \cup B \qquad \text{is equivalent to} \qquad (a \in A) \vee (a \in B).$$

Similarly,

$$a \in A \cap B \qquad \text{is equivalent to} \qquad (a \in A) \wedge (a \in B);$$

$$a \in A \triangle B \qquad \text{is equivalent to} \qquad (a \in A) \not\equiv (a \in B);$$

$$a \in \bar{A} \qquad \text{is equivalent to} \qquad \sim(a \in A); \text{ and}$$

$$A \subseteq B \qquad \text{is equivalent to} \qquad (a \in A) \supset (a \in B).$$

Thus we have completed the correspondence between set operators, Boolean operators, and logical operators. In practical terms, this correspondence provides a variety of equivalent approaches to many data manipulation problems. For example, in writing an information retrieval program based on keywords, we may think and write in terms of the sets of documents characterized by the various keywords. Or, if we prefer, we may think and write in terms of logical combinations of keywords. Either approach will produce the same results.

We referred in Chapter 1 to the representation of sets as "wordsets," each position in a computer word denoting a particular set element. At the time we mentioned the use of logical operators in the computer to implement the set operators. Now, in the light of the correspondence between these two types of operators, the validity of this use is clear.

EXERCISE

1. Find set operators equivalent to the logical operators $\mid$, $\downarrow$, and $\equiv$.

REFERENCE

1. A. Newell, J. C. Shaw, and H. A. Simon. Empirical explorations with the logic theory machine: a case study in heuristics, *in* "Computers and Thought" (E. A. Feigenbaum and J. Feldmen, eds.), pp. 109–133. McGraw-Hill, New York, 1963.

CHAPTER 10

Combinatorics

1. INTRODUCTION

One process which is fundamental to all computational work is that of *counting* or *enumerating* objects. This process is basic to determining program size, storage requirements, time estimates, error calculations, and the sequencing of operations. The study of enumerating techniques is known as *combinatorics* or *combinatorial analysis*.

In this chapter we study a variety of techniques, beginning with the counting of simple permutations and combinations. The concept of a generating function is discussed, as a means of easily computing rather complex counts. Finally, partitions and compositions of integers, which give rise to frequently recurring formulas, are introduced. A more detailed exposition of combinatorics may be found in Berge [1] or Riordan [2].

2. PERMUTATIONS OF OBJECTS

In Chapter 5 we viewed a permutation as a transformation on a set of objects. Here we wish to examine the results of such transformations. Let

S be a set of n distinct objects. An ordered selection of r elements from S is called an *r-permutation of n objects*. The number of such permutations is denoted by $P(n, r)$ or by $(n)_r$. Note the emphasis here on *order*. The sequence in which the objects are chosen is important to this concept.

To compute $P(n, r)$, we observe that the first object chosen may be any one of the n objects in S. Having chosen that, the next object may be any one of the $n - 1$ remaining. This pattern continues until r objects have been chosen. Thus

$$P(n, r) = n(n - 1)(n - 2) \cdots (n - r + 1).$$

(This is an instance of the *rule of products*: if A can be chosen in m ways, and then B chosen in n ways, then A and B can be chosen *in that order* in $m \cdot n$ ways.) Note that $P(n, r) = n!/(n - r)!$. For example, Fig. 2.1 illustrates the choice of three objects from four. Note that there are four possible first choices; for each first choice, three possible second choices; and for each first and second choice, two possible third choices.

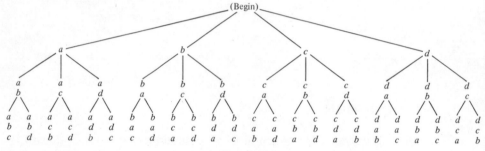

Figure 2.1

The development of enumerating procedures depends heavily upon *recurrence relations*, defining one value of a function in terms of previously computed values. These are often derived by considering one particular object. For example, consider a particular object A in the set S. Among all r-permutations, some will contain A and some will not. Those containing A have A in one of the r possible positions. The remaining $r - 1$ positions are filled by objects chosen from the $n - 1$ elements of $S - \{A\}$. That is, once the object A has been assigned to one of the r possible positions, there are $P(n - 1, r - 1)$ ways of filling of the remaining positions. Hence there are $rP(n - 1, r - 1)$ r-permutations containing A. Those not containing A are, in effect, r-permutations of $n - 1$ objects; there are $P(n - 1, r)$ of these. Adding, we have the relation

$$P(n, r) = rP(n - 1, r - 1) + P(n - 1, r).$$

We observe that $P(n, 1) = n$ since there are only n objects to choose, and $P(n, n) = n!$ From these boundary conditions and the recurrence relation we can construct a table of values by the process diagrammed in Fig. 2.2. See Table 2.1.

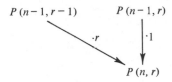

Figure 2.2

TABLE 2.1

n \ r	1	2	3	4	5	6	7	8
1	1							
2	2	2						
3	3	6	6					
4	4	12	24	24				
5	5	20	60	120	120			
6	6	30	120	360	720	720		
7	7	42	210	840	2520	5040	5040	
8	8	56	336	1680	6720	20160	40320	40320

Observe that the value of $P(n, r)$ is ultimately a function of the $n-r$ values $P(2, 1), \ldots, P(n-r+1, 1)$ and the $r-1$ values $P(2, 2), \ldots, P(r, r)$. For example,

$$
\begin{aligned}
P(7, 4) &= 144P(2, 1) + 72P(3, 1) + 24P(4, 1) \\
&\quad + 72P(2, 2) + 12P(3, 3) + 1P(4, 4) \\
&= 144 \cdot 2 + 72 \cdot 3 + 24 \cdot 4 + 72 \cdot 2 + 12 \cdot 6 + 1 \cdot 24 \\
&= 840.
\end{aligned}
$$

This development is shown in Fig. 2.3, where the arc numbers show the coefficient development.

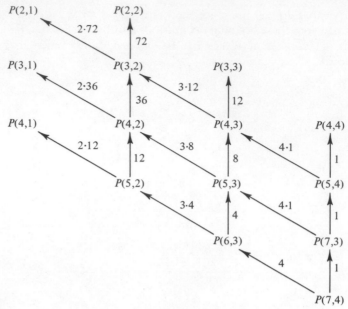

Figure 2.3

Now suppose that the set of objects contains p objects of one kind, q of another kind, r of another, and so on. We consider only the problem of counting n-permutations, that is, permutations of the entire set. Let x be the number of such permutations. If all n objects were distinct, there would be $P(n,n) = n!$ permutations. However, not all of the objects are distinct. The p objects of the first kind can be arranged in $p!$ ways, independently of the objects of the other kinds. (Think of them as labeled.) Similarly, the objects of the second kind have $q!$ arrangements. Hence

$$xp!\,q!\,r! \cdots = n!, \qquad \text{where} \quad p + q + \cdots = n.$$

Thus

$$x = \frac{n!}{p!\,q!\,r! \cdots}, \qquad \text{where} \quad p + q + r + \cdots = n.$$

Example 2.1

Suppose a knight must slay three indistinguishable tigers and two indistinguishable dragons. In how many arrangements may he slay the beasts?

Solution.   For the moment, label the tigers T_1, T_2, and T_3, and the dragons D_1 and D_2. There are now five distinct beasts, and hence $5! = 120$ arrangements. But if the tigers are indistinguishable, then for each arrangement, say $T_1 T_2 T_3 D_1 D_2$, there will be $3! - 1$ additional arrangements indistinguishable from it, $T_1 T_3 T_2 D_1 D_2$, $T_2 T_1 T_3 D_1 D_2$, $T_2 T_3 T_1 D_1 D_2$, $T_3 T_1 T_2 D_1 D_2$, and $T_3 T_2 T_1 D_1 D_2$. Thus the total number of distinct arrangements is reduced by a factor $3!$. Similarly, since the two dragons are indistinguishable, the total number of distinct arrangements is further reduced by a factor of $2!$. Thus there are

$$\frac{5!}{3! 2!} = \frac{120}{6 \cdot 2} = 10$$

distinct arrangements:

$TTTDD$,	$TTDTD$,	$TDTTD$,	$DTTTD$,	$TTDDT$,
$TDTDT$,	$DTTDT$,	$TDDTT$,	$DTDTT$,	$DDTTT$.

ANSWER.   Ten arrangements.

Example 2.2

Suppose 50 programs are submitted from three computer science courses—24 from one course, 14 from the seond, and 12 from the third. If the programs in any one course are indistinguishable, determine the number of distinct arrangements of the programs.

Solution.   For this example $n = 50$, $p = 24$, $q = 14$, and $r = 12$. Hence the number of distinct arrangements is

$$\frac{50!}{24! 14! 12!}.$$

ANSWER.

11 73880 49403 25745 10000.

Another type of situation that occurs is selection with unlimited repetition. Suppose that r objects are to be selected in order from a set of n objects. With unlimited repetition, the choice each time is from the entire set of n objects. Hence the number of permutations is

$$U(n, r) = n^r.$$

Selection with unlimited repetition is generally associated with sampling with repetition. That is, an object is selected from a population, studied, and replaced. Another object is then similarly selected, as is a third object, and so forth. Since the first object has been replaced, it may well be chosen again and again in the course of the study. If this is undesirable, then each object selected must be either permanently withdrawn, or marked in some way so that it can be recognized and rejected if chosen again.

Summary:

r-permutations of n objects, without repetition

$$P(n,r) = \frac{n!}{(n-r)!}$$

r-permutations of n objects, unlimited repetition

$$U(n,r) = n^r$$

n-permutations of n objects, in classes of $p, q, \cdots$ objects, $p + q + \cdots = n$,

$$\frac{n}{p! \, q! \cdots}$$

EXERCISES

Determine the number of permutations of r objects from a set of n objects, both without repetition and with unlimited repetition:

1. $r = 3$, $n = 7$.   2. $r = 6$, $n = 8$.

3. $r = 13$, $n = 52$.   4. $r = 12$, $n = 10$.

Determine the number of distinct arrangements of n objects that are divided into k classes of objects containing $p_1, p_2, ..., p_k$ objects.

5. $n = 10$, $p_1 = 3$, $p_2 = 7$.

6. $n = 12$, $p_1 = 3$, $p_2 = 5$, $p_3 = 4$.

7. $n = 4$, $p_1 = p_2 = p_3 = p_4 = 1$.

8. $n = 100$, $p_1 = 50$, $p_2 = 25$, $p_3 = 12$, $p_4 = 6$, $p_5 = 3$, $p_6 = 2$, $p_7 = 2$.

*9. Suppose n objects are divided into k classes containing $p_1, p_2, ..., p_k$ objects, with $p_1 \leqslant p_2 \leqslant \cdots \leqslant p_k$. Determine the number of distinct arrangements of r objects chosen from these n, when $r \leqslant p_1$. *Hint:* All objects chosen may or may not be from the same class; hence consider different cases. Try with small values of r and k first, then generalize.

*10. (cont.) Now determine the number of distinct arrangements of r objects, when $r \leqslant n$.

11. (cont.) Suppose that unlimited repetition is allowed. What is the counting formula now, and why?

12. Write a program to generate all r-permutations (without repetition) of n objects, given r and n as input, $r, n \leqslant 20$.

13. Write a program to generate all r-permutations (unlimited repetition) of n objects, given r and n as input, $r, n \leqslant 20$.

14. Write a program to generate all distinct arrangements of n objects, in classes of $p_1, ..., p_k$ objects, given n, k, and $p_1, ..., p_k$ as input, $n, k \leqslant 20$.

*15. Write a program to generate all distinct r-permutations of n objects, in classes of $p_1, ..., p_k$ objects, given r, n, k, and $p_1, ..., p_k$ as input, $r, n, k \leqslant 20$.

16. Let A be an $n \times n$ matrix. Recall the definition of the determinant,

$$\det(A) = \sum \text{sgn}(\pi) a_{1,\pi(1)} a_{2,\pi(2)} \cdots a_{n,\pi(n)},$$

where the sum is over all permutations of $1, ..., n$. (See p. 35.) Show that there are $n!$ terms in the sum for $\det(A)$.

17. (cont.)
   (a) Show that if B is derived from A by multiplying one row (or column) of A by a constant k, then $\det(B) = k \cdot \det(A)$.
   (b) Show that if two rows (or columns) of A are identical, then $\det(A) = 0$.
   (c) Let $C^{(p,q;k)}$ be an $n \times n$ matrix identical to A except for the qth row, which is defined by

$$C^{(p,q,k)}_{qj} = a_{qj} + k \cdot a_{pj}.$$

   Show that $\det(C^{(p,q;k)}) = \det(A)$.

18. (cont.)
   (a) Show that it is possible to transform A into a matrix A^* with the properties
   (1) one column (or row) of A^* contains only one nonzero entry;
   (2) $\det(A^*) = \det(A)$.
   (b) Determine the number of operations necessary to compute $\det(A^*)$, given A; and compare this with the number of operations required to compute $\det(A)$ directly from the definition.

3. COMBINATIONS OF OBJECTS

While the order of selection is often important, in some instances it is not. In fact, one often wishes to identify r-samples that contain the same elements, regardless of order. Let S be a set of n distinct objects. An unordered selection of r elements from S is called an r-combination of n objects. The number of such combinations is denoted by $C(n, r)$ or $\binom{n}{r}$.

We compute $C(n, r)$ by noting that each r-combination corresponds to $r!$ r-permutations. Hence $r! C(n, r) = P(n, r)$, or

$$C(n, r) = \frac{P(n, r)}{r!} = \frac{n!}{(n-r)! r!} = \binom{n}{r}.$$

By definition, for $n < 0$, and $r > n$ or $r < 0$, $C(n, r) = 0$.

The recurrence relation for $C(n, r)$ is developed by the standard trick of including or excluding a specific element A. If A is included, then the remaining $r-1$ objects are chosen from $n-1$ available objects. There are $C(n-1, r-1)$ such choices. If A is excluded, then there are $C(n-1, r)$ choices of the r objects from the remaining $n-1$ objects. Hence

$$C(n, r) = C(n-1, r-1) + C(n-1, r).$$

In contrast to the recurrence relation for $P(n,r)$, there is no multiplier r on the term $C(n-1,r-1)$ since the position of A, if it is included, is immaterial. This relation may also be established by manipulating the factorial expressions:

$$C(n-1,r-1) + C(n-1,r) = \frac{(n-1)!}{(n-r)!(r-1)!} + \frac{(n-1)!}{(n-1-r)!r!}$$

$$= \frac{(n-1)!}{(n-r-1)!(r-1)!}\left(\frac{1}{n-r}+\frac{1}{r}\right)$$

$$= \frac{n!}{(n-r)!r!}$$

$$= C(n,r).$$

Expansion of either the first term or the last term of the recurrence, iterated, yields the useful identities

$$C(n,r) = C(n-1,r) + C(n-2,r-1) + \cdots + C(n-1-r,0), \tag{1}$$

$$C(n,r) = C(n-1,r-1) + C(n-2,r-1) + \cdots + C(r-1,r-1). \tag{2}$$

As with the relation for r-permutations, the recurrence relation for combinations may be used to develop a table by the scheme shown in Fig. 3.1.

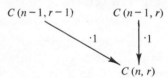

Figure 3.1

See Table 3.1, the classical *Pascal's triangle*. Identities (1) and (2) develop the value of $C(n,r)$ from a diagonal and a column, respectively of this table (Figs. 3.2 and 3.3).

The development of the formula for r-combinations with unlimited repetition (corresponding to $U(n,r)$ for r-permutations) involves the same trick of looking at a particular element A. Call the number of such r-combinations $f(n,r)$. If A is included as one element of an r-combination, it may also be included among the remaining $r-1$ elements because of the allowed repetition. Hence there are $f(n,r-1)$ r-combinations containing A at least once. In addition, there are $f(n-1,r)$ r-combinations not containing A. Thus

$$f(n,r) = f(n,r-1) + f(n-1,r).$$

TABLE 3.1

n \ r	0	1	2	3	4	5	6	7	8
0	1								
1	1	1							
2	1	2	1						
3	1	3	3	1					
4	1	4	6	4	1				
5	1	5	10	10	5	1			
6	1	6	15	20	15	6	1		
7	1	7	21	35	35	21	7	1	
8	1	8	28	56	70	56	28	8	1

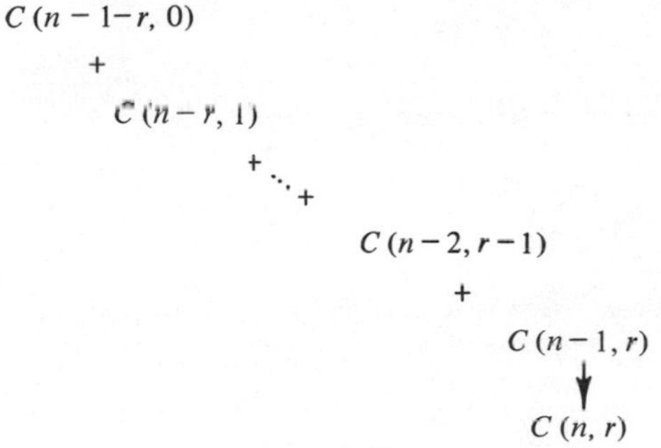

$$C(n-1-r, 0)$$
$$+$$
$$C(n-r, 1)$$
$$+ \cdots +$$
$$C(n-2, r-1)$$
$$+$$
$$C(n-1, r)$$
$$\downarrow$$
$$C(n, r)$$

Figure 3.2

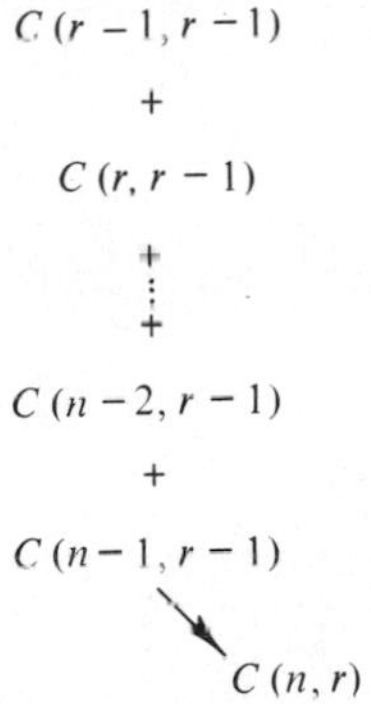

$$C(r-1, r-1)$$
$$+$$
$$C(r, r-1)$$
$$+$$
$$\vdots$$
$$+$$
$$C(n-2, r-1)$$
$$+$$
$$C(n-1, r-1)$$
$$\searrow$$
$$C(n, r)$$

Figure 3.3

We determine f exactly by calculating the boundary values $f(n, 1)$ and $f(1, r)$, and then experimenting with the recurrence relation. Now $f(n, 1) = n$ since any one of the n elements may be chosen; and $f(1, r) = 1$ since there is only one element from which to choose.

Let us calculate $f(n, 2)$.

$$
\begin{aligned}
f(n, 2) &= f(n, 1) + f(n-1, 2) \\
&= f(n, 1) + f(n-1, 1) + f(n-2, 2) \\
&= f(n, 1) + f(n-1, 1) + f(n-2, 1) + f(n-3, 2) \\
&\vdots \\
&= f(n, 1) + f(n-1, 1) + \cdots + f(2, 1) + f(1, 2) \\
&= n + (n-1) + (n-2) + \cdots + 2 + 1 \\
&= \frac{n(n+1)}{2} = \binom{n+1}{2}.
\end{aligned}
$$

How about $f(n, 3)$?

$$f(n, 3) = f(n, 2) + f(n-1, 3)$$
$$= f(n, 2) + f(n-1, 2) + f(n-2, 3)$$
$$\vdots$$
$$= f(n, 2) + f(n-1, 2) + \cdots + f(2, 2) + f(1, 3)$$
$$= \binom{n+1}{2} + \binom{n}{2} + \cdots + \binom{3}{2} + 1.$$

Now use identity (2) with $r = 3$, noting that $1 = \binom{2}{2}$:

$$f(n, 3) = \binom{n+2}{3}.$$

Recall

$$f(n, 2) = \binom{n+1}{2}, \qquad f(n, 1) = \binom{n}{1}.$$

We may now conjecture that

$$f(n, r) = \binom{n+r-1}{r}.$$

We leave it as as an exercise to show that this is the desired function. Summary:

r-combinations of n objects, without repetition

$$C(n, r) = \binom{n}{r} = \frac{n!}{r!(n-r)!}$$

r-combinations of n objects, unlimited repetition

$$f(n, r) = \binom{n+r-1}{r}$$

EXERCISES

1. Show that

$$\binom{n}{r} = \binom{n-1}{r} + \binom{n-2}{r-1} + \cdots + \binom{n-1-r}{0}.$$

2. Show that

$$\binom{n}{r} = \binom{n-1}{r-1} + \binom{n-2}{r-1} + \cdots + \binom{r-1}{r-1}.$$

3. Show that

$$f(n,r) = \binom{n+r-1}{r}$$

satisfies the boundary conditions $f(n,1) = n$, $f(1,r) = 1$, and the recurrence relation

$$f(n,r) = f(n,r-1) + f(n-1,r).$$

4. Write a program to generate all r-combinations (without repetition) of n objects, given r and n as input, $r, n \leqslant 20$.
5. Write a program to generate all r-combinations (unlimited repetition) of n objects, given r and n as input, $r, n \leqslant 20$.

4. ENUMERATORS FOR COMBINATIONS

While the technique of looking at a particular element is useful, it does have its limitations. Another technique of wider applicability is the use of *enumerating generating functions*, or *enumerators*.

A generating function in general is a finite or infinite series whose coefficients have values equal to the values of some function under study. In our case, the coefficients should count or enumerate objects by some method.

Consider, for example, three objects represented by x_1, x_2, x_3. We compute a polynomial:

$$(1+x_1 t)(1+x_2 t)(1+x_3 t)$$
$$= 1 + (x_1+x_2+x_3)t + (x_1 x_2+x_2 x_3+x_3 x_1)t^2 + (x_1 x_2 x_3)t^3$$
$$= a_0 + a_1 t + a_2 t^2 + a_3 t^3.$$

Observe that each a_r is a function of x_1, x_2, and x_3:

$$(1+x_1 t)(1+x_2 t)(1+x_3 t) = \sum_{r=0}^{3} a_r(x_1, x_2, x_3)t^r.$$

In particular, a_r $(r > 0)$ contains one term corresponding to each r-combination chosen from x_1, x_2, x_3. Thus by setting $x_1 = x_2 = x_3 = 1$ we can count the r-combinations:

$$a_r(1,1,1) = \binom{3}{r}.$$

(Note that by convention this also holds for $r = 0$.) Hence

$$(1+t)^3 = \sum_{r=0}^{3} a_r(1,1,1)t^r = \sum_{r=0}^{3} \binom{3}{r}t^r.$$

The enumerator for r-combinations of three objects is $(1+t)^3$. Similarly, for any n, the enumerator of r-combinations is

$$(1+t)^n = \sum_{r=0}^{n} a_r(1,1,\ldots,1)\, t^r = \sum_{r=0}^{n} \binom{n}{r} t^r.$$

(Because of their appearance in the expansion of binomials such as $(1+t)^n$, the numbers $\binom{n}{r}$ are called *binomial coefficients*.)

We are interested only in the coefficients. The general coefficients $a_r(x_1, x_2, \ldots, x_n)$ are called the *elementary symmetric functions* on n variables. They are important throughout algebra, number theory, and combinatorics.

The importance of the enumerators lies in their ease of manipulation. They are formal objects, so that we may manipulate them without the analytical worries of convergence, and so on.

Example 4.1

Show that

$$\sum_{r=0}^{\lfloor n/2 \rfloor} \binom{n}{2r} = \sum_{r=0}^{\lfloor n/2 \rfloor} \binom{n}{2r+1} = 2^{n-1}.$$

Solution. Consider $(1+t)^n = \sum_{r=0}^{n} \binom{n}{r} t^r$. First set $t = 1$:

$$2^n = \sum_{r=0}^{n} \binom{n}{r}.$$

Then set $t = -1$:

$$0 = \sum_{r=0}^{n} (-1)^r \binom{n}{r}.$$

Add these two equations and divide by 2:

$$\sum_{r=0}^{\lfloor n/2 \rfloor} \binom{n}{2r} = 2^{n-1}.$$

Similarly, by subtracting

$$\sum_{r=0}^{\lfloor n/2 \rfloor} \binom{n}{2r+1} = 2^{n-1}.$$

Example 4.2

Show that

$$\binom{n}{r} = \sum_{k=0}^{r} \binom{n-m}{k}\binom{m}{r-k}$$

Solution. Set $n = (n-m)+m$, thus decomposing $(1+t)^n$:

$$(1+t)^n = (1+t)^{n-m}(1+t)^m.$$

On the left, the coefficient of t^r is $\binom{n}{r}$. On the right, the t^r term is the sum of all terms in $t^k t^{r-k}, k = 0, 1, \ldots, r$, where the t^k comes from $(1+t)^{n-m}$ and t^{r-k} comes from $(1+t)^m$. Hence the coefficient on the right is the sum of the coefficients of $t^k t^{r-k}$. Thus

$$\binom{n}{r} = \sum_{k=0}^{r} \binom{n-m}{k}\binom{m}{r-k}.$$

Many combinatorial identities can be established by similar reasoning.

The idea that the term in t^r corresponds to r-combinations provides the germ for generalizing this work to the case of nondistinct or recurring elements. The term $1 + x_k t$ allows for object x_k to occur once or not at all. By analogy, the term

$$1 + x_k t + x_k^2 t^2 + \cdots + x_k^j t^j$$

allows for $0, 1, 2, \ldots, j$ occurrences of x_k. Similarly, the term

$$1 + x_k^2 t^2 + x_k^4 t^4 + \cdots + x_k^{2i} t^{2i}$$

allows for an even number of occurrences of x_k, up to a maximum of $2i$ occurrences.

Example 4.3

Determine the enumerator for combinations of n objects if unlimited repetition is allowed.

Solution. In this case the term for each x_k is $1 + x_k t + x_k^2 t^2 + \cdots$, an infinite series. Setting each of the n coefficients x_k equal to 1, the enumerator is

$$(1 + t + t^2 + t^3 + \cdots)^n.$$

However, $1 + t + t^2 + t^3 + \cdots = 1/(1-t)$. Thus the enumerator is

$$(1-t)^{-n} = \sum_{r=0}^{\infty} \binom{-n}{r}(-t)^r.$$

Even though $\binom{-n}{r}$ has no direct meaning in counting, it can be established formally that

$$\binom{-n}{r} = (-1)^r \binom{n+r-1}{r}.$$

(See Exercise 11.) Hence

$$(1-t)^{-n} = \sum_{r=0}^{\infty} \binom{n+r-1}{r} t^r.$$

ANSWER.

$$(1-t)^{-n} = \sum_{r=0}^{\infty} \binom{n+r-1}{r} t^r.$$

It is satisfying that the coefficients of this expansion agree with the values previously computed for $f(n,r)$. (See Section 3.)

Example 4.4

Determine the enumerator for combinations of n objects with unlimited repetition, if each object must occur at least once.

Solution.   Observe that the coefficients for $r < n$ must all be zero. (Why?) An object which may occur with unlimited repetition corresponds to a factor

$$(1 + t + t^2 + t^3 + \cdots).$$

However, the first term in this factor corresponds to no occurrence of the object. If the object must occur, this term is eliminated and the factor becomes

$$(t + t^2 + t^3 + \cdots).$$

The enumerator is

$$(t + t^2 + t^3 + \cdots)^n = t^n (1-t)^{-n}$$

$$= t^n \sum_{r=0}^{\infty} \binom{n+r-1}{r} t^r$$

$$= \sum_{r=0}^{\infty} \binom{n+r-1}{r} t^{n+r}$$

$$= \sum_{r=n}^{\infty} \binom{r-1}{r-n} t^r$$

$$= \sum_{r=n}^{\infty} \binom{r-1}{n-1} t^r.$$

ANSWER.

$$\left(\frac{t}{1-t}\right)^n = \sum_{r=n}^{\infty} \binom{r-1}{n-1} t^r.$$

Note that the coefficients are zero for $r < n$, as they should be.

Example 4.5

Determine the enumerator for combinations of n objects with unlimited repetition if each object occurs on even number of times.

Solution. Zero is, of course, an even number. Each object corresponds to a factor

$$(1 + t^2 + t^4 + t^6 + \cdots).$$

$$(1 + t^2 + t^4 + \cdots)^n = (1 - t^2)^{-n} = \sum_{r=0}^{\infty} \binom{n+r-1}{r} t^{2r}.$$

ANSWER.

$$(1 - t^2)^{-n} = \sum_{r=0}^{\infty} \binom{n+r-1}{r} t^{2r}.$$

Note that $(1 - t^2)^{-n} = (1+t)^{-n}(1-t)^{-n}$. Hence we can establish another combinatorial identity, namely

$$\binom{n+r-1}{r} = \sum_{k=0}^{2r} \binom{n+2r-k-1}{2r-k} \binom{n+k-1}{k} (-1)^k.$$

From these examples it is clear that the use of enumerators allows us to compute rapidly the number of r-combinations of n objects under a wide variety of restrictions. While these values can be computed by other means, generally the computation is not as simple and clean-cut as the method that we have demonstrated. The enumerating function method is so efficient that it often pays to generalize from a specific problem to a general one in order to solve the specific problem.

Example 4.6

In a programming course students are allowed a maximum of ten runs per assignment. The professor notes that a total of 38 runs were made for one particular assignment. Assuming that there are 15 students enrolled, and that each student makes at least one run for this assignment, determine the number of combinations of runs which total to the observed number.

Solution. Generalize the problem to consider a class of n students, and an arbitrary observed total number of runs. Since each student completed at least one and at most ten runs, the generating function for combinations of runs in a class of n students is

$$(t + t^2 + \cdots + t^{10})^n.$$

Rewrite this as a series:

$$(t + t^2 + \cdots + t^{10})^n = [(t + t^2 + \cdots) - (t^{11} + t^{12} + \cdots)]^n$$

$$= [t(1 + t + \cdots) - t^{11}(1 + t + \cdots)]^n$$

$$= (t - t^{11})^n (1 + t + t^2 + \cdots)^n$$

$$= t^n (1 - t^{10})^n \frac{1}{(1 - t)^n}$$

$$= t^n \sum_{k=0}^{n} \binom{n}{k} (-1)^k t^{10k} \sum_{r=0}^{\infty} \binom{n + r - 1}{r} t^r$$

$$= \sum_{r=0}^{\infty} \sum_{k=0}^{n} \binom{n + r - 1}{r} \binom{n}{k} (-1)^k t^{n + r + 10k}.$$

The coefficient of $t^{n+r+10k}$ is the number of combinations producing a total of $n + r + 10k$ runs. The specified total number is 38. Hence

$$n + r + 10k = 38.$$

Since $n = 15$,

$$r + 10k = 23.$$

Only three solutions are possible:

$$r = 23, \quad k = 0; \qquad r = 13, \quad k = 1; \qquad r = 3, \quad k = 2.$$

Hence the number of combinations is the sum of the three coefficients corresponding to these values,

$$\binom{37}{23}\binom{15}{0} - \binom{27}{13}\binom{15}{1} + \binom{17}{3}\binom{15}{2}.$$

ANSWER.   58062 83700.

Note that in solving this problem we have obtained a formula that is applicable to a wide range of problems having various parameter values.

EXERCISES

Determine the enumerator for combinations of n objects with the specified conditions:

1. Each object occurs at most twice.

2. Each object occurs at most k times, k finite.

3. Each object occurs at least twice.

4. Each object occurs at least once and at most k times, $1 < k < \infty$.

5. Each object occurs either not at all, or at least k times.

6. The kth object occurs at most k times, $k = 1, \ldots, n$.

7. The kth object occurs at least k times, $k = 1, \ldots, n$.

8. Show that p a's and q b's may be placed in a row so that no two b's occur together in $\binom{p+1}{q}$ ways. From this, derive a combinatorial formula for the number of labeled, cyclically 3-connected narrow-banded cubic graphs on n vertices. (See Chapter 2, Section 10.)

9. Show that

$$\sum_{r=1}^{n} \frac{(-1)^{r+1}}{r+1} C(n,r) = \frac{n}{n+1}$$

and

$$\sum_{r=0}^{n} \frac{1}{r+1} C(n,r) = \frac{2^{n+1}-1}{n+1}.$$

Hint: Integrate the generating function $(1+t)^n$.

10. Show that

$$\binom{n+r-1}{r} = \sum_{k=0}^{2r} \binom{n+2r-k-1}{2r-k}\binom{n+k-1}{k}(-1)^k.$$

11. By substituting $-n$ for n, show that

$$\binom{-n}{r} = (-1)^r \binom{n+r-1}{r}.$$

12. The answer obtained for Example 4.6 is rather surprising in view of the fact that if the total were 15 rather than 38, then the number of combinations would be 1. Write a program to compute the coefficients

$$\sum_{k=0}^{n} \binom{n+r-1}{r}\binom{n}{k}(-1)^k$$

given n and $n+r+10k$ as input. Remember that these coefficients are integers and should be computed as such, not as floating point numbers. Test your program by computing these coefficients for $n = 15$ and $n+r+10k = 15, 16, \ldots, 38$.

13. (cont.) Verify by direct enumeration the correctness of your answer to Exercise 12 for $n+r+10k \leqslant 19$.

5. ENUMERATORS FOR PERMUTATIONS

The enumerator for combinations of n distinct objects is

$$(1+t)^n = \sum_{r=0}^{n} \binom{n}{r} t^r = \sum_{r=0}^{n} C(n,r) t^r.$$

Since we know that $P(n,r) = r!\, C(n,r)$, we may rewrite this equation, showing $P(n,r)$ as coefficients of $t^r/r!$:

$$(1+t)^n = \sum_{r=0}^{n} P(n,r) \frac{t^r}{r!}.$$

This indicates that whereas the polynomials $1+t+t^2+\cdots+t^k$ yield enumerators for combinations, the polynomials

$$1 + t + \frac{t^2}{2!} + \frac{t^3}{3!} + \cdots + \frac{t^k}{k!}$$

will yield enumerators for permutations.

Because of the exponential series

$$e^{at} = \sum_{r=0}^{\infty} a^r \frac{t^r}{r!}$$

these generating functions are called *exponential* generating functions.

Example 5.1

Determine the enumerator for r-permutations of n objects in classes containing $p, q, \cdots$ objects, $p+q+\cdots = n$.

Solution. Objects from the first class may be chosen at most p times. This choice is represented by the polynomial

$$1 + t + \frac{t^2}{2!} + \frac{t^3}{3!} + \cdots + \frac{t^p}{p!}.$$

Similarly, the choice from the second class is represented by

$$1 + t + \frac{t^2}{2!} + \cdots + \frac{t^q}{q!}.$$

Thus the generating function is

$$\left(1+t+\frac{t^2}{2!}+\cdots+\frac{t^p}{p!}\right)\left(1+t+\frac{t^2}{2!}+\cdots+\frac{t^q}{q!}\right)\cdots,$$

where $p+q+\cdots = n$. Now collect the coefficient of t^n.

ANSWER.
$$\sum_{n=0}^{\infty} \frac{n!}{p!\,q!\cdots}\, t^{n},$$

where $\qquad p + q + \cdots = n.$

Note that this is in agreement with our previous result. Observe also that using this approach, the answers to Exercises 9 and 10, Section 2 are relatively easy to obtain.

Example 5.2

Determine the enumerator for permutations of n objects with unlimited repetition.

Solution. With unlimited repetition, the choice function for one object is

$$1 + t + \frac{t^2}{2!} + \frac{t^3}{3!} + \cdots.$$

Hence the enumerator is

$$\left(1 + t + \frac{t^2}{2!} + \frac{t^3}{3!} + \cdots\right)^{n}.$$

Note, however, that

$$1 + t + \frac{t^2}{2!} + \frac{t^3}{3!} + \cdots = e^{t}.$$

Hence the enumerator is

$$e^{nt} = \sum_{r=0}^{\infty} \frac{(nt)^{r}}{r!} = \sum_{r=0}^{\infty} n^{r} \frac{t^{r}}{r!}.$$

ANSWER.
$$e^{nt} = \sum_{r=0}^{\infty} n^{r} \frac{t^{r}}{r!}.$$

Once again, the coefficients agree with our previous results.

Example 5.3

Determine the enumerator for permutations of n objects with unlimited repetition, but with the condition that each object must appear at least once.

Solution. The choice function is

$$t + \frac{t^2}{2!} + \frac{t^3}{3!} + \cdots,$$

so the enumerator is

$$\left(t + \frac{t^2}{2!} + \frac{t^3}{3!} + \cdots\right)^n.$$

When the analogous expression occurred in our work with combinations, we factored out t. (See Example 4.4.) Here this will not help. (Why?) Rather, consider the series as $e^t - 1$, and expand using the binomial theorem:

$$\left(t + \frac{t^2}{2!} + \frac{t^3}{3!} + \cdots\right)^n = (e^t - 1)^n$$

$$= \sum_{j=0}^{n} \binom{n}{j} (-1)^j e^{(n-j)t}$$

$$= \sum_{j=0}^{n} \binom{n}{j} (-1)^j \sum_{r=0}^{\infty} (n-j)^r \frac{t^r}{r!}$$

$$= \sum_{r=0}^{\infty} \frac{t^r}{r!} \sum_{j=0}^{n} \binom{n}{j} (-1)^j (n-j)^r.$$

ANSWER.

$$(e^t - 1)^n = \sum_{r=0}^{\infty} d_r \frac{t^r}{r!}, \qquad \text{where}$$

$$d_r = \sum_{j=0}^{n} \binom{n}{j} (-1)^j (n-j)^r.$$

Note the ease of obtaining quite complicated coefficients.

EXERCISES

1–7. Determine the enumerator for permutations of n objects, using the conditions specified in Exercises 1–7 of Section 4, p. 297.

8. Write a program to compute the coefficients d_r of Example 5.3, for $n = 1, \ldots, 10$ and $r = 1, \ldots, 10$. Remember that these coefficients are integers and should be computed as such, not as floating point numbers.

9. (cont.) Verify for $n = 3$ and $r = 0, 1, \ldots, 5$ that the coefficients d_r correctly count the permutations.

6. STIRLING NUMBERS

Various well-defined sets of numbers occur with sufficient frequency in combinatorial work that they are singled out, identified by name, and studied with some thoroughness. The binomial coefficients $\binom{n}{r}$ are the best known of these numbers. The Stirling numbers form two more well-known sets, which arise in the finite difference calculus and in the problem of dividing a set into classes. They are the numbers relating "factorial" expansion to powers. The *factorial expansions* of t are defined as follows:

$$(t)_0 = 1$$
$$(t)_1 = t$$
$$(t)_2 = t(t-1)$$
$$\vdots$$
$$(t)_n = t(t-1)(t-2)\cdots(t-n+1).$$

(*Note:* These are falling factorials. One can analogously define rising factorials $(t)^n$.)

Thus each $(t)_n$ is a polynomial in t:

$$(t)_n = \sum_{k=0}^{n} s(n,k)\, t^k.$$

The coefficients in this polynomial are the *Stirling numbers of the first kind*. (Thus the polynomial $(t)_n$ is a generating function for the numbers $s(n,k)$.) Let us compute a few of the polynomials:

$$(t)_0 = 1$$
$$(t)_1 = t$$
$$(t)_2 = t^2 - t$$
$$(t)_3 = t^3 - 3t^2 + 2t$$
$$(t)_4 = t^4 - 6t^3 + 11t^2 - 6t$$
$$(t)_5 = t^5 - 10t^4 + 35t^3 - 50t^2 + 24t.$$

From these few polynomials it appears that

$$
\begin{aligned}
s(n,0) &= 0, & n &\geqslant 1 \\
s(n,1) &= (n-1)!, & n &\geqslant 1 \\
s(n,n-1) &= -\binom{n}{2}, & n &\geqslant 2 \\
s(n,n) &= 1, & n &\geqslant 1
\end{aligned}
\tag{1}
$$

and

$$s(n,k) = 0, \qquad k > n.$$

(See Exercise 2.)

We may also express the powers of t in terms of the falling factorials:

$$t^0 = 1$$

$$t^1 = t = (t)_1$$

$$t^2 = (t)_2 + t = (t)_2 + (t)_1$$

$$t^3 = (t)_3 + 3t^2 - 2t = (t)_3 + 3(t)_2 + (t)_1$$

$$t^4 = (t)_4 + 6(t)_3 + 7(t)_2 + (t)_1$$

$$t^5 = (t)_5 + 10(t)_4 + 25(t)_3 + 15(t)_2 + (t)_1.$$

In general,

$$t^n = \sum_{k=0}^{n} S(n,k)(t)_k.$$

These coefficients are known as the *Stirling numbers of the second kind*. The relationships suggested by these polynomials are

$$S(n,0) = 0, \qquad n \geqslant 1 \qquad S(n,n-1) = \binom{n}{2}, \quad n \geqslant 2$$

$$S(n,1) = 1, \qquad n \geqslant 1 \qquad S(n,n) = 1, \quad n \geqslant 1 \qquad (2)$$

$$S(n,2) = 2^{n-1} - 1, \quad n \geqslant 2 \qquad S(n,k) = 0, \quad k > n.$$

(See Exercise 3.)

Tables of Stirling numbers are computed from the recurrence relations

$$s(n+1,k) = s(n,k-1) - ns(n,k), \qquad S(n+1,k) = S(n,k-1) + kS(n,k).$$

See Tables 6.1 and 6.2.

TABLE 6.1 $s(n,k)$

n \ k	1	2	3	4	5	6	7
1	1						
2	-1	1					
3	2	-3	1				
4	-6	11	-6	1			
5	24	-50	35	-10	1		
6	-120	274	-225	85	-15	1	
7	720	-1764	1624	-735	175	-21	1

TABLE 6.2 $S(n, k)$

n \ k	1	2	3	4	5	6	7
1	1						
2	1	1					
3	1	3	1				
4	1	7	6	1			
5	1	15	25	10	1		
6	1	31	90	65	15	1	
7	1	63	301	350	140	21	1

EXERCISES

1. Verify the recurrence relations for the Stirling numbers.

2. Establish relationships (1) for Stirling numbers of the first kind.

3. Establish relationships (2) for Stirling numbers of the second kind.

4. Write programs to compute $s(n, k)$ and $S(n, k)$.

5. Show that

$$S(n + 1, m) = \sum_{i=0}^{n} \binom{n}{k} S(k, m - 1).$$

6. Another set of numbers of use in combinatorics is the set of *Bell numbers*, defined by

$$B_n = \sum_{k=1}^{n} S(n, k).$$

Show that

$$B_{n+1} = \sum_{k=0}^{n} \binom{n}{k} B_k.$$

Hint: Recall that $S(n, k) = 0$ for $k > n$.

7. CYCLE CLASSES OF PERMUTATIONS

Recall the cycle notation for permutations. Each permutation of n objects consists of a number of cycles of various lengths. The lengths of these cycles are useful in classification of the permutations. The *cycle class* $(k) = (k_1, k_2, \ldots, k_n)$ consists of all permutations on n objects that have exactly

k_1 1-cycles, k_2 2-cycles, ..., k_n n-cycles. Another notation for this class is

$$1^{k_1} 2^{k_2} 3^{k_3} \cdots n^{k_n},$$

where the $k_i = 1$ are omitted, and j^{k_j} is omitted for $k_j = 0$.

Example 7.1

Determine the cycle class of each permutation on three objects.

	Permutations	Class (two notations)	
ANSWER.	(123), (132)	(001)	3
	(12)(3), (13)(2), (23)(1)	(110)	12
	(1)(2)(3)	(300)	1^3

Note that the k_i are related by

$$\sum_{i=1}^{n} i k_i = n.$$

The basic problem is to compute $C(k_1, k_2, ..., k_n)$, the number of permutations of class $(k_1, k_2, ..., k_n)$ on n objects. To do this, pick one such permutation, and consider all possible permutations of its elements. There are $n!$ of them, not all distinct.

One reason for lack of distinctness is that the cyclic order within any one cycle is immaterial. For example, (1432), (2143), (3214), and (4321) are all the same cycle. Each r-cycle has r equivalents in this way, so that the number of permutations is reduced by the factor

$$1^{k_1} 2^{k_2} \cdots n^{k_n}.$$

The other reason for lack of distinctness is that the order of the cycles is immaterial since they are disjoint. For example, (12)(34) and (34)(12) are the same permutation. Since the k_r r-cycles have $k_r!$ possible orders, the factor due to this effect is

$$k_1! k_2! \cdots k_n!.$$

Thus the number of distinct permutations of class $(k_1, k_2, ..., k_n)$ is

$$C(k_1, k_2, ..., k_n) = \frac{n!}{1^{k_1} k_1! 2^{k_2} k_2! \cdots n^{k_n} k_n!}.$$

The generating function for cycle classes, called the *cycle indicator*, involves n variables, one for each possible cycle length. It is

$$C_n(t_1, t_2, \ldots, t_n) = \sum C(k_1, k_2, \ldots, k_n) t_1^{k_1} t_2^{k_2} \cdots t_n^{k_n}$$

$$= \sum \frac{n!}{k_1! \cdots k_n!} \left(\frac{t_1}{1}\right)^{k_1} \left(\frac{t_2}{2}\right)^{k_2} \cdots \left(\frac{t_n}{n}\right)^{k_n},$$

where the sum is over all $k_1, \ldots, k_n$ such that $k_1 + 2k_2 + \cdots + nk_n = n$.

The first few cycle indicators are

$$C_1 = t_1$$
$$C_2 = t_1^2 + t_2$$
$$C_3 = t_1^3 + 3t_1 t_2 + 2t_3$$
$$C_4 = t_1^4 + 6t_1^2 t_2 + 3t_2^2 + 8t_1 t_3 + 6t_4.$$

For example, from C_3 we find that there are one permutation of class 1^3, three permutations of class 1^2, and two permutations of class 3. Note that $C_n(1, 1, \ldots, 1) = n!$, so that

$$\sum \frac{1}{1^{k_1} k_1! \, 2^{k_2} k_2! \cdots n^{k_n} k_n!} = 1,$$

a result due to Cauchy.

EXERCISES

1. Classify the permutations on four objects.

2. Classify the permutations on five objects.

It may be argued that our derivation of $C(k_1, \ldots, k_n)$ does not take into account the inter-mixing of cycle lengths. For example, it does not account for both (12)(34)(5) and (12)(5)(34). This series of four exercises shows that intermixing is of no consequence:

3. Show that any particular ordering of n elements produces

$$\frac{(\sum_{i=1}^n k_i)!}{\prod_{i=1}^n k_i!}$$

permutations of class $1^{k_1} 2^{k_2} \cdots n^{k_n}$.

(Example: 14235 produces three permutations of class $1^1 2^2$: (14)(23)(5), (14)(2)(35), (1)(42)(35)).

4. (cont.) Hence, show that the total number of permutations of class $1^{k_1} 2^{k_2} \cdots n^{k_n}$ produced by all orderings of n elements is

$$\frac{(\sum_{i=1}^n k_i)!}{\prod_{i=1}^n k_i!} n!$$

5. (cont.) Show that given any particular permutation of class $1^{k_1} 2^{k_2} \cdots n^{k_n}$, the number of duplicate permutations due solely to intermixing cycles of various lengths is

$$\frac{(\sum_{i=1}^{n} k_i)!}{\prod_{i=1}^{n} k_i!}.$$

6. (cont.) Hence, show that

$$C(k_1, k_2, \ldots, k_n) = \frac{n!}{\prod_{i=1}^{n} i^{k_i} k_i!}.$$

7. Write a program to generate $C_n(t_1, \ldots, t_n)$, given n as input. Test your program on $n = 1, \ldots, 5$.

8. PARTITIONS AND COMPOSITIONS

In Section 7 we considered the problem of organizing a set of n elements into classes of objects (cycles), with the condition that the k_i objects of the ith class have length i. This led, in the cycle indicator, to the rather interesting condition that the sum be "over all $k_1, \ldots, k_n$ such that $k_1 + 2k_2 + \cdots + nk_n = n$." In this section we consider the general problem of organizing a set of r elements into classes, which is important in many applications.

Let $p_1, p_2, \ldots, p_m$ be positive integers summing to n. For example, if $n = 4$, $p_1, \ldots, p_m$ could be any of the following:

$$m = 1: \quad 4; \qquad\qquad m = 2: \quad 1 + 3, \, 2 + 2;$$

$$m = 3: \quad 1 + 1 + 2; \qquad m = 4: \quad 1 + 1 + 1 + 1.$$

Such a division of n into a sum is called a *partition* of n if the order in which the p_i are summed is unimportant, or a *composition* of n if the order of summation is important. Thus $n = 4$ has five partitions, and eight compositions since $1 + 3$ and $3 + 1$ are distinct compositions, as are $1 + 1 + 2$, $1 + 2 + 1$, and $2 + 1 + 1$. We refer to m as the number of *parts* of the partition or composition. Since the order is unimportant for partitions, we assume a standard order $p_1 \geqslant p_2 \geqslant \cdots \geqslant p_m$ for work with partitions. Table 8.1 gives

TABLE 8.1

n	Partitions	Compositions
1	1	1
2	$2, 1^2$	$2, 1^2$
3	$3, 21, 1^3$	$3, 21, 12, 1^3$
4	$4, 31, 2^2, 21^2, 1^4$	$4, 31, 13, 2^2, 21^2, 121, 1^2 2, 1^4$
5	$5, 41, 32, 31^2, 2^2 1, 21^3, 1^5$	$5, 41, 14, 32, 23, 31^2, 131, 1^2 3, 2^2 1, 212, 12^2, 21^3, 121^2, 1^2 21, 1^3 2, 1^5$

the partitions and compositions for $n = 1, ..., 5$, where the notation 121^2, for example, represents the composition $1 + 2 + 1 + 1$.

(Unfortunately, a P–C mnemonic is reversed here. Partitions are unordered, as are combinations; while compositions are ordered, as are permutations.)

Partitions and compositions are often listed by the number of parts, the smallest part, and so on. They may be displayed by means of the *Ferrers graph* and the *zigzag graph*, for partitions and compositions respectively. For example, the Ferrers graph of the partition 532 is

$$\cdot \quad \cdot \quad \cdot \quad \cdot \quad \cdot \tag{5}$$

$$\cdot \quad \cdot \quad \cdot \tag{3}$$

$$\cdot \quad \cdot \tag{2}$$

The *conjugate partition* may be read from the columns of this graph; for this example it is $3^2 21^2$.

The zigzag graph of the composition 532 is

$$\cdot \quad \cdot \quad \cdot \quad \cdot \quad \cdot$$
$$\cdot \quad \cdot \quad \cdot$$
$$\cdot \quad \cdot$$

For the composition 352, the graph is

$$\cdot \quad \cdot \quad \cdot$$
$$\cdot \quad \cdot \quad \cdot \quad \cdot \quad \cdot$$
$$\cdot \quad \cdot$$

Again, conjugate compositions are read from the columns: $1^4 2121$ for 532, and $1^2 21^3 21$ for 352.

Because of the order involved in compositions, it is possible to associate each composition of n with a subset of $\{1, 2, ..., n\}$. In particular, given a composition $c_1, ..., c_m$, take $x_0 = 0$, and let $x_i = x_{i-1} + c_i, i = 1, 2, ..., m$. Note that $x_m = x_{m-1} + c_m = \sum_{i=1}^{m} c_i = n$. Then associate with the composition $c_1, ..., c_m$ the set $\{x_1, ..., x_m\}$. Observe that since $x_m = n$ for any composition, this association, in fact, establishes a one-to-one correspondence between the compositions of n and the subsets of $\{1, 2, ..., n-1\}$. We leave it as an exercise to establish the following count of compositions:

There are $\binom{n-1}{m-1}$ m-part compositions of n, and 2^{n-1} compositions of n in all.

Obviously, use of wordsets provides a very compact computer representation of compositions. Observe that since $x_m = n$, the position of the left-most 1 in the wordset corresponds to the number whose composition is being represented. (Remember that wordsets read from right to left.) However, by storing n separately we could shorten the word length necessary to represent a composition, and simplify the right-to-left scan of the word since we would know a priori when we had scanned the last 1.

Example 8.1

Represent as wordsets the compositions $512^2 3$ and 12532 of 13.

Solution.   Let $x_0 = 0$. For the first composition, $x_1 = 5$, $x_2 = 6$, $x_3 = 8$, and $x_4 = 10$. For the second, $x_1 = 1$, $x_2 = 3$, $x_3 = 8$, and $x_4 = 11$. For each of these compositions, $x_5 = 13$.

ANSWER.

$$512^2 3: \quad 1001010110000;$$
$$12532: \quad 1010010000101.$$

For partitions, a similar correspondence can be established. However, because of the standard order which we have imposed, not every wordset corresponds to a partition. Another useful bit representation of partitions is the *rim representation*, suggested by Comét [3]. This is obtained by traversing the rim of the Ferrers graph for the partition from bottom to top, using 1 to denote a horizontal increment, and 0 to denote a vertical increment. Because of the manner of traverse, the number of zeros in the rim representation equals the number of parts, and the number of ones to the right of the ith zero from the left (in wordset notation) is the value of p_i. The number whose partition is represented is generally given separately. (See Exercise 11.)

Example 8.2

Give the rim representation of the partition $643^2 21^2$ of 20.

Solution.   The Ferrers graph is given in Fig. 8.1, with the traverse and the bits indicated.

ANSWER. $0110100101001 \, (n = 20)$.

We have noted that the number of compositions of n is 2^{n-1}. Counting partitions is more difficult, and best accomplished by a generating function. The parts of size k contribute a factor

$$1 + t^k + t^{2k} + \cdots,$$

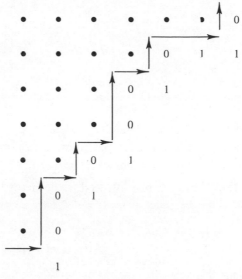

Figure 8.1

since there may be $0, 1, 2, \ldots$ parts of this size. Thus the generating function for unrestricted partitions of n is

$$p(t) = (1 + t + t^2 + \cdots)(1 + t^2 + t^4 + \cdots) \cdots (1 + t^k + t^{2k} + \cdots) \cdots$$

$$= \frac{1}{(1-t)(1-t^2)(1-t^3) \cdots (1-t^k) \cdots}.$$

Hence

$$p(t) = p_0 + p_1 t + p_2 t^2 + \cdots + p_n t^n + \cdots,$$

where p_n is the number of unrestricted partitions of n. The first few terms of this series are

$$p(t) = 1 + t + 2t^2 + 3t^3 + 5t^4 + 7t^5 + 11t^6 + 15t^7 + \cdots.$$

Various restrictions on the partitions yield other enumerators.

Example 8.3

Determine the enumerator for partitions with no parts larger than k.

Solution. If the largest part has size k, then the factors $(1 + t^m + t^{2m} + \cdots)$ with $m > k$ are not used.

ANSWER.

$$p^k(t) = \sum_{n=0}^{\infty} p_{nk} t^n$$

$$= \frac{1}{(1-t)(1-t^2)\cdots(1-t^k)}.$$

Example 8.4

Determine the enumerator for partitions with no repeated parts (that is, with unequal parts).

Solution. Since repetition is not allowed, each factor $(1 + t^m + t^{2m} + \cdots)$ reduces to $1 + t^m$.

ANSWER.

$$u(t) = (1+t)(1+t^2)(1+t^3)\cdots.$$

Example 8.5

Determine the enumerator for partitions with only odd parts.

Solution. For this, only the factors $(1 + t^m + t^{2m} + \cdots)$ with m odd are used.

ANSWER.

$$\sigma(t) = \frac{1}{(1-t)(1-t^3)(1-t^5)\cdots}.$$

EXERCISES

1. Show that $u(t) = \sigma(t)$. (Example for $n = 6$. Partitions with unequal parts: $6, 51, 42, 321$; partitions with only odd parts: $51, 3^2, 31^2, 1^6$—four of each.)

2. Show that the number of partitions of n whose largest part is m equals the number of m-part partitions of n. (Example for $n = 7$, $m = 3$. Partitions with largest part 3: $3^2 1, 32^2, 321^2, 31^4$; partitions into 3 parts: $51^2, 421, 3^2 1, 32^2$—four of each.) *Hint:* Use the Ferrers graph.

3. To each m-part partition of n there correspond one or more partitions of a set of n elements into nonempty subsets. For example, to the partition 21 of $n = 3$ there correspond the set partitions $\{1, 2\} \cup \{3\}$, $\{1, 3\} \cup \{2\}$, and $\{2, 3\} \cup \{1\}$ of the set $\{1, 2, 3\}$. For $n = 1, 2, \ldots, 5$ verify that the number of partitions of a set of n elements into m nonempty subsets is $S(n, m)$, a Stirling number of the second kind.

4. (cont.) Let $P(n, m)$ be the number of partitions of a set of n elements into m nonempty subsets. Show that $P(n, m) = S(n, m)$ by proving

(a) $P(n, 1) = 1$;

(b) $P(n, n) = 1$;

(c) $P(n+1, k) = P(n, k-1) + kP(n, k)$.

5. (cont.) Let $p_1^{i_1} p_2^{i_2} \cdots p_m^{i_m}$ be a partition of n having i_j parts of size $p_j, j = 1, 2, \ldots, m$. Show that the number of partitions of a set of n elements corresponding to this partition of n is

$$\frac{n!}{(p_i!)^{i_1} (p_2!)^{i_2} \cdots (p_m!)^{i_m} i_1! i_2! \cdots i_m!}.$$

6. Determine the enumerator for compositions with no parts larger than k.

*7. Determine the enumerator for compositions with unequal parts.

8. Determine the enumerator for compositions with only odd parts.

9. Write programs to compute the compositions and partitions of n, given n as input. Test your program with $n = 1, 2, \ldots, 5$.

10. Write a program to convert compositions of n from vector notation $(c_1, c_2, \ldots, c_m)$ to wordset notation, and back. Test your programs with the compositions of $n = 1, 2, \ldots, 5$, and those of Example 8.1.

11. Write programs to convert partitions of n from vector notation $(p_1, p_2, \ldots, p_m)$ to rim representation and back. Test your programs with the partitions of $n = 1, 2, \ldots, 5$, and that of Example 8.2.

*12. Write a program to compute the number of partitions of n, given n as input. *Hint:* Use the enumerator $p(t)$. Use your program to compute the number of partitions for $n = 1, 2, \ldots, 5$, and for $n = 10, n = 15$.

REFERENCES

1. C. Berge, "Principles of Combinatorics." Academic Press, New York, 1971.
2. J. Riordan, "An Introduction to Combinatorial Analysis." Wiley, New York, 1958.
3. S. Comét, Notations for partitions. *Math. Tables and Aids to Computation*, **9** (1955) 143–146.

CHAPTER 11

Systems of
Distinct Representatives

1. INTRODUCTION AND HISTORY

Throughout this book one theme has been the interplay between discrete mathematics and computing. In this chapter we examine one example of this interplay; this will bring together many of the concepts that we have discussed.

The idea of using individuals to represent a set of objects or people is a common one. It is the basis of elections, committee selections, quality control procedures and other sampling techniques, and many mathematical theorems. Such individuals may or may not be selected at random. In this chapter we are concerned with a particular set of questions about this selection process. We shall discuss some of the history of these questions, and a few of their applications, and describe the interplay of the computer and combinatorics in studying these questions. The questions are these:

1. Suppose that we have n nonempty sets of objects. Under what circumstances can we choose n distinct objects, one from each set? Note that the sets may have elements in common. For example, we may have 25 classes of students, and wish to set up a committee of 25 members, one representing each class.

2.   Now suppose that the same collection of objects referred to in question 1 is classified into n nonempty sets in another (distinct from the first) way. Under what circumstances can we choose n distinct objects so that simultaneously one is from each set of the first classification, and one is from each set of the second? For example, suppose our students reside in 25 housing units. When can we select a committee of 25 members having one member from each class, and one from each housing unit?

3.   The question is the same, except that there are three (four, five, ...) simultaneous classifications.

Historically these questions were asked as far back as 1911, with the sets being the cosets of a group. If G is a group, H is a subgroup, and a is a particular element of G, then the sets $aH = \{ah\,|\,h\in H\}$ and $Ha = \{ha\,|\,h\in H\}$ are called the *right* and *left cosets* of G, respectively. (Some authors call aH a left coset and Ha a right coset. It does not matter which definition is used, as long as the distinction between aH and Ha is maintained.) It is easy to show that the right cosets of a group G partition G, as do the left cosets, and that the right and left cosets are generally different, unless G is Abelian, but equinumerous. (See Exercises 9, 10, Section 1, Chapter 6.) In group theory one often picks elements $a_1, a_2, \ldots$ to represent the right cosets of a group, each element being from a different right coset. Because of the partitioning, these elements are distinct. Similarly, one can pick distinct elements $b_1, b_2, \ldots$ to represent the left cosets. The question asked was, can one pick distinct elements $c_1, c_2, \ldots$ that simultaneously represent both right and left cosets? (The answer is affirmative.)

This same type of question appeared in various contexts over the next twenty years. Our first question was answered for general sets in 1935 by an English mathematician, Philip Hall [1]. Hall proved such a set of individuals, called a *system of distinct representatives* (*SDR*), can be chosen if and only if each subcollection of k of the sets contains at least k distinct elements, for $k = 1, 2, \ldots, n$.

Example 1.1

Select an SDR for the following collection:

$$S_1 = \{1,2\}, \qquad S_2 = \{1,3,4\},$$

$$S_3 = \{2,3\}, \qquad S_4 = \{3,4\}.$$

Solution.   By Hall's criterion, an SDR can be selected if and only if (a) each set contains at least one element, (b) any two sets have (in their union) at least two elements, (c) any three sets have at least three elements, and (d) the four sets together contain at least four elements. Since this set of conditions is satisfied, an SDR can be found. Thorough investigation reveals three SDRs, shown in Table 1.1.

TABLE 1.1

Set represented	Element		
S_1	1	1	2
S_2	3	4	1
S_3	2	2	3
S_4	4	3	4

ANSWER.   From S_1, 1; from S_2, 3; from S_3, 2; from S_4, 4.

Example 1.2

Select an SDR for the following collection:

$$S_1 = \{1, 2, 3\}, \qquad S_2 = \{2, 4, 5\}, \qquad S_3 = \{1, 3\},$$

$$S_4 = \{2, 3\}, \qquad S_5 = \{3\}.$$

Solution.   Hall's criterion holds for $k = 1$, 2, and 3, but fails for $k = 4$ since $S_1 \cup S_3 \cup S_4 \cup S_5 = \{1, 2, 3\}$.

ANSWER. There is no SDR.

Note in Example 1.2 that 3 must represent S_5, and hence 1 represents S_3, while 2 represents S_4. Thus there is no element left to represent S_1.

The second question, answered in a special case by Hall in 1935, was answered in general by L. R. Ford, Jr. and D. R. Fulkerson in 1958 [2]. Before discussing the answer, we shall need some notation.

Let $A = \{a_1, a_2, \ldots, a_s\}$ be a finite set, and let $\mathscr{S}_1, \ldots, \mathscr{S}_m$ be sequences of n (not necessarily distinct) subsets of A, $\mathscr{S}_i = \langle S_{i1}, \ldots, S_{in} \rangle$ $(i = 1, \ldots, m)$. An element $a_k \in A$ *represents* the sets $S_{1j_1}, S_{2j_2}, \ldots, S_{mj_m}$ if $a_k \in \bigcap_{i=1}^{m} S_{ij_i}$. The system of sequences $\mathscr{S}_1, \mathscr{S}_2, \ldots, \mathscr{S}_m$ has a (*common*) *system of distinct representatives* (SDR) if there exist distinct elements $a_{k_1}, \ldots, a_{k_n}$ and a reordering of the sets within each sequence such that a_{k_j} represents $S_{1j}, \ldots, S_{mj}$ $(j = 1, \ldots, n)$. Recall that for any set S, we let $|S|$ denote the cardinality or number of elements in S. Further, we let $X_1, X_2, \ldots$ denote nonempty subsets of $\{1, 2, \ldots, n\}$, and $\mathscr{X}_i$ denote the subsequence S_{ij_1}, $S_{ij_2}, \ldots,$ of $\mathscr{S}_i$ when $X_i = \{j_1, j_2, \ldots\}$. Finally, we let

$$I_i = I(\mathscr{X}_i) = \left\{ i \,\middle|\, a_i \in \bigcup_{j \in X_i} S_{ij} \right\}.$$

Example 1.3

Given the sequences $\mathcal{S}_1$ and $\mathcal{S}_2$ described below and $X_1 = \{1,3\}$, $X_2 = \{1,2,4\}$, determine I_1 and I_2.

$$\mathcal{S}_1: \quad \begin{aligned} S_{11} &= \{a_1, a_2, a_3\} \\ S_{12} &= \{a_2, a_4, a_5\} \\ S_{13} &= \{a_1, a_3\} \\ S_{14} &= \{a_2, a_3\} \\ S_{15} &= \{a_3, a_4\} \end{aligned} \qquad \mathcal{S}_2: \quad \begin{aligned} S_{21} &= \{a_2, a_3, a_5\} \\ S_{22} &= \{a_1, a_4\} \\ S_{23} &= \{a_2, a_3, a_4\} \\ S_{24} &= \{a_4, a_5\} \\ S_{25} &= \{a_1, a_3\} \end{aligned}$$

Solution. Since $X_1 = \{1,3\}$, $\mathcal{X}_1 = S_{11}, S_{13}$. Hence $I_1 = \{i \,|\, a_i \in S_{11} \cup S_{13}\} = \{1,2,3\}$. Similarly $\mathcal{X}_2 = S_{21}, S_{22}, S_{24}$, and hence $I_2 = \{i \,|\, a_i \in S_{21} \cup S_{22} \cup S_{24}\} = \{1,2,3,4,5\}$.

ANSWER. $I_1 = \{1,2,3\}$;
$\qquad\qquad I_2 = \{1,2,3,4,5\}$.

Thus X_i is a set of indices determining a subsequence of $\mathcal{S}_i$, and I_i is the set of indices of elements in that subsequence.

In this notation, Hall's result is that for $m = 1$ (that is, one sequence of sets), there is an SDR if and only if for any $\mathcal{X}_1 \subseteq \mathcal{S}_1$, $|X_1| \leq |I_1|$.

The Ford and Fulkerson result is but little more complex: for $m = 2$, there is a common SDR if and only if for any $\mathcal{X}_1 \subseteq \mathcal{S}_1$ and any $\mathcal{X}_2 \subseteq S_2$,

$$|X_1| + |X_2| \leq n + |I_1 \cap I_2|.$$

Example 1.4

Find a common SDR for $\mathcal{S}_1$ and $\mathcal{S}_2$ of Example 1.3.

Solution. Since $n = 5$, obviously $1 \leq |X_i| \leq 5$, for $i = 1,2$. Rewrite the Ford and Fulkerson condition:

$$|I_1 \cap I_2| \geq |X_1| + |X_2| - n.$$

For various values of the cardinalities, this inequality reduces to the inequalities shown in Table 1.2. Observe that the first four inequalities are trivially satisfied, while the last is true since it involves all sets, so that $I_1 = I_2 = \{1,2,3,4,5\}$. The remaining four must be tested. For example, if $|X_1| + |X_2| = 6$, there are the 210 possibilities in Table 1.3. Similarly, for $|X_1| + |X_2| = 7$, 8, or 9 there are 120, 45, or 10 combinations, respectively. Actually, it is not necessary to test all of these. In fact, a small problem like this can be solved by inspection. Note that just five elements are involved, so that if there is a common SDR, it must consist of the five elements. Try each sequence separately: for each there is an SDR consisting of $\{a_1, \ldots, a_5\}$. Hence there is a common SDR.

TABLE 1.2

$|X_1| + |X_2|$	Inequality
2	$|I_1 \cap I_2| \geqslant -3$
3	$|I_1 \cap I_2| \geqslant -2$
4	$|I_1 \cap I_2| \geqslant -1$
5	$|I_1 \cap I_2| \geqslant 0$
6	$|I_1 \cap I_2| \geqslant 1$
7	$|I_1 \cap I_2| \geqslant 2$
8	$|I_1 \cap I_2| \geqslant 3$
9	$|I_1 \cap I_2| \geqslant 4$
10	$|I_1 \cap I_2| \geqslant 5$

TABLE 1.3

$|X_1|$	$|X_2|$	Number of combinations of sets
1	5	$5 \cdot 1 = 5$
2	4	$\binom{5}{2} \cdot \binom{5}{4} = 50$
3	3	$\binom{5}{3} \cdot \binom{5}{3} = 100$
4	2	$\binom{5}{4} \cdot \binom{5}{2} = 50$
5	1	$1 \cdot 5 = 5$
		210 total

ANSWER. a_1 representing S_{11}, S_{22}
a_2 representing S_{14}, S_{21}
a_3 representing S_{13}, S_{25}
a_4 representing S_{15}, S_{23}
a_5 representing S_{12}, S_{24}.

Despite further efforts, the third question remains unanswered. We shall discuss the difficulties which it presents in Sections 2 and 3.

It is interesting to observe that, while Hall approached the problem via set theory, Ford and Fulkerson used network theory to answer the second question. Consider the network schematic in Fig. 1.1, for the one sequence

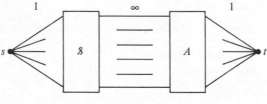

Figure 1.1

case, $m = 1$. The blocks labeled $\mathscr{S}$ and A denote vertices representing the sets and the elements, respectively. From the source s an arc of capacity 1 leads to each vertex in $\mathscr{S}$. From each set vertex S an arc of infinite capacity leads to each element vertex a such that $a \in S$. And from each element vertex an arc of capacity 1 leads to the sink t. An SDR exists if and only if the maximal flow in this network equals n, the number of set vertices. This is demonstrated in Figs. 1.2 and 1.3, corresponding to Examples 1.1 and 1.2, respectively. The network in each figure has a maximal flow of 4.

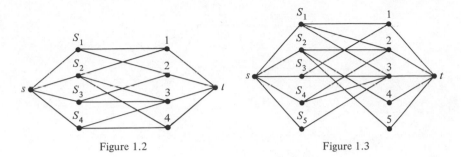

Figure 1.2   Figure 1.3

We can often show by inspection of a network that no SDR is possible. For example, in Fig. 1.3 a flow of value 5 is necessary if there is an SDR corresponding to the network. Since there must be flows of value 1 out of each of the vertices 4 and 5, there must be flows of value 1 into these vertices. That is, the flow out of vertex S_2 must be at least 2. But this is impossible since the flow into S_2 is at most 1. Hence there is no SDR corresponding to this network.

Ford and Fulkerson arrived at their result for $m = 2$ by considering the type of network represented in Fig. 1.4. The vertex sets A_1 and A_2

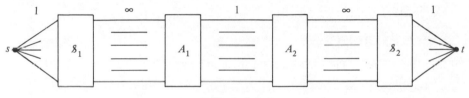

Figure 1.4

are copies of vertices representing the set A. These sets are connected by arcs of capacity 1, joining vertices representing the same element. This is necessary to prevent double use of an element as a representative. Other arcs and capacities are as for the case $m = 1$. The network corresponding to Examples 1.3 and 1.4 is given in Fig. 1.5. It has capacity 5. Ford and Fulkerson established their inequality by examining possible flows through such a network, relating the maximal flow to the matching between sets and elements.

Work on systems of representatives is important in problems of matching. For $m = 1$, a classic example is pairing off for a dance. Suppose each set S_{1i} is the set of girls that the ith boy knows. It is possible for all boys to select dance partners simultaneously? For $m = 2$, let the sequences $\mathscr{S}_1$ and $\mathscr{S}_2$ denote classifications of the elements of A in different ways. For example, A may be a set of thread samples, classified by thread size ($\mathscr{S}_1$) and

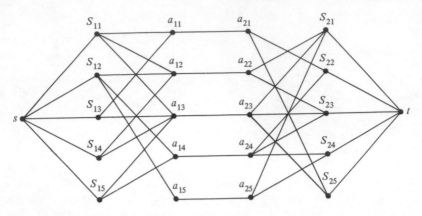

Figure 1.5

color ($\mathscr{S}_2$). If there are n sizes and n colors, can a salesman select a set of n thread samples that will contain all sizes and all colors?

This brief sketch of the SDR problem is highly incomplete. Not only have other approaches to this problem been used, but also variations and generalizations of the problem have been investigated. The interested reader will find more information in Ford and Fulkerson's book [3], and in many books on operations research.

EXERCISES

1. In the network of Fig. 1.3, find a minimal-valued cut.

2. Let S_1, S_2, ..., S_n be sets having s_1, s_2, ..., s_n elements respectively, and suppose that Hall's condition (the union of any k sets contains at least k elements, $k = 1, ..., n$) is satisfied. What is the minimum number of SDRs this sequence could have? What is the maximum number? Give an example in each case.

3. Write a program to take as input the integer n ($\leqslant 10$) and n sets of integers with at most ten elements each, and print out all SDRs, up to a maximum of 20. For each SDR, print each element used, together with the set it represents.

4. Write a program to take as input the integer n ($\leqslant 10$) and two sequences of n sets of integers each, with at most ten elements in each set, and print out all SDRs, up to a maximum of 20. For each SDR, print each element used, together with the sets it represents.

5. (cont.) Generalize your programs of Exercises 3 and 4 to work for $m(\geqslant 3)$ input sequences, with m as an input parameter.

2. THE THIRD QUESTION, GENERAL CASE

From the discussion of the cases $m = 1$ and $m = 2$ several observations can be made. First, the solutions suggest the following conjecture: There is some function $f(m,n)$ such that for any m, a common SDR exists if and only if for any $\mathscr{X}_i \subseteq \mathscr{S}_i \, (i = 1, \ldots, m)$

$$\sum_{i=1}^{m} |X_i| \leqslant f(m,n) + \left| \bigcap_{i=1}^{m} I_i \right|.$$

This is certainly true for $m = 1$ with $f(1,n) = 0$, and $m = 2$ with $f(2,n) = n$. Unfortunately it is false for $m \geqslant 3$. (See Exercise 1.)

Second, while the stated criteria are "for all $\mathscr{X}_i$," not all are needed: for example, the Ford and Fulkerson inequality automatically holds if $|X_1| + |X_2| \leqslant n$.

Third, even so, there are many specific sets of subsequences to be investigated (385 for the very simple case of Example 1.4) to determine if the condition is satisfied.

What *are* necessary and sufficient conditions for the existence of an SDR for given $\mathscr{S}_1, \ldots, \mathscr{S}_m$? Two experimental approaches suggest themselves. One is a classical mathematical approach. By contemplation or inspiration, make a conjecture about a necessary or sufficient condition. Since networks were successful for $m = 2$, this may involve trying networks for $m = 3$, or it may involve examining simple forms for $f(m,n)$. If it is not immediately obvious that the conjecture is false, try to find a counterexample, or to prove the conjecture correct. In this way, it is easy to show the following:

For any m and n, if there is an SDR, then for any $\mathscr{X}_i \subseteq \mathscr{S}_i \, (i = 1, \ldots, m)$

$$\sum_{i=1}^{m} |X_i| \leqslant (m-1)n + \left| \bigcap_{i=1}^{m} I_i \right|.$$

Also, if for $n \geqslant 2$, any m, and any $\mathscr{X}_i \subseteq \mathscr{S}_i \, (i = 1, \ldots, m)$

$$\sum_{i=1}^{m} |X_i| \leqslant m - n + 1 + \left| \bigcap_{i=1}^{m} I_i \right|,$$

then there is an SDR.

The first approach relies heavily on the mathematician's intuition and experience in making good conjectures and finding relevant examples. The second approach is to use a computer to generate examples before trying to state any conjecture. This second approach is highly systematic, and less reliant on intuition to produce relevant examples. However, left to itself, the computer may prove to be too systematic. To take one of the simplest situations, suppose $m = 3$, $n = 2$, and $|A| = 4$, and suppose that the sets in each sequence are pairwise disjoint. There are eleven distinct (nonisomorphic) sets of nonempty sequences satisfying these stated conditions, shown in Table 2.1. For n and $|A|$ a little larger, the number grows to hundreds. We

TABLE 2.1

$\mathcal{S}_1$	$\mathcal{S}_2$	$\mathcal{S}_3$
{1}	{1}	{1}
{2, 3, 4}	{2, 3, 4}	{2, 3, 4}
{1}	{1}	{1, 2, 3}
{2, 3, 4}	{2, 3, 4}	{4}
{1}	{1, 2, 3}	{1, 2, 4}
{2, 3, 4}	{4}	{3}
{1}	{1}	{1, 2}
{2, 3, 4}	{2, 3, 4}	{3, 4}
{1}	{1, 2, 3}	{1, 2}
{2, 3, 4}	{4}	{3, 4}
{1}	{1, 2, 3}	{1, 4}
{2, 3, 4}	{4}	{2, 3}
{1}	{1, 2}	{1, 2}
{2, 3, 4}	{3, 4}	{3, 4}
{1}	{1, 2}	{1, 3}
{2, 3, 4}	{3, 4}	{2, 4}
{1, 2}	{1, 2}	{1, 2}
{3, 4}	{3, 4}	{3, 4}
{1, 2}	{1, 2}	{1, 3}
{3, 4}	{3, 4}	{2, 4}
{1, 2}	{1, 3}	{1, 4}
{3, 4}	{2, 4}	{2, 3}

are thus faced with the problem of generating and analyzing immense quantities of data. And the analysis itself is not short: remember the 385 sets of subsequences for Example 1.4.

We make three observations: (1) There are enough potential data that intuition alone is likely to miss some relevant data. (2). There are enough potential data that we would do well to identify and avoid generating large amounts of data having low relevance. (3) The analysis of each datum is sufficiently lengthy that some way should be found to reduce the analysis. These observations suggest that we use a computer to generate and analyze data, but monitor the process closely and modify it whenever possible.

In addition to the need to analyze large quantities of data, the SDR problem exhibits another trait common to many combinatorial problems.

This is a dramatic sensitivity of the data analysis to small parameter changes. Suppose, for example, that you are generating data for $m = 3$, and outputting only those data for which

$$\sum_{i=1}^{3} |X_i| - \left| \bigcap_{i=1}^{3} I_i \right|$$

exceeds a certain value. An increase of one in this value can increase computing time by a factor of 200 or more. Thus your computer may appear to be stuck in a tight loop for half an hour or more. It is, and it should be—at least in the sense that it is still correctly performing the desired calculation. This sensitivity has a double impact. First, a modest computer budget is very quickly depleted by half-hour runs. Second, if your operator is nervous about tight loops, he may abort the run after twenty minutes or so, before there is any output. This is a waste of both money and computer time, and frustrating to the user.

As a general rule, by closely monitoring computer runs and carefully analyzing the output, the effects of this sensitivity can be minimized. A carefully designed series of computer experiments can lead to modifications in the program to counterbalance the sensitivity effects. However, if extremely long runs with little or no intermediate output cannot be avoided, at least warn the operator so that he will let the program run to completion.

EXERCISES

1. Consider these two examples with $m = n = 3$, $A = \{1, 2, 3, 4, 5, 6\}$.

    I: $S_{11} = \{1, 2\}$ $S_{21} = \{1, 3\}$ $S_{31} = \{1, 6\}$
        $S_{12} = \{3, 4\}$ $S_{22} = \{2, 5\}$ $S_{32} = \{2, 4\}$
        $S_{13} = \{5, 6\}$ $S_{23} = \{4, 6\}$ $S_{33} = \{3, 5\}$

    II: $S_{11} = \{1, 2\}$ $S_{21} = \{1, 3\}$ $S_{31} = \{1, 6\}$
        $S_{12} = \{3, 4\}$ $S_{22} = \{2, 5\}$ $S_{32} = \{2, 3\}$
        $S_{13} = \{5, 6\}$ $S_{23} = \{4, 6\}$ $S_{33} - \{4, 5\}$

    (a) Show that I has two SDRs, while II has none.
    (b) Show that both satisfy the inequality

    $$\sum_{i=1}^{3} |X_i| - \left| \bigcap_{i=1}^{3} I_i \right| \leqslant 6$$

    for all $\mathcal{X}_i \subseteq \mathcal{S}_i$ $(i = 1, 2, 3)$, and that the value 6 cannot be lowered.

2. Show that for any m and n, if there is an SDR, then for any $\mathcal{X}_i \subseteq \mathcal{S}_i$ $(i = 1, \ldots, m)$

    $$\sum_{i=1}^{m} |X_i| \leqslant (m-1)n + \left| \bigcap_{i=1}^{m} I_i \right|.$$

3.   Show that for if $n \geqslant 2$, any m, and any $\mathscr{X}_i \subseteq \mathscr{S}_i (i = 1, \ldots, m)$

$$\sum_{i=1}^{m} |X_i| \leqslant m - n + 1 + \left| \bigcap_{i=1}^{m} I_i \right|,$$

then there is an SDR.

4.   Determine the number of nonisomorphic sets of sequences of pairwise disjoint sets for

   (a) $m = 3, n = 3, |A| = 6$;
   (b) $m = 3, n = 4, |A| = 8$.

5.   Determine for arbitrary m and n the number of possible ways to choose the $\mathscr{X}_i$. Write a program to tabulate values of this for $m + n \leqslant 11$.

3. THE THIRD QUESTION, PARTITION CASE

As one generates and studies data, patterns begin to emerge. Some patterns prove to be false leads—they do not generalize in a helpful way. Others are quite useful. Fairly quickly it becomes apparent that the study of SDRs is simpler if the sets in each sequence partition A. We will call this condition the *partition hypothesis*.

The effect of the partition hypothesis becomes apparent if we examine Fig. 3.1. Let

$$y = \max \sum_{i=1}^{m} |X_i| - \left| \bigcap_{i=1}^{m} I_i \right|,$$

where the maximum is taken over all choices of $\mathscr{X}_i$. In the figure, the lines $y = (m-1)n$ and $y = m - n + 1$ correspond respectively to the necessary and the sufficient conditions that we have noted. That is, no system for which the y value falls in region A will have an SDR; any system for which the y value falls in region D will have an SDR. If the y value for a system of sets falls in either region B or C, the system may or may not have an SDR.

The effect of the partition hypothesis is apparent in the line $y = m - 1$. If this hypothesis holds, then any system for which the y value falls in region C will have an SDR.

For any m and n, if the sets in each sequence partition A, and if for any $\mathscr{X}_i \subseteq \mathscr{S}$ $(i = 1, \ldots, m)$

$$\sum_{i=1}^{m} |X_i| \leqslant m - 1 + \left| \bigcap_{i=1}^{m} I_i \right|,$$

then there is an SDR.

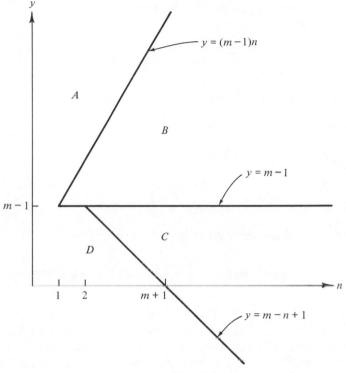

$y = (m-1)n$

A

B

$y = m - 1$

$m - 1$

C

D

$1 \quad 2 \quad\quad m + 1$

n

$y = m - n + 1$

Figure 3.1

A more striking consequence of the partition hypothesis is its effect on the computation to determine y. In a sense which we shall describe, it eliminates "almost all" of the computation. Study of region B of Fig. 3.1 suggests that the lines $y = m + k$ ($k = -1, 0, 1, \ldots$) and $y = (m-1)n - k$ ($k = 0, 1, 2, \ldots$) may yield insights into the conditions for the existence of an SDR. While the former lines do not prove significant, the latter yield this result.

For any m and n, if the set in each sequence $\mathscr{S}_i$ partitions A, and if for some $k = 0, 1, \ldots$

$$\sum_{i=1}^{m} |X_i| \leqslant (m-1)n - k + \left| \bigcap_{i=1}^{m} I_i \right|$$

for all choices of $\mathscr{X}_i$ such that $\sum_{i=1}^{m} |X_i| = (m-1)n - k + 1$, then

$$\sum_{i=1}^{m} |X_i| \leqslant (m-1)n - k + \left| \bigcap_{i=1}^{m} I_i \right|$$

for all possible choices of $\mathscr{X}_i$.

Thus, if the partition hypothesis holds, then to determine y we need examine only those $\mathscr{X}_i$ such that

$$\sum_{i=1}^{m} |X_i| = (m-n)n - k + 1,$$

rather than all possible $\mathscr{X}_i$. The latter number of cases is

$$T = (2^n - 1)^m;$$

a lengthy combinatorial argument shows the former to be

$$N = \sum \frac{(n!)^m m!}{\prod_{v=1}^{n} [v!(n-v)!]^{p_v} p_v!},$$

where the sum is over all nonnegative integers p_v such that

$$\sum_{v=1}^{n} p_v = m \qquad \text{and} \qquad \sum_{v=1}^{n} v p_v = (m-1)n - k + 1.$$

The argument leading to N is based on an examination of m-part compositions of $(m-1)n-k+1$. Compositions rather than partitions are needed because the elements within individual sets are involved in the computation of y. Thus the choice of three sets, for example, from one sequence may yield different results than would the choice of three sets from another sequence. The number of such compositions is

$$\sum \frac{m!}{p_1! p_2! \cdots p_n!},$$

where the p_i satisfy the above two conditions. Then for each such composition it is necessary to determine the number of ways in which the subsequences can be chosen.

While the expression for N is complex, it has been shown for $m = 2$ and $k = 0$ that

$$\lim_{n \to \infty} \frac{N}{T} = 0.$$

Evidence indicates that this same limit value holds for all m and k. Thus while the value of N is large, it is in this sense—that the ratio N/T becomes arbitrarily small as we increase n—that "almost all" computation for y is eliminated. Table 3.1 shows values of N and T for small m and n, with k selected to maximize N.

TABLE 3.1

n:		2	3	4	5	6
$m = 2$	k:	0	1	1	1	1
	N:	4	18	68	250	922
	T:	9	49	225	961	3969
$m = 3$	k:	1	2	3	3	4
	N:	12	108	840	6300	47960
	T:	27	343	3375	29791	250047
$m = 4$	k:	2	3	5	6	7
	N:	32	648	10896	172750	2629906
	T:	81	2401	50625	923521	15752961
$m = 5$	k:	3	5	6	8	10
	N:	80	4050	146240	4813750	148588160
	T:	243	16807	759375	28629151	992436543

EXERCISES

1. Determine the formula for N in the special case $m = 2$, $k = 0$.

2. (cont.) Show that in this case

$$N = \sum_{k=1}^{n} \binom{n}{k}\binom{n}{n+1-k}.$$

3. (cont.) Show that

$$\sum_{k=1}^{n} \binom{n}{k}\binom{n}{n+1-k} = \binom{2n}{n+1}.$$

4. (cont.) Let $N(t)$ and $T(t)$ be N and T evaluated for $m = 2$, $k = 0$, and $n = t$. Let $R(t) = N(t)/T(t)$. Show that

$$\frac{R(t+1)}{R(t)} < 1,$$

that is, that R is a monotone decreasing function of n.

4. SUMMARY

The attack on the SDR problem is an example of the interplay between much of the mathematics and computing that we have discussed. The problem arose from group theory, graph theory, and set theory. The solution for the case $m = 2$ was achieved by use of a network model. Work on the general case has involved extensive use of computers to generate and analyze examples. The results arising from this analysis involve combinatorial formulas. In particular, the conditions on the summation for N are formulas related to compositions of integers. And we are not through. While the ratio $R(n)$ is monotone decreasing, this does not imply that $\lim_{n \to \infty} R(n) = 0$. We are able to show this using an approximation formula which is discussed in Chapter 12. Awareness of this type of interplay, and the consequent ability to call on a wide variety of mathematical and computational techniques is useful in solving many problems, of which the SDR problem is only one example.

REFERENCES

1. P. Hall, On representatives of subsets, *J. London Math. Soc.* **10** (1953) 26–30.
2. L. R. Ford, Jr. and D. R. Fulkerson, Network flow and systems of representatives, *Canad. J. Math.* **10** (1958) 78–85.
3. L. R. Ford, Jr. and D. R. Fulkerson, "Flows in Networks." Princeton Univ. Press, Princeton, New Jersey, 1962.

CHAPTER 12

Discrete Probability

1. PROBABILITIES ON A DISCRETE SET

Suppose that we are observing some event, and have decided upon a variable to measure at the conclusion of the event. Suppose further that the variable is quantized, so that there are only a discrete number of values that it can assume. For example, suppose that the event is the dealing of a poker hand, and we have decided to measure (count) the number of aces dealt. There are only five possible values. Or suppose that the event is a human life, and we are measuring age (in years) at death. Again there are only finitely many possible values—but it is not clear how many. If we were to measure age in seconds, then there would be billions of possibilities. Let us ask a question: Of all the possible outcomes, what is the likelihood that one particular outcome will occur?

Let us denote the set of possible outcomes by S, with elements x_i, $i = 1, 2, 3, \ldots$. With each outcome x_i we associate a nonnegative number $p(x_i)$ having the property that

$$p(x_1) + p(x_2) + \cdots + p(x_n) + \cdots = 1.$$

This number is called the *probability* of the outcome x_i. Notice that we have not yet specified *how* to determine the probability values. Notice also that we allow $p(x_i)$ to be zero. This is convenient in problems like the age-at-death problem. We can probably obtain agreement that a maximum possible age is 300 years, and that from some number of years N on, the probability of a person surviving to that age is zero, $p(N) = p(N+1) = \cdots = p(300) = 0$. That is, nobody will live N years or more. The argument will come in determining the value of N.

Since the probabilities $p(x_i)$ are nonnegative numbers summing to 1, it is clear that each $p(x_i)$ is bounded by 0 and 1, $0 \leqslant p(x_i) \leqslant 1$. The intuitive interpretation of the extreme values is that if $p(x_i) = 0$, then outcome x_i will certainly not occur; if $p(x_i) = 1$, then outcome x_i will certainly occur. Values between the extremes indicate, in some sense, the proportion of the time we expect outcome x_i to occur.

For example, suppose the event is the toss of a coin, with the measured outcome either a head (H) or a tail (T). If the coin has two heads then $p(H) = 1$ and $p(T) = 0$. (We assume that it will not balance on the edge.) That is, the outcome will certainly be a head, and never a tail. However, if the coin is "fair," there is no reason to expect more or fewer heads than tails in a set of repetitions of the event. That is, we expect $p(H) = p(T)$. Since $p(H) + p(T) = 1$, this implies $p(H) = p(T) = \frac{1}{2}$.

Now suppose the event is the roll of two dice, and the measured outcome is the value rolled. Thus the outcome has 11 possible values. However, some outcomes may happen in more than one way. For example, the value 4 is obtained in three ways $(3+1, 2+2, 1+3)$. This calls for a more thorough investigation.

Example 1.1

Compute the probabilities of the various values that can be rolled with two dice.

Solution.   Consider for the moment just one die. Its roll has six possible values, all equally likely (as heads and tails are equally likely for a coin). Thus for one die,

$$p(1) = p(2) = \cdots = p(6) = \frac{1}{6}.$$

Now take two dice. Assume as intuitively reasonable that the roll of one die does not affect the roll of the other. Since there are six possible rolls for each, there are 36 ($=6 \cdot 6$) possible rolls in all (Table 1.1). Arrange these 36 possibilities by value, and observe, for example, that five of the 36 yield a value of 6 (Table 1.2). Thus in the long run, you would expect $\frac{5}{36}$ of a

TABLE 1.1

1,1	1,2	1,3	1,4	1,5	1,6
2,1	2,2	2,3	2,4	2,5	2,6
3,1	3,2	3,3	3,4	3,5	3,6
4,1	4,2	4,3	4,4	4,5	4,6
5,1	5,2	5,3	5,4	5,5	5,6
6,1	6,2	6,3	6,4	6,5	6,6

TABLE 1.2

Value							Number of ways
2	1,1						1
3	2,1	1,2					2
4	3,1	2,2	1,3				3
5	4,1	3,2	2,3	1,4			4
6	5,1	4,2	3,3	2,4	1,5		5
7	6,1	5,2	4,3	3,4	2,5	1,6	6
8		6,2	5,3	4,4	3,5	2,6	5
9			6,3	5,4	4,5	3,6	4
10				6,4	5,5	4,6	3
11					6,5	5,6	2
12						6,6	1

series of rolls of the dice to have value 6. Take this as the probability of rolling 6. The other probabilities are similarly calculated.

ANSWER. See Table 1.3.

TABLE 1.3

$p(2) = \frac{1}{36}$	$p(8) = \frac{5}{36}$
$p(3) = \frac{1}{18}$	$p(9) = \frac{1}{9}$
$p(4) = \frac{1}{12}$	$p(10) = \frac{1}{12}$
$p(5) = \frac{1}{9}$	$p(11) = \frac{1}{18}$
$p(6) = \frac{5}{36}$	$p(12) = \frac{1}{36}$
$p(7) = \frac{1}{6}$	

There are several observations to be made from Example 1.1. First, the computed numbers *are* probabilities, nonnegative numbers summing to one. Second, consider any one roll of the dice, say 3 on the first die and 4 on the second. Since this is one of 36 equally likely possibilities, the probability of this specific roll is $p(\langle 3,4 \rangle) = \frac{1}{36}$. Observe that $\frac{1}{36} = \frac{1}{6} \cdot \frac{1}{6}$, and that, turning to a single die, $p(3) = p(4) = \frac{1}{6}$. This is an instance of the product rule.

The Product Rule. If events with outcomes $A_1, A_2, ..., A_n$ occur independently, and the outcomes have probabilities $p(A_i)$, $i = 1, 2, ..., n$, then the probability of the joint occurrence of these events having the outcome $\langle A_1, A_2, ..., A_n \rangle$ is the product of the individual probabilities:

$$p(\langle A_1, A_2, ..., A_n \rangle) = p(A_1) p(A_2) \cdots p(A_n).$$

Third, consider the probability of obtaining a specific value, say 6, on a roll of the dice. Since there are k ways to obtain this value (in this instance $k = 5$), the probability of obtaining this value is $k/36$ (in this instance $\frac{5}{36}$). Observe that this is the sum of $\frac{1}{36}$ taken k times. This is an instance of the sum rule.

The Sum Rule. If the outcome E of an event is composed of several independent alternatives, $A_1, A_2, ..., A_n$, then the probability of the outcome E is the sum of the individual probabilities:

$$p(E) = p(A_1) + p(A_2) + \cdots + p(A_n).$$

Example 1.2

Two pair of dice Π_1 and Π_2 are rolled. What is the probability of rolling 6 on Π_1 and 9 on Π_2?

Solution. The probability of rolling 6 on Π_1 is $p(6) = \frac{5}{36}$; the probability of rolling 9 on Π_2 is $p(9) = \frac{1}{9}$. Since the rolls of the two pair of dice are independent of each other, the probabilities follow the product rule:

$$p(6 \wedge 9) = p(6) \cdot p(9).$$

ANSWER. $\frac{5}{324}$.

We may verify this result by direct counting. Since there are 36 possible rolls for each pair of dice, there are 36^2 possible pairs of rolls. Visualize these in a tableau (Fig. 1.1). Observe that only those roll combinations in the doubly shaded rectangle have both a 6 on Π_1 and a 9 on Π_2.

There are 20 such combinations. Hence the probability of obtaining one of these combinations is $20/36^2 = 5/324$.

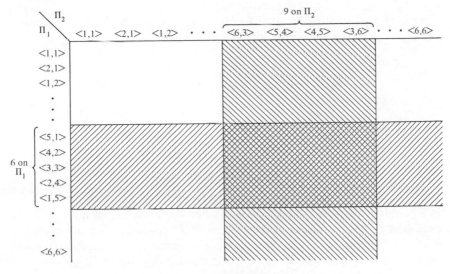

Figure 1.1

Example 1.3

Using Π_1 and Π_2 as defined in Example 1.2, what is the probability of rolling 6 on Π_1 or 9 on Π_2?

Solution. To argue that the rolls are independent and that hence the sum rule applies is wrong. Applying the sum rule gives the result

$$p = p(6) + p(9) = \frac{5}{36} + \frac{1}{9} = \frac{1}{4}.$$

But observe Fig. 1.1. Within the horizontal band corresponding to 6 on Π_1 lie $\frac{5}{36}$ of the points, and within the vertical band corresponding to 9 on Π_2 lie $\frac{1}{9}$ of the points. But simply summing these numbers counts those points in the doubly shaded rectangle twice. Thus you must subtract those points, so that they are only counted once:

$$p(6 \vee 9) = p(6) + p(9) - p(6 \wedge 9).$$

ANSWER. $\frac{19}{81}$.

This example illustrates the principle of inclusion and exclusion which is applied to account for co-occurrences of independent events.

Principle of Inclusion and Exclusion.   If $E_1, E_2, ..., E_n$ are out-comes of independent events, then

$$p(E_1 \lor E_2 \lor \cdots \lor E_n)$$
$$= p(E_1) + p(E_2) + \cdots + p(E_n)$$
$$- p(E_1 \land E_2) - p(E_1 \land E_3) - \cdots - p(E_{n-1} \land E_n)$$
$$+ p(E_1 \land E_2 \land E_3) + \cdots + p(E_{n-2} \land E_{n-1} \land E_n)$$
$$- \cdots$$
$$+ (-1)^{n-1} p(E_1 \land E_2 \land E_3 \land \cdots \land E_{n-1} \land E_n).$$

When does the sum rule apply, and when the principle of inclusion and exclusion.? The answer lies in the possibility of co-occurrence of outcomes. The sum rule applies only when the outcomes cannot co-occur. For example, while a pair of dice may total to six, this total cannot simultaneously be $1+5$ and $4+2$. If outcomes can co-occur, as in Example 1.3, the principle of inclusion and exclusion is used. See Exercise 1.

EXERCISES

1. Show that the sum rule is a special case of the principle of inclusion and exclusion.

TABLE 1.4[a]

e	1231	r	603	p	229	v	93
t	959	h	514	f	228	k	52
a	805	l	403	m	225	q	20
o	794	d	365	w	203	x	20
n	719	c	320	y	188	j	10
i	718	u	310	b	162	z	9
s	659			g	161		

[a] Data for this and the probability tables in Section 2 taken from Gaines [1].

Table 1.4 shows the frequency of occurrence of letters per 10,000 words of English text. The following five exercises relate to this table.

2. (a) If the occurrences of all letters were equally likely, what would be the probability of occurrence of each letter?
   (b) From Table 1.4, how would you estimate the probability of occurrence of each letter? Why is this only an estimate?

3. Write a program to count occurrences of letters in a string of letters. Use it to obtain frequency counts for roughly the first 1000 letter occurrences in each of Chapters 1, 5, and 10 of this text. In each text segment, punch periods to denote sentences, and keep track of the total number of letters counted. (Ignore numerals, formulas, figure references, and so on.)

From the total of the three runs estimate the probability of occurrence of each letter in English text. Compare your estimates with those obtained from Table 1.4, and explain any discrepancies. Save your data and results for future use.

4. From Table 1.4 estimate the probability of occurrence of the letter pairs *an*, *th*, *ht*, *qu*, *qb*, *ea*, *nt*, *rs*, *ck*, and *be*.

5. (cont.) Write a program to count occurrences of letter pairs in a string of letters. Apply your program to the same data as you used for Exercise 3. Remember that blanks break letter pairs, so that "the work" has five letter pairs, not six. Save your results for future use.

6. (cont.) Compare the results of Exercises 4 and 5. How do you explain the discrepancy?

2. CONDITIONAL PROBABILITY AND INDEPENDENCE

In Exercises 4–6 of Section 1 you observed that the product of the probabilities of two individual letter occurrences does not match the probability of occurrence of the pair of letters. Your results should have been similar to those given in Table 2.1. In explaining these results you probably

TABLE 2.1

	Product	Actual		Product	Actual
an	0.0058	0.0172	ea	0.0099	0.0131
th	0.0049	0.0315	nt	0.0069	0.0110
ht	0.0049	0.0022	rs	0.0040	0.0039
qu	0.000062	0.0020	ck	0.00017	0.0008
qb	0.000032	0	be	0.0020	0.0058

noted that in English q is always followed by u. Thus $p(qu)$ should be relatively high, but $p(qb) = 0$. Also, although the pairs *th* and *ht* both occur, *th* seems to be more common. Thus you would expect $p(th)$ to be higher than $p(ht)$. And certainly $p(th)$ intuitively should be higher than $p(qh)$, $p(zh)$, or $p(jh)$.

Suppose that we have examined one letter occurrence, say b, and we ask for the probability that the next letter is an h. Table 2.2 gives probability

TABLE 2.2

ah	0	hh	0.0005	nh	0.0009	uh	0.0002
bh	0	ih	0	oh	0.0003	vh	0
ch	0.0046	jh	0	ph	0.0007	wh	0.0048
dh	0.0003	kh	0	qh	0	xh	0
eh	0.0015	lh	0	rh	0.0003	yh	0.0005
fh	0.0006	mh	0.0001	sh	0.0030	zh	0
gh	0.0016			th	0.0315		

estimates for each pair of letters αh. From the table it is clear that this probability is not $p(h)$, for then the probability of the pair bh would be $p(bh) = p(b)p(h) = 0.00083$. The probability that the next letter is h depends on the letter that we have just examined. The *conditional probability* of outcome B given that outcome A has occurred is

$$p(B|A) = \frac{p(A \wedge B)}{p(A)}.$$

Thus,

$$p(h|b) = \frac{p(bh)}{p(b)}.$$

The conditional probabilities for h, given different preceding letters α, are given in Table 2.3, along with the probability products $p(\alpha)p(h)$ for

TABLE 2.3

α	$p(h|\alpha)$	$p(\alpha)p(h)$	α	$p(h|\alpha)$	$p(\alpha)p(h)$
a	0	0.0041	n	0.013	0.0037
b	0	0.00083	o	0.0038	0.0041
c	0.14	0.0016	p	0.031	0.0012
d	0.0082	0.0019	q	0	0.00010
e	0.012	0.0063	r	0.0050	0.0031
f	0.026	0.0012	s	0.046	0.0034
g	0.099	0.00083	t	0.33	0.0049
h	0.0097	0.0026	u	0.0065	0.0016
i	0	0.0037	v	0	0.00048
j	0	0.000051	w	0.24	0.0010
k	0	0.00027	x	0	0.00010
l	0	0.0021	y	0.027	0.00097
m	0.0044	0.0012	z	0	0.000046

comparisons with the probabilities $p(\alpha h)$ of Table 2.2. For example, from Table 2.2 $p(ch) = 0.0046$, and from Table 1.4, $p(c) = 320/10000 = 0.0320$. Hence $p(h|c) = 0.0046/0.0320 = 0.14$.

Example 2.1

Determine the conditional probabilities for the value of a roll of two dice, given the value rolled on the first die.

Solution.   It is clear that the probabilities for the total rolled depend on the value rolled on the first die. For example, if a 4 is rolled first, then the total t cannot be less than 5 nor more than 10. Thus

$$p(t|4) = 0 \qquad \text{if} \quad t = 2, 3, 4, 11, \text{ or } 12.$$

However, all rolls of the second die are equally probable. Thus the totals 5, 6, 7, 8, 9, and 10 are equally probable, or

$$p(t|4) = \frac{1}{6} \quad \text{if} \quad t = 5, 6, 7, 8, 9, \text{ or } 10.$$

A similar analysis holds for any first value f.

ANSWER.

$$p(t|f) = \begin{cases} \frac{1}{6} & \text{if } f+1 \leqslant t \leqslant f+6 \\ 0 & \text{otherwise.} \end{cases}$$

Ignoring the conditional relationships between outcomes can lead to erroneous conclusions. For example, let us determine the probabilities in two cases:

Case 1.   First roll is 4 ($f = 4$) and total is 3 ($t = 3$);

Case 2.   First roll is 4 ($f = 4$) and total is 5 ($t = 5$).

If we ignore the conditional probabilities, then $p(f = 4) = \frac{1}{6}$, $p(t = 3) = \frac{1}{18}$, and $p(t = 5) = \frac{1}{9}$. Hence it seems that

$$p(f = 4 \wedge t = 3) = \frac{1}{6} \cdot \frac{1}{18} = \frac{1}{108} \quad \text{and} \quad p(f = 4 \wedge t = 5) = \frac{1}{6} \cdot \frac{1}{9} = \frac{1}{54}.$$

But from the formula for conditional probabilities,

$$p(A \wedge B) = p(A)p(B|A).$$

Thus since $p(t = 3 | f = 4) = 0$ and $p(t = 5 | f = 4) = \frac{1}{6}$,

$$p(f = 4 \wedge t = 3) = \frac{1}{6} \cdot 0 = 0 \quad \text{and} \quad p(f = 4 \wedge t = 5) = \frac{1}{6} \cdot \frac{1}{6} = \frac{1}{36}.$$

Ignoring the conditional probabilities yields an answer that is too high in the first case, too low in the second. Using the conditional probabilities yields the correct answer: there is no way to roll a total of 3 if the first roll is 4; and only one out of the 36 possible rolls combines the properties of 4 on the first die, and 5 for total.

We say that two outcomes A and B are *statistically independent* if

$$p(B|A) = p(B),$$

that is, if knowledge that outcome A has or has not occurred does not affect our expectations for the occurrence of B. Note that if A and B are statistically independent, then

$$p(A \wedge B) = p(A)p(B).$$

It is a common practice in experimental work to assume statistical independence since this greatly simplifies any statistical calculations. Nevertheless, in many applications such as information retrieval there are clear dependencies that invalidate such an assumption. For example, one would expect the word *digital* to occur quite frequently in sentences containing the word *computer*, but rather infrequently in sentences containing the word *daisy*. Similarly, the order of occurrence of words can be expected to affect the probability of co-occurrence, as is illustrated by *green car* and *car green*. Because of such dependencies, if statistical independence is assumed, any conclusions reached are open to question on the basis that the effect of the assumption is not negligible. Assumption of statistical independence can often yield results that are "good enough" from a practical viewpoint. However, if the assumption is false, then some additional work should be done to determine whether making the false assumption leads to erroneous conclusions.

EXERCISES

These exercises refer to the data and results of the Exercises in Section 1.

1. From your frequency counts for Section 1 determine for each ordered pair $\alpha\beta$ of letters the conditional probabilities $p(\beta|\alpha)$ and $p(\alpha|\beta)$.

Prepare additional data consisting of roughly the first 1000 letter occurrences in each of Chapters 3, 7, and 12 of this text, as before. Use this data together with that previously prepared for the following nine exercises:

2. Recompute the probabilities of single letters and letter pairs. Compare your results with those previously obtained.

3. Modify the letter pair program to ignore blanks (so that "*ew*" in "*the word*" will count as a pair), and recompute the letter pair probabilities. Compare with your previous results.

4. Compute the probabilities of occurrence for the initial letters of words.

5. Compute the probabilities of occurrence for the final letters of words.

6. Compute the probabilities of occurrence for triples of letters, both recognizing and ignoring blanks.

7. Compute the probabilities of occurrence for words of n letters, $n = 1, 2, 3, \ldots$.

8. Compute the probabilities of occurrence for sentences of n words, $n = 1, 2, 3, \ldots$.

For Exercises 9 and 10 ignore occurrences of words in this list:

a	all	was	them
an	and	who	they
as	any	even	this
at	are	from	were
be	but	have	what
by	can	into	will
in	for	like	with
is	has	many	about
it	its	more	first
no	may	most	other
of	not	only	their
on	now	said	there
or	one	some	these
so	our	such	where
to	the	than	which
we	two	that	would

*9. Compute the probabilities of co-occurrence of two words within a sentence. (For example, the sentence "Boys sleep well." has three co-occurrences, (boys, sleep), (boys, well), and (sleep, well).) Print out only those co-occurrences that occur more than once, but note that you will also need the remaining co-occurrences for Exercise 10.

10. (cont.) Compute the conditional probability of occurrence of a word α in a sentence, given that a word β occurs in the sentence.

3. COMPUTATION OF BINOMIAL COEFFICIENTS

The computation of large binomial coefficients and factorials is a recurring problem in combinatorial work. Obviously one does not compute

$$\binom{150}{65} = \left(\frac{150!}{65! \, 85!} \right)$$

directly. The numbers involved are so large and require so many multiplications that even with double precision arithmetic the result is not very accurate. We present three simple methods for computing such numbers.

First, suppose that complete accuracy is necessary. Consider $\binom{150}{65}$. To compute the three factorials directly and then divide involves extensive multiple precision arithmetic. For example, 150! is a number of some 875 bits—25 words on a typical computer. Prime power factorization provides a simple way to reduce the amount of computation. Since there are 35 primes less than 150, each integer involved can be represented by a vector of 35 components, the value of each component being the exponent of that prime in the factorization. For example, since $150 = 2 \cdot 3 \cdot 5^2$, it is represented by $\langle 1, 1, 2, 0, 0, ..., 0 \rangle$; while 149, a prime, is represented by $\langle 0, 0, ..., 0, 1 \rangle$.

Since the entries are exponents, multiplication is achieved by adding vectors, and division by subtracting vectors. Thus although 35 words are needed for the developing vector, most of the arithmetic operations are simple and single precision.

Example 3.1

Use prime power vectors to compute $\binom{40}{18}$.

Solution. Since there are only 12 primes less than 40, vectors of twelve components are used. Thus

$$40 = \langle 3,0,1,0,0,0,0,0,0,0,0,0 \rangle,$$
$$39 = \langle 0,1,0,0,0,1,0,0,0,0,0,0 \rangle,$$
$$38 = \langle 1,0,0,0,0,0,0,0,1,0,0,0,0 \rangle,$$
$$37 = \langle 0,0,0,0,0,0,0,0,0,0,0,1 \rangle,$$
$$\vdots$$
$$2 = \langle 1,0,0,0,0,0,0,0,0,0,0,0 \rangle.$$

Adding these,

$$40!: \qquad V_1 = \langle 38,18,9,5,3,3,2,2,1,1,1,1 \rangle.$$

Similarly

$$18!: \qquad V_2 = \langle 16,8,3,2,1,1,1,0,0,0,0,0 \rangle,$$

and

$$22!: \qquad V_3 = \langle 19,9,4,3,2,1,1,1,0,0,0,0 \rangle.$$

Thus $\binom{40}{18}$ is represented by

$$V_1 - V_2 - V_3 = \langle 3,1,2,0,0,1,0,1,1,1,1,1 \rangle.$$

That is,

$$\binom{40}{18} = 2^3 \cdot 3 \cdot 5^2 \cdot 13 \cdot 19 \cdot 23 \cdot 29 \cdot 31 \cdot 37.$$

There is still much multiplication to do, but far less than that involved in computing 40!.

ANSWER. $\binom{40}{18} = 11\,33802\,61800.$

Now suppose that complete accuracy is not necessary; a good estimate will suffice. In this case you can use *Stirling's approximation,*

$$n! \approx (2\pi)^{1/2} n^{n+1/2} e^{-n}.$$

This estimate will be slightly low, and another estimate,

$$n! \approx (2\pi)^{1/2} n^{n+1/2} e^{-n+1/(12n)},$$

is slightly high. While these formulas appear awesome, in practice they work very well.

Example 3.2

Estimate $\binom{40}{18}$ using Stirling's approximation.

Solution.   Estimate the factorials:

$$40! = (2\pi)^{1/2} 40^{40.5} e^{-40},$$

$$18! \approx (2\pi)^{1/2} 18^{18.5} e^{-18},$$

$$22! \approx (2\pi)^{1/2} 22^{22.5} e^{-22}.$$

Thus

$$\binom{40}{18} = \frac{40!}{18!\,22!}$$

$$\approx \frac{(2\pi)^{1/2} 40^{40.5} e^{-40}}{(2\pi) 18^{18.5} 22^{22.5} e^{-40}}$$

$$= (2\pi)^{-1/2} \frac{40^{40.5}}{18^{18.5} 22^{22.5}}.$$

This value is easily computed using logarithms.

ANSWER. $\binom{40}{18} \approx 1.1409 \times 10^{11}.$

The second approximation formula corrects this by a factor $e^{-301/47520}$, yielding

$$\binom{40}{18} \approx 1.1335 \times 10^{11}.$$

Note that whereas the first Stirling formula underestimates $n!$, it over-estimates binomial coefficients. For the second formula the situation is reversed. In each case, the relative error is very small for large numbers. In our example, the first estimate is 0.63 percent too high, while the second is 0.079 percent too low.

The third procedure is to revert to a "brute force" approach. Notice that we do not need to compute all three factorials $n!$, $m!$, and $(n-m)!$ in order to determine $\binom{n}{m}$. One of the factorials in the denominator directly cancels a portion of $n!$ For example,

$$\binom{150}{65} = \frac{150 \cdot 149 \cdot 148 \cdot \cdots \cdot 87 \cdot 86}{1 \cdot 2 \cdot 3 \cdot \cdots \cdot 64 \cdot 65}.$$

This computation may be done as

$$\left(\frac{150}{1}\right)\left(\frac{149}{2}\right)\left(\frac{148}{3}\right) \cdots \left(\frac{87}{64}\right)\left(\frac{86}{65}\right),$$

or as

$$\left(\frac{150}{65}\right)\left(\frac{149}{64}\right)\left(\frac{148}{63}\right) \cdots \left(\frac{87}{2}\right)\left(\frac{86}{1}\right).$$

While this procedure is too tedious for hand computation, computer results are surprisingly accurate. Even for $\binom{150}{65}$ the result, 2.4590×10^{43}, is more accurate than Stirling's approximation.

Example 3.3

Compute $\binom{40}{18}$ as

$$\left(\frac{40}{18}\right)\left(\frac{39}{17}\right) \cdots \left(\frac{23}{2}\right)\left(\frac{22}{1}\right).$$

ANSWER. Using Fortran on a CDC Cyber 72,

$$\binom{40}{18} = 1.13380261800E11.$$

Note: Such accuracy should not be expected from computers with shorter word lengths.

EXERCISES

1. Write a program to compute exactly factorials and binomial coefficients. The input will be either a single number m for which $m!$ is to be computed, or a pair of numbers $\langle m, n \rangle$, for which $\binom{m}{n}$ is to be computed. Assume that the output will not have more than 20 decimal digits. Compute a table of factorials up to 20!, and the binomial coefficients $\binom{40}{18}$, $\binom{52}{25}$, and $\binom{39}{37}$.

2. With the same input and output assumptions as in Exercise 1, write a program to compute factorials and binomial coefficients by the two Stirling's approximations. Compute the same numbers as in Exercise 1, and compare your results.

3. With the same input and output assumptions as in Exercise 1, write a "brute force" program to compute factorials and binomial coefficients. Compute the same numbers as in Exercise 1, and compare your results.

4. Modify your programs of Exercises 1, 2, and 3 to compute $\binom{150}{65}$ by the three methods. Compare your results and run times.

5. Determine exactly the probability of obtaining n aces, $n = 0, 1, 2, 3, 4$, in a poker hand. *Hint:* Determine the number of possible hands, and the number of hands with exactly n aces.

6. Recall the formulas developed in Section 3 of Chapter 11 for SDRs in the case $m = 2$, $k = 0$. Specifically,

$$N(n) = \binom{2n}{n+1}, \qquad T(n) = (2^n - 1)^2, \qquad \text{and} \qquad R(n) = \frac{N(n)}{T(n)}.$$

Use Stirling's approximation to estimate $R(n)$ for $n = 10^2, 10^4, 10^6$.

*7. (cont.) Use Stirling's approximation to show that

$$\lim_{n \to \infty} R(n) = 0.$$

4. DISTRIBUTIONS

In measuring various phenomena it is quickly obvious that the values obtained are not all equal, but vary from one measurement to the next. This is true whether we are measuring age at death, height, sentence lengths, or properties of some manufactured item. It is also true, but slightly less obvious, that the values obtained for any one type of measurement fall into some pattern of distribution. It is a task of the statistician to detect and characterize these distributions for some useful purpose. In quality control, variations from the distribution may indicate a malfunction in an assembly line. In insurance, the distribution is used to set insurance rates. In stylistic analysis, word and sentence distributions are used to determine authorship.

While the variety of distributions is infinite, most natural distributions can be closely approximated by a few standard types of distribution. In this section we describe the more important of these.

Binomial Distribution

Suppose that we "run a series of experiments" that are statistically independent and which each have only two possible outcomes, success and failure. Suppose also that the probability of success (and of failure) remains fixed throughout the series. For example, we may elect to roll seven dice in each experiment, counting as a success only those rolls for which either three or four 5's appear on the dice. Or we may define the experiment to be selecting ten widgets from an assembly line, with success defined to mean at most one defective widget in the group of ten. Under ideal circumstances these "experiments" satisfy the conditions specified, and are called *Bernoulli trials*.

Suppose that we are interested in the total number of successes in n Bernoulli trials, and not in their order of occurrence. Then among the n trials, k successes can be distributed in $\binom{n}{k}$ ways. If the probability of success is p, and of failure is $q = 1 - p$, then each way for k successes to occur has probability $p^k q^{n-k}$. The function that characterizes this distribution of successes is the *binomial distribution function*.

In n Bernoulli trials with success probability p and failure probability $q = 1 - p$, the probability of k successes and $n - k$ failures ($0 \leqslant k \leqslant n$) is

$$b(k; n, p) = \binom{n}{k} p^k q^{n-k}.$$

Suppose, for example, that we are manufacturing widgets and expect 10 percent of the widgets to be defective. If our quality control procedure consists of counting the number of defective widgets in a sample of size 35, then after 400 applications of this procedure we might find that our experience is reflected in Table 4.1. While we do not expect exactly 10 percent defects every time, is this experience consistent with our belief that 10 percent are defective?

TABLE 4.1

Number of defective widgets in sample	Number of samples	Fraction of samples
0	6	0.0150
1	47	0.1175
2	69	0.1725
3	107	0.2675
4	72	0.1800
5	56	0.1400
6	24	0.0600
7	14	0.0350
8	3	0.0075
9	2	0.0050
10	0	0.0000
	400	1.0000

This situation is typical of those fitting a binomial distribution. "Success" is finding a defective widget, and we are counting the number of successes per 35 trials. Since the "success" probability is $p = 0.1$ the distribution should be the binomial distribution

$$b(k; 35, 0.1) = \binom{35}{k}(0.1)^k(0.9)^{35-k}.$$

The agreement between data and theory is shown in Table 4.2 and Fig. 4.1. Four hundred is a relatively small sample size, and the agreement is fairly good. A statistician would apply further tests to determine exactly how good.

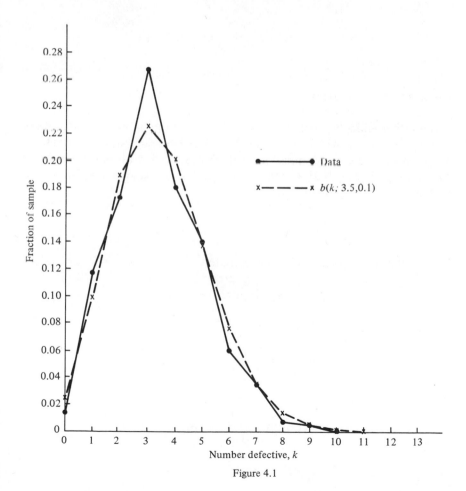

Figure 4.1

TABLE 4.2

k = number defective	Data	$b(k; 35, 0.1)$
0	0.0150	0.0250
1	0.1175	0.0974
2	0.1725	0.1839
3	0.2675	0.2248
4	0.1800	0.1998
5	0.1400	0.1376
6	0.0600	0.0765
7	0.0350	0.0352
8	0.0075	0.0137
9	0.0050	0.0046
10	0.0000	0.0013
11	0.0000	0.0003
12	0.0000	0.0000

Poisson Distribution

It is rather difficult to compute a single value of the binomial distribution, although successive values are easily computed by recursion. Fortunately, a good and simple approximation to the binomial distribution is provided by the Poisson distribution. Let $\lambda = np$, and suppose that n is large and p small, so λ is fairly small. Then

$$b(0; n, p) = (1 - p)^n = \left(1 - \frac{\lambda}{n} \right)^n.$$

For n sufficiently large, this is approximately $e^{-\lambda}$:

$$b(0; n, p) \approx e^{-\lambda}.$$

The binomial distribution satisfies a recurrence relation,

$$b(k + 1; n, p) = \frac{n - k}{k + 1} \cdot \frac{p}{1 - p} \cdot b(k; n, p).$$

Thus

$$b(1; n, p) = \frac{\lambda}{1 - \lambda/n} b(0; n, p)$$

$$\approx \lambda e^{-\lambda};$$

and

$$b(2; n, p) = \frac{n-1}{2} \frac{p}{1-p} b(1; n, p)$$

$$\approx \frac{\lambda^2}{2} e^{-\lambda}.$$

In general,

$$b(k; n, p) \approx \frac{\lambda^k}{k!} e^{-\lambda},$$

where the approximation is valid if λ^2 is small compared to n.

The *Poisson distribution function* is

$$p(k; \lambda) = \frac{\lambda^k}{k!} e^{-\lambda}.$$

For n sufficiently large, this approximates the binomial distribution function $b(k; n, \lambda/n)$.

Note that since λ is a fixed parameter, $p(k; \lambda)$ is quite simple to compute.

Example 4.1

Fit a Poisson distribution to the data of Table 4.1.

Solution. These data were assumed to have $p = 0.1$. Hence $\lambda = np = 3.5$. Thus the distribution desired is

$$p(k; 3.5) = \frac{3.5^k}{k!} e^{-3.5}.$$

ANSWER.   See Table 4.3 and Fig. 4.2.

While we have introduced the Poisson distribution as an approximation to the binomial distribution, it has its own uses as well. In particular, if we are conducting an experiment over a period of time and observing the number of successes in a time interval of fixed length, then the Poisson distribution is appropriate provided that the experimental conditions remain constant and the numbers of successes in nonoverlapping time intervals are

TABLE 4.3

k = number defective	Data	$p(k; 3.5)$
0	0.0150	0.0302
1	0.1175	0.1057
2	0.1725	0.1850
3	0.2675	0.2158
4	0.1800	0.1888
5	0.1400	0.1322
6	0.0600	0.0771
7	0.0350	0.0385
8	0.0075	0.0169
9	0.0050	0.0066
10	0.0000	0.0023
11	0.0000	0.0007
12	0.0000	0.0002

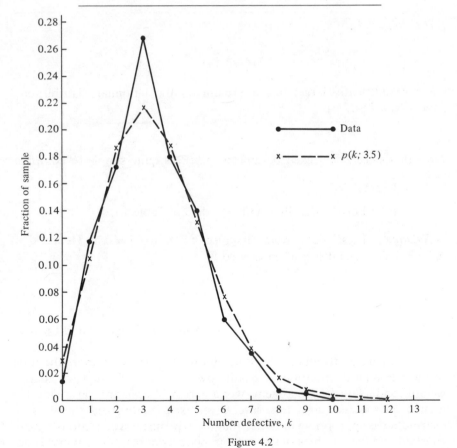

Figure 4.2

statistically independent. Under these conditions, the probability of exactly k successes in an interval of length t is

$$p(k; \lambda t) = \frac{(\lambda t)^k}{k!} e^{-\lambda t},$$

where $\lambda = \lim_{n \to \infty} n p_n$, and p_n is the probability of at least one success in an interval of length $1/n$. The Poisson distribution fits radioactive decay patterns, traffic accident patterns, and certain queuing problems such as waiting lines at computer terminals.

Normal Distribution

One of the earliest distributions studied, the normal distribution is appropriate in many situations where there is no boundary, or where the boundary is sufficiently distant that it has a negligible effect. For example, the distribution of hits around a bull's-eye, or of actual lengths of bars having nominal length 5 meters is normal. In the first case there is no real boundary; in the second case the boundary (length 0 meters) is sufficiently distant that it has no measurable influence on the distribution of bar lengths. The *normal density function*

$$\phi(x) = \frac{1}{\sqrt{2\pi}} \exp\left(-\frac{x^2}{2}\right)$$

and its integral, the *normal distribution function*,

$$\Phi(x) = \frac{1}{\sqrt{2\pi}} \int_{-\infty}^{x} \exp\left(\frac{-y^2}{2}\right) dy$$

are fairly difficult to compute. Fortunately they have been extensively tabulated, and thus are relatively easy to use. The graph of the normal density function is the familiar bellshaped curve of Fig. 4.3.

For statistical uses, the normal density function is written in a parameterized form,

$$\phi(x; m, \sigma) = \frac{1}{\sigma\sqrt{2\pi}} \exp\left[-\frac{1}{2} \frac{(x-m)^2}{\sigma}\right],$$

with the *mean* m and *standard deviation* σ as parameters. The mean determines the value $x = m$ for which the normal density function attains its maximum, and the standard deviation determines the spread of the bell. These parameters may be estimated from experimental data by the formulas

$$m \approx \frac{1}{n} \sum_{i=1}^{h} x_i f_i$$

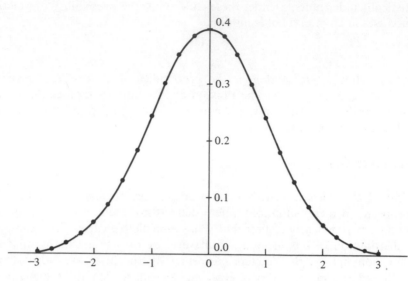

Figure 4.3

and

$$\sigma \approx \left(\frac{1}{n}\sum_{i=1}^{h}(x_i-m)^2 f_i\right)^{\frac{1}{2}},$$

where x_i is the value of the variable measured or counted, f_i is the count for x_i, h is the number of different values studied, and n is the total count, $\sum f_i$.

Example 4.2

Estimate m and σ for the data of Table 4.1.

Solution. For this, x is the number of defective widgets in the sample, f_i is the number of samples with x_i defective widgets, and $n = 400$ is the total number of samples. Since there are at most nine defective widgets in a sample, $h = 10$. Then

$$m = \frac{1}{400}\sum_{i=1}^{10} x_i f_i$$

$$= \frac{1}{400}(0\cdot6+1\cdot47+2\cdot69+\cdots+8\cdot3+9\cdot2)$$

$$= 3.395;$$

and

$$\sigma \approx \left(\frac{1}{400}\sum_{i=1}^{10}(x_i-3.395)^2 f_i\right)^{\frac{1}{2}}.$$

ANSWER. $m \approx 3.395$; $\sigma \approx 1.682$.

Note that m and σ are independent of any supposed distribution.

Observing that the binomial distribution has a rather bell-shaped curve, we might suggest that the normal distribution could approximate the binomial distribution. This is indeed possible. For n sufficiently large, the normal distribution with $m = np$ and $\sigma = \sqrt{npq}$ approximates the binomial distribution $b(k;n,p)$. That is, if $x_k = (k-np)/\sqrt{npq}$, then

$$b(k;n,p) \approx \frac{1}{\sigma}\phi(x_k).$$

Example 4.3

Approximate the binomial distribution of Table 4.2 by normal distributions based on the mean and standard deviation for the sample, and on $m = np$ and $\sigma = \sqrt{npq}$.

Solution. For each approximation, calculate the x_k, as shown in Table 4.4, and then determine the $\phi(x_k)$ from tabulated values.

TABLE 4.4

k	x_k (m = 3.395)	x_k (m = 3.5)
0	−2.0184	−1.9718
1	−1.4239	−1.4085
2	−0.8293	−0.8451
3	−0.2348	−0.2817
4	−0.3597	0.2817
5	0.9542	0.8451
6	1.5488	1.4085
7	2.1433	1.9718
8	2.7378	2.5352
9	3.3323	3.0986
10	3.9269	3.6620

ANSWER. See Table 4.5.

Despite a small value of n, the distributions computed in Example 4.3 approximate the binomial distribution reasonably well. Note that using the sample mean and standard deviation yields a generally better approximation.

TABLE 4.5

k	Data	$m = 3.395$ $\sigma = 1.682$	$m = 3.5$ $\sigma = 1.775$
0	0.0150	0.0309	0.0323
1	0.1175	0.1866	0.0832
2	0.1725	0.1681	0.1566
3	0.2675	0.2310	0.2161
4	0.1800	0.2137	0.2161
5	0.1400	0.1511	0.1566
6	0.0600	0.0713	0.0832
7	0.0350	0.0240	0.0323
8	0.0075	0.0055	0.0089
9	0.0050	0.0010	0.0019
10	0.0000	0.0001	0.0003

To complete the picture, since the Poisson distribution also approximates the binomial distribution, the normal distribution can be used to approximate the Poisson distribution, using $m = np = \lambda$ and $\sigma = \sqrt{npq} = \sqrt{\lambda q}$. All of these approximations improve for larger values of n. But particularly for small n the validity of any approximations should be checked by a statistician.

Uniform Distribution

In a uniform distribution all outcomes are equally probable. That is, if an event has n possible outcomes uniformly distributed, then the probability of each outcome is $1/n$. This is the case for fair coins ($p = \frac{1}{2}$), fair dice ($p = \frac{1}{6}$), and fair draws of cards ($p = \frac{1}{52}$). Because of its simplicity the uniform distribution is used for many examples in probability texts; but the other distributions that we have discussed are more important in practice.

There are of course many other statistical distributions. But those discussed in this chapter are widely used and fit, at least approximately, many data sets.

EXERCISES

1. Write a program to compute the binomial distribution $b(k; n, p)$, given n, p, and a range of k as input. Check your program against Table 4.2.

2. Write a program to compute the Poisson distribution $p(k; \lambda)$, given λ and a range of k as input. Check your program against Table 4.3.

3. The normal approximations in Table 4.5 suggest that the data used do not conform to $p = 0.1$, but rather to $p = 0.097$. Use this value of p to recompute the binomial and Poisson distributions for the data of Table 4.1, and compare the results with Tables 4.2 and 4.3. Do you think this change in p is significant?

TABLE 4.6

Number of pages	Number of samples	Fraction of samples	Number of pages	Number of samples	Fraction of samples
1	7	0.01072	24	7	0.01072
2	16	0.02450	25	12	0.01838
3	19	0.02910	26	4	0.00613
4	22	0.03369	27	5	0.00766
5	39	0.05972	28	3	0.00459
6	38	0.05819	29	4	0.00613
7	30	0.04594	30	2	0.00306
8	53	0.08116	31	1	0.00153
9	40	0.06126	32	2	0.00306
10	47	0.07198	33	1	0.00153
11	39	0.05972	34	0	0.00000
12	41	0.06279	35	1	0.00153
13	39	0.05972	36	1	0.00153
14	31	0.04747	37	0	0.00000
15	35	0.05360	38	1	0.00153
16	19	0.02910	39	2	0.00306
17	10	0.01531	40	1	0.00153
18	21	0.03216	41	1	0.00153
19	19	0.02910	—		
20	13	0.01991	43	1	0.00153
21	9	0.01378	50	1	0.00153
22	10	0.01531	61	1	0.00153
23	5	0.00766	Total	653	0.99998

4. Consider the data of Table 4.6, representing page lengths of papers in the *J. ACM* from January 1960 through July 1972. Compute the mean and standard deviation for this data.

5. (cont.) Determine the value of q from the equations $m = np$, $\sigma = \sqrt{npq}$. What does this tell you about fitting a binomial or Poisson distribution to the data of Table 4.6?

6. Fit a normal distribution to the data of Table 4.6.

5. RANDOM NUMBERS

There are many reasons for choosing things "at random" from a space. For example, quality control is often maintained by choosing parts "at random" from an assembly line and subjecting them to tests. Politicians base campaign strategies on polls of a "random" sample of the electorate. Monte Carlo techniques use values chosen "at random" to estimate area and other quantities. Large systems are tested by simulating their performance for conditions chosen "at random," rather than for all possible combinations of conditions.

For some purposes the sample of parts, voters, or conditions is left unstructured, while for other purposes it is structured or stratified. For example, the politician may request a poll of certain ethnic groups, or ask that the results of a farm poll be grouped by farm size. But within each structured block, or for the whole sample if it is unstratified, a general policy is that the sample be randomly chosen.

The basis for choice in many sampling situations is a *random number generator*. This is an algorithm designed to produce a sequence of numbers that satisfy certain statistical tests embodying the intuitive (and to some extent mathematical) concept of randomness.

Since randomness, like beauty, is in the mind of the beholder, there is no simple characterization of the concept. It is generally agreed that randomness excludes simple patterns of sampling. For example, choosing every tenth name in a phone book does not yield a population sample that most statisticians would regard as random. In practice, judgement on randomness is a posteriori. That is, after a sample is selected it is subjected to certain tests of statistical properties to determine whether or not it is random.

Among the properties that are desirable, the most obvious one is uniformity. If numbers are to be chosen from a uniform distribution, then each number should have equal probability of being chosen. Thus, in the decimal system, each digit should have probability 0.1, each two-digit numeral probability 0.01, and so forth. In practice, some slight variation from the ideal is to be expected. For example, the data for the widget example in Section 4 was derived by counting zeros in a table of 14,000 random digits. In theory there should be 1400 zeros. In this particular table there are 1358 zeros. This much variation in a sample of 14,000 is enough to lead one to question the randomness of the table.

Among other patterns that are easily checked are these: number of consecutive occurrences of a digit; the gaps between successive occurrences of a digit; the pattern of runs, for example, 2345; the number pairs of the form $(n, n+1)$; and the number of digits taken to obtain at least one of each digit (the "coupon collector's test"). These are empirical tests whose results can be interpreted by means of standard statistical techniques. In addition more sophisticated tests are available if needed. In using random number tables or generators that are available in the literature and in program libraries you should verify that the numbers generated satisfy your concept of randomness, and not blindly assume that the author or programmer's concept of randomness agrees with yours.

A variety of algorithms have been used to generate random numbers. Most of the simple techniques yield poor results, and hence are only of use in desperate situations. The most successful generator to date is the *linear congruential generator*,

$$x_{n+1} = (ax_n + c) \bmod m.$$

It can be shown that appropriate choices for the initial value x_0 and the parameters a, c, and m will yield sequences satisfying the standard tests of randomness. From the initial value, each succeeding value is chosen by reducing $ax_n + c$ modulo m, that is, taking the remainder of $ax_n + c$ on division by m. For example, choosing $x_0 = 6$, $a = 5$, $c = 7$, and $m = 11$ leads to the sequence

$$x_0 = 6, \qquad x_1 = 4, \qquad 5, 10, 2, 6, 4, 5, \cdots.$$

Note that this sequence is periodic. In fact all linear congruential generators produce periodic sequences. The length of the period is bounded by m since there are only m distinct residues (remainders) modulo m. However, the initial value x_0 and the parameters a and c also affect period length. Thus the choice $x_0 = 7$, $a = 7$, $b = 3$, and $m = 11$ leads to a sequence of period 10. It is desirable to generate sequences of large period. The first of the above examples, with $x_0 = 6$, generates only four of the eleven possible values $0, 1, \ldots, 10$ modulo 11; whereas the second example, with $x_0 = 7$, generates ten of the possible values. In some sense the second example makes better use of the modulus. In running tests, for example, it would provide us with more variety in the values to be tested.

Under certain conditions we can guarantee that the period of the sequence generated is m. These conditions, which are both necessary and sufficient, are that

  (i) $a - 1$ is a multiple of p, for each prime factor, p, of m;
  (ii) $a - 1$ is a multiple of 4 if m is a multiple of 4; and
  (iii) c and m are relatively prime.

Thus to ensure long sequences in a computer of word length w and base b arithmetic, choose $m = b^w$, or $m = b^w \pm 1$, and the remaining parameters in accordance with the above values. The choice of $m = b^w \pm 1$ results in the low order digits of the x_n being more randomly distributed than those generated using $m = b^w$, if that is important; but the use of $m = b^w + 1$ involves the overflow register to signal $x_n = b^w$. Note that the choice of x_0 is immaterial since all m residues are generated.

Example 5.1

Determine maximal cycle length random number generators for a binary computer with $w = 3$ bits.

Solution. For $w = 3$ and $b = 2$, m should be chosen as 7, 8, or 9. Choose $x_0 = 0$.

Consider first $m = 7$. To achieve a cycle length of 7, choose $a = 1$, $c = 2$. The cycle is 0, 2, 4, 6, 1, 3, 5, 0, ...

For $m = 8$, an appropriate parameter choice is $a = 5$, $c = 3$. This yields the cycle 0, 3, 2, 5, 4, 7, 6, 1, 0, ...

For $m = 9$, choose $a = 4$, $c = 2$, yielding the cycle 0, 2, 1, 6, 8, 7, 3, 5, 4, 0, ...

$$\text{ANSWER.} \quad x_{n+1} = (x_n + 2) \bmod 7;$$
$$x_{n+1} = (5x_n + 3) \bmod 8;$$
$$x_{n+1} = (4x_n + 2) \bmod 9.$$

It is instructive to consider these generators in a little more detail. The patterning for the first generator ($m = 7$) is quite obvious. Hence the sequence produced is not really "random." For $m = 8$, the pattern becomes obvious if the bit pattern is examined:

000,   011,   010,   101,   100,   111,   110,   001,   000, ...

Only for $m = 9$ is the pattern fairly complex:

000,   010,   001,   110,   1000 (overflow),   111,   011,   101,   100,   000, ...

Note that although 000 appears twice in the register, the two appearances are distinguished by the overflow indicator.

Warning: Higher level languages such as Fortran often automatically handle overflow and sign bits in such a way that use of $b^w + 1$ as modulus will not yield the proper congruences.

Finally, observe that use of b^n in a base b computer, either as the modulus m or the multiplier a, has the desirable feature that multiplication and division are accomplished by shift instructions.

EXERCISES

1. Show that for any m, the choices $a = 0$ and $a = 1$ yield poor random number generators.

2. Suppose $m = p_1 p_2 \cdots p_n$, where the p_i are distinct primes. Show that any generator with this modulus and cycle length m produces a sequence that is too patterned to be "random."

3. Find appropriate constants a and c for linear congruential generators with modulus and period $m = 2^n + i$, $i = -1, 0, 1$, where $n = 5, 10, 15$. If possible, choose a value for a other than 0 or 1.

4. Show that for given m, only $0 \leqslant a \leqslant m$ and $0 \leqslant c \leqslant m$ need be considered in developing linear congruential generators.

5. Write a subroutine to generate the congruences

$$x_{n+1} = (ax_n + c) \bmod m,$$

given x_0, a, b, and m as input. Use it to investigate thoroughly the linear congruential generators for $m = 11$.

6. Suppose a ternary $(b = 3)$ computer has ten-digit words. Develop and investigate linear congruential generators with cycle lengths $3^{10} + i$, $i = -1, 0, 1$.

REFERENCE

1. H. F. Gaines, "Elementary Cryptanalysis." Dover, New York, 1956.

Answers and Hints for Selected Exercises

CHAPTER 1

Section 3

1. Proof by induction. For $n = 1$, $1 = 1 f_1$. Suppose that such a representation holds for all integers less than n. Let $n = 1 + n_1$, where $n_1 = \sum_{i=1}^{\infty} a_i f_i = a_1 f_1 + a_2 f_2 + a_3 f_3 + \cdots$. Then $n = (1 + a_1) f_1 + a_2 f_2 + a_3 f_3 + \cdots$. If $1 + a_1 = 1$ and $a_2 = 0$, we have the representation for n. If $1 + a_1 = 1$ and $a_2 = 1$, then by definition of n_1, $a_3 = 0$. But $f_3 = f_1 + f_2$. Hence $n = f_3 + a_4 f_4 + \cdots$. If $a_4 = 0$, we are done; if not, this process continues until some $a_i = 0$ is reached. Finally, if $1 + a_1 = 2$, then $a_1 = 1$ and $a_2 = 0$. Hence $n = 2f_1 + a_3 f_3 + \cdots = f_2 + a_3 f_3 + \cdots$. Finish as above.

7. $(1 - t - t^2)^{-1} = \sum_{i=0}^{\infty} f_i t^i$, where the f_i are the Fibonacci numbers. $(1 + t - t^2)^{-1} = (1 - (-t) - (-t)^2)^{-1} = \sum_{i=0}^{\infty} (-1)^i f_i t^i$.

Section 6

1. (a) $\overline{A \cup B} = \bar{A} \cap \bar{B}$ (b) $A \cap B$ (c) $(\bar{A} \cap \bar{B}) \cup (A \cap B)$,
   (d) $A \triangle B = \bar{A} \triangle \bar{B}$ (e) $\overline{A \cap B} = \bar{A} \cup \bar{B}$ (f) $A \cup B$.

Section 7

2. $A \subseteq A \cup (B \bigtriangleup C)$;   but   since $A \subseteq A \cup B$ and $A \subseteq A \cup C$, $A \nsubseteq (A \cup B) \bigtriangleup (A \cup C)$.

Section 8

2. Suppose $|A| = m$ and $|B| = n$, $m \leqslant n$. If the sets are unordered, up to mn comparisons must be made. If the sets are ordered, let $A \cup B = \{d_1, d_2, \ldots, d_k\}$ ($k \leqslant m+n$) in this order, and let $i_1 < i_2 < \cdots < i_m$ be the subscripts of A in the union. The number of comparisons needed is $i_1 + (i_2 - i_1) + (i_3 - i_2) + \cdots + (i_m - i_{m-1}) = i_m \leqslant k \leqslant m+n$.

Section 9

3. If A and B are disjoint, then $|A \oplus B| = |A \otimes B|$. If $A = B$, then

$$|A \oplus B| = n + \frac{n(n-1)}{2} = \frac{1}{2}n(n+1),$$

whereas $|A \otimes B| = n^2$. ($|A| = n$.)

Section 10

5. $R \circ S = \{\langle x, y \rangle \mid y = x^4 - (2i-1)x^2 + i^2 - i + 1, i \geqslant 7\}$.
6. $R \circ S = I \otimes I$, where I is the set of integers. (For any x we can choose z arbitrarily, and find a y such that $xy \leqslant 100$ and $yz \leqslant 5$. Thus there is a composition $x \to y \to z$ for any pair $\langle x, z \rangle$.)

Section 11

2. The proposition is true. Within any one rectangle all points on a given horizontal line represent the same equivalence: if $\langle x_1, y \rangle \in R$ and $\langle x_2, y \rangle \in R$, then $\langle x_1, x_2 \rangle \in R$. Similarly all points on a given vertical line represent the same equivalence. Thus, since any two such lines intersect, all points within the rectangle are equivalent. However, if the rectangle is not on the diagonal $x = y$, then there will be a mirror-image rectangle representing the same set of equivalent points. Thus the converse of the proposition is not true.

Section 13

1.

$$AB^T = \begin{bmatrix} 0 & 4 & 7 \\ 1 & 12 & 24 \\ 4 & 0 & 12 \end{bmatrix}, \qquad BA^T = \begin{bmatrix} 0 & 1 & 4 \\ 4 & 12 & 0 \\ 7 & 24 & 12 \end{bmatrix}$$

$$B^T A = \begin{bmatrix} 10 & 28 \\ 1 & 14 \end{bmatrix}, \qquad A^T B = \begin{bmatrix} 10 & 1 \\ 28 & 14 \end{bmatrix}.$$

2.

$$A^{-1} = \begin{bmatrix} -2 & 1 \\ \frac{3}{2} & -\frac{1}{2} \end{bmatrix}.$$

3.

$$A^{-1} = \begin{bmatrix} -\frac{5}{18} & \frac{1}{18} & \frac{7}{18} \\ \frac{1}{18} & \frac{7}{18} & -\frac{5}{18} \\ \frac{7}{18} & -\frac{5}{18} & \frac{1}{18} \end{bmatrix}.$$

CHAPTER 2

Section 2

1. $\varepsilon(G) \leqslant \deg(G)$: If $\deg(G) = 0$, then there are no lines, and $\varepsilon(G) = 0$. Suppose $\deg(G) = k > 0$, and let v be a vertex of degree k. Deletion of the k edges incident to v disconnects G. Hence $\varepsilon(G) \leqslant \deg(G)$. $v(G) \leqslant \varepsilon(G)$: If G is disconnected or consists of a single point, then $v(G) = 0$. Suppose $\varepsilon(G) = m > 0$, and let $e_i = (v_{i1}, v_{i2})$, $i = 1, \ldots, m$ be a set of m edges whose removal disconnects G. Then removal of the vertices v_{i1}, $i = 1, \ldots, m$ disconnects G (or reduces it to a single point). Hence

$$v(G) \leqslant |\{v_{i1}, \ldots, v_{m1}\}| \leqslant m = \varepsilon(G).$$

3. Let $p_1, p_2, \ldots, p_m$ be m paths from v to w, disjoint except for the end points, and assume that removal of these paths disconnects G. If the edge (v, w) is in G, let it be p_1. From edge p_i choose a vertex $v_i \neq v, w$, for $i = 1, \ldots, m$. (If $p_1 = (v, w)$. let $v_1 = v$.) Then $\{v_1, v_2, \ldots, v_m\}$ is a set of m vertices whose removal disconnects G. Hence if $v(G) = k$, then $m = |\{v_1, \ldots, v_m\}| \geqslant k$. Thus there are at least k disjoint paths from v to w. (This theorem together with its converse is a variant of Menger's theorem, a classical result in combinatorics and graph theory.)

Section 3

7. (1) If G had a cycle, removal of one of the edges in the cycle would not disconnect G. Hence G is acyclic.

(2) The existence of a path assures that G is connected. If G had a cycle, then there would be two paths between any two points on the cycle. Hence G is acyclic.

(3) By induction. Clearly G is a tree if $n = 1$ or 2. Suppose that the characterization holds for $n = k$, and let G be a graph of $k + 1$ vertices, k edges, and no cycles. Choose a vertex v of degree 0 or 1, and delete it from G, obtaining graph G'. If v were isolated, then G' would have k vertices and k edges. Hence G' would not be a tree. This implies that G would have a cycle (since no edges were removed). Thus v is not isolated, $\deg(v) = 1$, and removal of v removes an edge. Thus G' has k vertices, $k - 1$ edges, and no cycles. By the induction hypothesis G' is a tree. Since $\deg(v) = 1$, reinsertion of v does not create a cycle. Hence G is a tree.

10. If G has a spanning tree, then there is a path (through the tree) from any vertex to any other. Therefore G is connected. If G is connected, then either it is a tree (that is, its own spanning tree), or it has a cycle. If G has a cycle, find one and delete an edge, breaking the cycle. The new graph G_1 is either a tree or has a cycle. In the latter case repeat the process. Thus we obtain a finite sequence $G = G_0, G_1, \ldots, G_n$, $n \geqslant 0$, such that each G_i is a connected subgraph of G_{i-1}, $i = 1, \ldots, n$, and G_n is a tree. Furthermore, each G_i contains all vertices of G. Thus G_n is a spanning tree for G.

12. (a) 1,   (b) n, the number of vertices,

(c) $\begin{pmatrix} \binom{n}{2} \\ n-1 \end{pmatrix}$,   (d) $\begin{pmatrix} mn \\ m+n-1 \end{pmatrix}$.

13. Such a graph has $nk/2$ edges.

Section 4

For Exercises 5–8 the matrices given are for Fig. 4.2 with the edge numbering

1: (v_1, v_2) 2: (v_1, v_4) 3: (v_1, v_6)

4: (v_2, v_3) 5: (v_2, v_4) 6: (v_3, v_4)

7: (v_4, v_5) 8: (v_4, v_6) 9: (v_5, v_6)

5.
$$\begin{bmatrix}
1 & 1 & 0 & 0 & 1 & 0 & 0 & 0 & 0 \\
0 & 1 & 1 & 0 & 0 & 0 & 0 & 1 & 0 \\
0 & 0 & 0 & 1 & 1 & 1 & 0 & 0 & 0 \\
0 & 0 & 0 & 0 & 0 & 0 & 1 & 1 & 1 \\
1 & 1 & 0 & 1 & 0 & 1 & 0 & 0 & 0 \\
1 & 0 & 1 & 0 & 1 & 0 & 0 & 1 & 0 \\
0 & 1 & 1 & 0 & 0 & 0 & 1 & 0 & 1 \\
1 & 0 & 1 & 1 & 0 & 1 & 0 & 1 & 0 \\
1 & 0 & 1 & 0 & 1 & 0 & 1 & 0 & 1 \\
1 & 0 & 1 & 1 & 0 & 1 & 1 & 0 & 1
\end{bmatrix}$$

6.
$$\begin{bmatrix}
0 & 0 & 1 & 0 & 0 & 0 & 0 & 0 & 0 \\
0 & 1 & 0 & 0 & 0 & 0 & 0 & 1 & 0 \\
1 & 0 & 0 & 0 & 1 & 0 & 0 & 1 & 0 \\
0 & 1 & 0 & 0 & 0 & 0 & 1 & 0 & 1 \\
1 & 0 & 0 & 1 & 0 & 1 & 0 & 1 & 0 \\
1 & 0 & 0 & 0 & 1 & 0 & 1 & 0 & 1 \\
1 & 0 & 0 & 1 & 0 & 1 & 1 & 0 & 1
\end{bmatrix}$$

7. The full spanning tree matrix has 31 rows:
$$\begin{bmatrix}
1 & 1 & 1 & 1 & 0 & 0 & 1 & 0 & 0 \\
1 & 1 & 1 & 1 & 0 & 0 & 0 & 0 & 1 \\
1 & 1 & 1 & 0 & 0 & 1 & 1 & 0 & 0 \\
1 & 1 & 1 & 0 & 0 & 1 & 0 & 0 & 1 \\
1 & 1 & 0 & 1 & 0 & 0 & 1 & 1 & 0 \\
1 & 1 & 0 & 1 & 0 & 0 & 1 & 0 & 1 \\
 & & & & \vdots & & & & \\
1 & 0 & 0 & 0 & 1 & 1 & 0 & 1 & 1
\end{bmatrix}$$

8.
$$\begin{bmatrix} 1 & 1 & 0 & 0 \\ 1 & 0 & 1 & 0 \\ 0 & 0 & 1 & 0 \\ 1 & 1 & 1 & 1 \\ 0 & 0 & 0 & 1 \\ 0 & 1 & 0 & 1 \end{bmatrix}$$

9. (Fig. 4.3)
$$\begin{bmatrix} 1 & 0 & 0 & 1 & 0 \\ 1 & 1 & 0 & 0 & 1 \\ 0 & 0 & 1 & 0 & 1 \\ 1 & 1 & 0 & 1 & 1 \\ 1 & 0 & 1 & 0 & 1 \end{bmatrix}$$

or its transpose.

Section 6

	girth	radius	diameter	crossing number	chromatic number
1.	0	2	4	0	2
5.	3	2	3	0	3
10.	4	3	3	4	2

Section 7

4. (a) $n = \sum_{i=1}^{k} \beta_i.$

(b) The ith component, with a basis of β_i cycles, has at most $2^{\beta_i} - 1$ cycles.

(c) Suppose $\beta_1 \geqslant \beta_i$ for all $i = 1, \ldots, k$. Then

$$N = \sum_{i=1}^{k} (2^{\beta_i} - 1) \leqslant k(2^{\beta_1} - 1);$$

and $2^n - 1 = 2^{\beta_1}(2^{\beta_2 + \cdots + \beta_k}) - 1.$ Thus

$$\frac{N}{2^n - 1} \leqslant E_1 = \frac{k \cdot 2^{\beta_1} - k}{2^{\beta_2 + \cdots + \beta_k} \cdot 2^{\beta_1} - 1}.$$

For large n, this is approximately $E_2 = k/2^{\beta_2 + \cdots + \beta_k}$. For example, if $k = 3$, $\beta_1 = 4$, $\beta_2 = 3$, and $\beta_3 = 2$, then $N/(2^n - 1) \approx 0.0489$. The estimates are $E_1 \approx 0.0881$ and $E_2 \approx 0.0938$.

5. The answers are the same as those for Exercise 4.

6. (a) $n = \sum_{i=1}^{k} \beta_i + k - 1$.

   (b) A cycle crossing $G_i, G_{i+1}, \ldots, G_{i+j}$ could be derived from cycles in each of these $j + 1$ subgraphs, joined together. Thus there are at most

$$(2^{\beta_i} - 1) + (2^{\beta_{i+1}} - 1) + \cdots + (2^{\beta_{i+j}} - 1)$$

such cycles. Hence an upper bound on the number of cycles is

$$N = \sum_{j=0}^{k-1} \sum_{l=1}^{k-j} \sum_{m=0}^{j} (2^{\beta_{l+m}} - 1).$$

   (c) N is considerably smaller than $2^n - 1$, although not so small as in Exercises 4 and 5. For example, if $k = 3$, $\beta_1 = 4$, $\beta_2 = 3$, and $\beta_3 = 2$, then $N = 82$ and hence $N/(2^n - 1) \approx 0.160$.

Section 8

4.
$$\begin{bmatrix} 1 & 0 & 0 & 0 \\ 1 & 1 & 0 & 0 \\ 0 & 1 & 1 & 0 \\ 0 & 0 & 1 & 0 \\ 0 & 0 & 1 & 0 \\ 0 & 1 & 1 & 1 \\ 0 & 0 & 0 & 1 \end{bmatrix}$$

or some permutation of this matrix.

6. Let δ be the density.

   adjacency matrix: $\delta = 2/m$

   incidence matrix: $\delta = 2m/n^2$

   m.c.s.g. matrix: $\delta = (\sum_{i=2}^{m} i k_i)/(m \sum_{i=2}^{m} k_i)$

8. The complete graph K_i on i vertices has $\binom{i}{2}$ edges. Hence the inequality. Equality holds if G is a complete graph.

9. A gross upper bound is obtained by noting that for $i = 1, ..., m$, $k_i \leqslant m - i + 1$. Using this,

$$\sum_{i=2}^{m} \binom{i}{2} k_i \leqslant \frac{1}{24} m(m^2 - 1)(m + 2).$$

Section 9

For matrix A, with 24 words required for the complete matrix:

1. 24 words required—no savings.
2. 24 bits + 11 words required. If word length is at least 24 bits, 12 words are saved.
3. 36 words required—a loss of storage.
4. 24 words required—no savings.
5. 16 words required—eight words saved.
6. The answers depend on the distribution of the k nonzero elements. Assuming that they are uniformly distributed in rows, then there are at most $\lceil k/m \rceil$ elements per row, where $\lceil k/m \rceil$ is the least integer i such that $i \geqslant k/m$. With this assumption the storage requirement estimates are:

full matrix	mn
position value matrices	$2m\lceil k/m \rceil$
bit map, value vector	$\lceil k/w \rceil + k$
position-value pair (coord.)	$3k$
position-value pair (linear)	$2k$
bandwidth	at least $m\lceil k/m \rceil$

Section 10

2. The best situation occurs when $l = k + 1$ and $m = k + 2$. Then operation "b" joins vertices k and $k + 2$. This leaves a new $\langle k, l, m \rangle$ triple, $\langle k + 1, k + 3, k + 5 \rangle$. Another operation "$b$" now joins vertices $k + 1$ and $k + 5$. Hence the bandwidth is at least $(k + 5) - (k + 1) = 4$. However, an operation "a" at this point leaves $\langle k, l, m \rangle$ as $\langle k + 4, k + 5, k + 6 \rangle$, that is, as $\langle k', k' + 1, k' + 2 \rangle$, thus repeating the initial situation. Thus the operation sequence "ba" (and similarly "aa" and "ab") preserves bandwidth. Since the initial triangle has bandwidth 3, the bandwidth of the final graph is 3 if and only if no sequence "bb" occurs.

3. Let $f_{n,a}$ and $f_{n,b}$ denote the number of sequences of length n ending in a and b respectively. Then $f_n = f_{n,a} + f_{n,b} = f_{n-1,a} + f_{n-1,b} + f_{n-1,a} = f_{n-1} + f_{n-1,a}$. But $f_{n-1,a} = f_{n-2,a} + f_{n-2,b} = f_{n-2}$. Hence $f_n = f_{n-1} + f_{n-2}$. The two sequences a and b correspond to numberings of the bandwidth-3 graph on six vertices. The three sequences aa, ab, and ba correspond to numberings of the bandwidth-3 graph on eight vertices. Thus $f_3 = 2$ and $f_4 = 3$.

4. It is generally a good rule to give vertices of high degree numbers in the middle of the range.

Section 11

2. Bandwidth $= 3$.

$$
\begin{bmatrix}
3 & 2 & 1 & 0 & 0 & 0 & 0 & 0 & 0 \\
1 & 3 & 0 & 2 & 0 & 0 & 0 & 0 & 0 \\
0 & 1 & 0 & 0 & 1 & 0 & 0 & 0 & 0 \\
0 & 0 & -3 & 1 & 1 & 2 & 0 & 0 & 0 \\
0 & 0 & 0 & 0 & 2 & 1 & -1 & 0 & 0 \\
0 & 0 & 0 & 0 & 0 & 1 & 2 & -1 & 2 \\
0 & 0 & 0 & 0 & 0 & 0 & 1 & -2 & 1
\end{bmatrix}
$$

Storage saved: 30 words = 47.6 percent.

Section 12

2. The proof for transformation of vertices of degree 2 is trivial. Now suppose v is a vertex of degree 4 or more, colored R, say. Replace v by a cycle of vertices surrounding a new region. Pick one of these and color it R. Now proceed around the cycle. Each vertex v' before the last one in the cycle is adjacent to two previously colored vertices— one in the cycle and one a vertex of the original graph. Hence there is always a choice of two colors available for v'. The last vertex in the cycle is adjacent to three previously colored vertices, so even here there is at worst a fourth color required.

CHAPTER 3

Section 2

2. The minimal tree domain D_0 is determined by using the smallest addresses allowable in the definition of a complete end point set from M_0. For example, if $M_0 = \{2.2\}$, the smallest complete end point set containing M_0 is $\{1, 2.1, 2.2\}$. Thus $D_0 = \{0, 1, 2, 2.1, 2.2\}$.

Section 3

1. If $b \in S\ a_1 \cdot a_2\ D$ then $b = a_1 \cdot a_2 \cdot c \in S\ a_1\ D = a_1 \cdot D_1$. Thus $a_2 \cdot c \in D_1$, or $a_2 \cdot c \in S\ a_2\ D_1$. Hence $b \in a_1 \cdot S\ a_2\ D_1$. If $b \in a_1 \cdot S\ a_2\ D_1$, then $b = a_1 \cdot c$, where $c \in S\ a_2\ D_1$. Thus $c = a_2 \cdot d$ for some d, or $b = a_1 \cdot a_2 \cdot d \in S\ a_1 \cdot a_2\ D$.

9. The definitions all are based on corresponding definitions in the tree domain D underlying T. For example, the scope of α in T consists of all elements of T whose addresses are in the scope of the address of α in D.

Section 4

9. The definition is the same as that for $<_{\mathscr{S}}$, except that (1) $a_1 < a_2$ is replaced by (1) $a_2 < a_1$.

Section 5

14. The fully eliminated tree permits d-introduction at addresses $0, 1, 2$, and 2.2. These d-introductions singly and in various combinations yield trees equivalent to the given tree.

Section 6

1. $26^1 < 2^5, 26^2 < 2^{10}, 26^3 < 2^{15}, 26^4 < 2^{19}$.

CHAPTER 4

Section 3

1. It is clear that the digraph need not be unilaterally connected since an arborescence itself is not (except in trivial cases). However, it is easy to show that "weakly connected" is not sufficient. A necessary and sufficient condition that a graph have a spanning arborescence is that it be "quasi strongly connected," that is, for any two vertices v and w there exists a vertex x such that there are paths from x to v and from x to w. See pp. 133–135 of "Programming, Games and Transportation Networks" by C. Berge and A. Ghouila-Houri (Wiley, New York, 1965).

3. For the set of $n-1$ vertices, the total number of arcs into the vertices equals the total number of arcs out. If for the remaining vertex the in-degree exceeds the out-degree, then for the entire graph there are more arcs leading into the vertices than leading out from them. This is absurd.

4. If G is separable, such paths do not exist.

Section 4

2. Assume G is an arborescence on n vertices, with a single source. These properties hold:

   (1) Every column of A except one has exactly one 1. That one column corresponds to the root, and consists entirely of zeros.
   (2) There exists an $m \leqslant n$ such that $A^m = 0$.
   (3) For an appropriate numbering of the vertices, A is an upper-triangular matrix.

   These properties characterize arborescences.

3. Properties 2 and 3 of Exercise 2 hold. Property 1 is replaced by

   (1') Every column of A except one has at least one 1. That one column corresponds to the root, and consists entirely of zeros.

   These properties characterize the given class of digraphs.

Section 5

1. *ABCEGI, ABDFHI, ABCEGHI, ABDFABCEGI, ABDFABDFHI, ABDFABCEGHI.*

CHAPTER 5

Section 2

4. Associativity fails since $(2^2)^3 \neq 2^{(2^3)}$.
5. If S consists of the finite subsets of T, then the answer to Exercise 2 does not change, but there is no identity for Exercise 3 since T is infinite. If S consists of the infinite subsets of T, then there is no identity for Exercise 2; and $\langle S, * \rangle$ is not a semigroup in Exercise 3 since closure fails.

Section 3

2. The second production yields infinitely many words.
3. Any d is easily eliminated. Any c occurs only in a substring of the form $acc \cdots c$. Repeated use of the fourth production eliminates any such string.
9. Let X be a new nonterminal symbol. Then $rpt \to rXt$, $Xt \to Xq$ (right context is empty), $rXq \to rtq$.

Section 4

1. With the obvious abbreviations, and $<$ as the precedence operator, the graph is derived from the following relations:

$$d < n < nfp < nf < tv,$$

$$n < ifp < if < tv, \qquad n < sfp < sf < tv,$$

$$n < rfp < rfb < rfc < rf < tv.$$

CHAPTER 6

Section 1

1. No identity.
4. Associativity fails. (Try cce.) Also inverse fails.
9. There are n distinct elements, since for any $x, y \in H$, if $x \neq y$, then $ax \neq ay$.

10. Suppose $aH \cap bH \neq \varnothing$, and let $d \in aH \cap bH$. Then for suitable $x, y \in H$, $d = ax = by$. Hence $a = byx^{-1}$. Now take $ah \in aH$. Then $ah = (byx^{-1})h = b(yx^{-1}h) = bh' \in bH$. Thus $aH \subseteq bH$. Similarly, $bH \subseteq aH$.

Section 2

1. A group of order 2.
4. Only the identity.

Section 4

1. The group is of order 18.
2. The group is of order 8.
3. The group is of order 24.
4. The group is of order 30.

Section 5

13. S_n has $n!$ elements.
14. A_n has $n!/2$ elements.

CHAPTER 7

Section 3

10. It is easy to show that any two chains of words obtained by successive d-eliminations from a given word will terminate, when all d's have been eliminated, in the same word. From this it is quite simple to build up a structure that appears to be a lattice. The difficulty comes in establishing the uniqueness of the meet and join.

Section 4

1. This is most easily established by contradiction. Assuming that the lattice is unbounded leads to an infinite chain of distinct elements.
5. If and only if n is a product of distinct primes.
8. Always. Use the technique of Example 3.2 to establish this.
9. $b = b \wedge 1 = b \wedge (c \vee a) = (b \wedge c) \vee (b \wedge a) = (b \wedge c) \vee 0 = b \wedge c$. Similarly $c = b \wedge c$. Hence $b = c$.

10. Prove this by contradiction, as Exercise 1 was done.
21. The lattice involves six elements. It is not modular.

CHAPTER 8

Section 3

1. The set of all subsets of the integers, with addition defined as set union, and multiplication defined as set intersection.
2. The set of all subsets of the integers, with addition defined as set intersection, and multiplication defined as set union. This produces infinite descending chains of elements, with no atoms.

Section 4

1. There are two functions.   **2.** Four functions.   **3.** One function.
4. No functions.   **5.** No functions.

Section 5

6. $(((xy)' + yz)'(xz))'$.
7. $(((xz)'x + y')'((x + z')'x)')'$.

Section 6

1. $yz + x'z + y'z'$, or $yz + x'y' + y'z'$.
3. $w + x'y'z'$.
7. $z + x'y$.
9. $vwxy' + vwx'y + w'xy + v'xy + x'y'z + vxz$.

CHAPTER 9

Section 3

1. Equivalent to $p \lor q$.
4. F if (p, r, s) has the value (F, F, F) or if (p, q, r, s) has the value (T, F, T, T); T otherwise.
10. $(p \equiv r) \supset (p \land q)$.
13. $(p \supset r) \downarrow ((q \lor r) \supset p)$.

Section 4

2. $((p) \supset (q)) \equiv \big((\sim(r)) \vee (\sim(((p) \wedge (q)) \vee (r)))\big).$
6. $p \vee q \equiv \sim(p \vee \sim(r \supset s)) \supset (\sim s \equiv \sim(p \wedge q)).$

Section 6

1. $ECApNqKpNrCNpAqr$; $pqNAprNKCpNqrACE.$
5. Well-formed.
8. Well-formed.
11. $((p \equiv q) \wedge \sim r) \vee ((p \wedge \sim r) \vee \sim q) \supset p.$
13. $\sim(p \vee (q \wedge r)) \equiv \big(((\sim p \equiv (q \equiv r)) \vee (p \wedge r)) \supset \sim p\big).$

CHAPTER 10

Section 2

9. Let $w(r,k)$ be the number of ways of choosing r objects from k classes. Consider the first class. If there is an object chosen from it, then the remaining $r-1$ objects may be chosen from k classes. If no object is chosen from the first class, then all r objects are chosen from the remaining $k-1$ classes. Hence $w(r,k) = w(r-1,k)+w(r,k-1)$. With the boundary conditions $w(1,k) = k$ and $w(r,1) = 1$, the solution is

$$w(r,k) = \binom{r+k-1}{k-1} = \binom{r+k-1}{r}.$$

Note that it is consistent to define $w(0,k) = 1$.

10. Since r is unrestricted it may not be possible to choose all r objects from one class. Let $g(r,k;i)$ be the number of ways of choosing r objects from k classes when $p_i < r \leqslant p_{i+1}$. Setting $p_0 = -1$, we have $g(r,k;0) = w(r,k)$. For $g(r,k;1)$ we may choose up to p_1 objects from the first class, with the remaining ones chosen freely from the other $k-1$ classes. Thus

$$g(r,k;1) = \sum_{j=0}^{p_1} g(j,1;0)g(r-j,k-1;0) = \sum_{j=0}^{p_1} w(r-j,k-1)$$
$$= w(r,k) - w(r-p_1,k).$$

Beyond this the expressions become more complex. For example,

$$g(r,k;2) = \sum_{j_0=0}^{p_1} g(j_0,2;0)g(r-j_0,k-2;0)$$
$$\sum_{j_1=p_1+1}^{p_2} g(j_1,2;1)g(r-j_1,k-2;0).$$

This evaluates to

$$g(r,k;2) = \sum_{j_0=0}^{p_1} w(j_0, 2) w(r-j_0, k-2)$$
$$- \sum_{j_1=p_1+1}^{p_2} w(j_1, 2) w(r-p_1-j_1, k-2).$$

11. The effect of allowing unlimited repetition is to make each class infinite. Thus in effect $r < p_1$ for all r, and the answer to Exercise 9 applies.

Section 4

1. $(1+t+t^2)^n$ or $\left(\dfrac{1-t^3}{1-t}\right)^n$.   2. $\left(\dfrac{1-t^{k+1}}{1-t}\right)^n$

3. $\left(\dfrac{t^2}{1-t}\right)^n$.   4. $\left(\dfrac{t(1-t^k)}{1-t}\right)^n$.   5. $\left(\dfrac{1-t+t^k}{1-t}\right)^n$

6. $\dfrac{(1-t^2)(1-t^3)\cdots(1-t^n)}{(1-t)^n}$.   7. $\dfrac{t^{n(n+1)/2}}{(1-t)^n}$

8. Let $r(p, q)$ be the desired number of ways, and let $r_a(p, q)$ and $r_b(p, q)$ represent the number of ways to arrange p a's and q b's in a row ending in a and b, respectively. Then $r_a(p, q) = r(p-1, q)$, and $r_b(p, q) = r_a(p, q-1) = r(p-1, q-1)$. Hence $r(p, q) = r(p, q-1) + r(p-1, q-1)$. The boundary conditions for this recurrence are $r(0, q) = r(p, 0) = 1$. The solution is

$$r(p, q) = \binom{p+1}{q}.$$

Observe that a sequence of length k represents a graph on $2(k+2)$ vertices. Thus, summing the recurrence for $p+q = k$ we establish that the number of sequences of length k is f_{k+2}, a Fibonacci number. Hence the number of the specified graphs on $2k$ vertices is f_k, the kth Fibonacci number.

9. From integrating the function and its expansion,

$$\sum_{r=0}^{n} C(n, r) \frac{t^{r+1}}{r+1} = \frac{(1+t)^{n+1}}{n+1} + C_n.$$

Then show that $C_n = -1/(n+1)$.

10. Use the result of Example 4.5 and expand $(1-t^2)^{-n}$ as $(1-t)^{-n}(1+t)^{-n}$.

Section 5

The functions $f(k)$ and $g(k)$ are those defined in the answers for Exercises 2 and 4.

1. $\left(1 + t + \dfrac{t^2}{2!}\right)^n.$ **2.** $\left(1 + t + \dfrac{t^2}{2!} + \cdots + \dfrac{t^k}{k!}\right)^n = f^n(k)$

3. $\left(\dfrac{t^2}{2!} + \dfrac{t^3}{3!} + \cdots\right)^n = (e^t - (1+t))^n.$

4. $\left(t + \dfrac{t^2}{2!} + \cdots + \dfrac{t^k}{k!}\right)^n = g^n(k) = (f(k) - 1)^n.$

5. $\left(1 + \dfrac{t^k}{k!} + \dfrac{t^{k+1}}{(k+1)!} + \cdots\right)^n = (e^t - g(k-1))^n.$

6. $(1 + t)\left(1 + t + \dfrac{t^2}{2!}\right) \cdots \left(1 + t + \dfrac{t^2}{2!} + \cdots + \dfrac{t^n}{n!}\right) = \displaystyle\prod_{k=1}^{n} f(k).$

7. $\left(t + \dfrac{t^2}{2!} + \cdots\right)\left(\dfrac{t^2}{2!} + \dfrac{t^3}{3!} + \cdots\right)\left(\dfrac{t^n}{n!} + \dfrac{t^{n+1}}{(n+1)!} + \cdots\right) = \displaystyle\prod_{k=0}^{n-1} (e^t - f(k)).$

Section 7

4. This is an application of the formula $n!/(p_1!p_2!\cdots)$, where $p_1 + p_2 + \cdots = n$, for choosing arrangements of objects from several classes. In this case the n of the general formula is $\sum k_i$.

6. The cycles of any one length may not be interchanged. For example, along with (12)(34)(5) we consider (12)(5)(34) and (5)(12)(34), but not (34)(12)(5). Thus the cycles of length i may be regarded as indistinguishable elements in a class of k_i objects. This again is an application of the general formula $n!/(p_1p_2!\cdots)$.

Section 8

1. This follows from the observation that $u(t)(1-t)(1-t^2)(1-t^3)\cdots = (1-t^2)(1-t^4)(1-t^6)\cdots.$

6. $c(t, k) = t\,\dfrac{1 + t + t^2 + \cdots + t^{k-1}}{1 - t - t^2 - \cdots - t^k} = t\,\dfrac{1 - t^k}{1 - 2t + t^{k+1}}.$

CHAPTER 11

Section 1

2. Assume $s_1 \leqslant s_2 \leqslant \cdots \leqslant s_n$. The minimum number is $\prod_{i=1}^{n}(s_i - i + 1)$, and the maximum number is $s_1 s_2 \cdots s_n$.

Section 2

2. Let $R = \{a_1, \ldots, a_n\}$ be an SDR, and for each set S_{ij} let S'_{ij} be the set consisting of the representative of S_{ij}. Then each collection $\mathscr{X}_i$ induces a collection $\mathscr{X}'_i$ such that $\cap I(\mathscr{X}'_i) \subseteq \cap I(\mathscr{X}_i)$. To establish the given inequality if suffices to establish it for the reduced system of sets. This is done by counting the elements omitted in choosing the $\mathscr{X}'_i$.

3. Choose each $\mathscr{X}_i$ to be a single set. If for some choice of $\mathscr{X}_i$, $|\cap I_i| \geqslant n$, then one can directly construct an SDR. If for all choices of $\mathscr{X}_i$, $|\cap I_i| < n$, then either the union of these intersections contains at least n elements, or it contains fewer than n elements. In the former case one can find an SDR by judicious choice; in the latter case the given inequality will not hold for any choice in which at least one $\mathscr{X}_i$ consists of two or more sets.

Section 3

1. The only solutions to the equations governing the sum are $p_i = p_{n-i+1} = 1$ for $i = 1, 2, \ldots, n/2$, and for n odd, $p_k = 2$, where $k = (n+1)/2$.

3. The sum represents a choice of $n+1$ objects from a set of $2n$ objects which contains n objects of each of two kinds.

CHAPTER 12

Section 3

6. $$R(n) \approx \frac{(2n)^{2n+1/2}}{\sqrt{2\pi}\,(n+1)^{n+3/2}\,(n-1)^{n-1/2}\,(2^n-1)^2}.$$

7. The approximation can be rewritten as

$$\frac{1}{\sqrt{\pi}} \cdot \left(\frac{n}{n^2-1}\right)^{1/2} \cdot \frac{n^{2n}}{(n+1)^2 (n^2-1)^{n-1}} \cdot \frac{2^{2n}}{(2^n-1)^2}.$$

The third and fourth factors tend to 1 while the second tends to 0 as $n \to \infty$.

Section 5

2. By the conditions given in the text choose $a = 1$. (Other choices of a will lead to shorter periods.) But then the generated values form an arithmetic sequence modulo m.

Index

A

Absorption laws
    lattice, 211
    set, 13
Algebra, 166
Algebraic system, 166
Alphabet, 168
And, 254
And-gate, 236
Arborescence, 134
Arc, 130
    antiparallel, 131
    parallel, 131
ARPA network, 160
Associativity
    group, 183
    lattice, 211
    semigroup, 167
    set, 13
Atom, 223
Automaton
    finite, 175
    linear bounded, 175
    push-down, 175
Automorphism, 186
Axiom, 273

B

Backus Naur form, 176
Backus normal form, 176
Banding (a matrix), 80
Bandwidth
    bipartite graph, 92
    cubic graph, 83
    graph, 83
Bandwidth numbering, 83
Bell numbers, 303
Bernoulli trial, 342
Bijection, 22
Binomial coefficient, 292, 337
Binomial distribution, 341
Bit map, 79
BNF, 176
Boolean addition, 226, 254
Boolean algebra, 226
Boolean function, 230
    minimization of, 239
Boolean multiplication, 226, 254
Branch and bound technique, 150

C

Cancellation, 184
Canonical form, 231

Capacity
    of an arc, 139
    of a cut, 140
Cardinality, 18
CDC Cyber, 72, 181
Chromatic number, 65
Closure
    group, 183
    semigroup, 167
Coarseness, 64
Combination, 287
Combinatorics, 281
Common SDR, 314
Commutativity
    group, 183
    lattice, 211
    semigroup, 167
    set, 13
Complement
    Boolean, 227
    graph, 44
    lattice, 218
    set, 7
Complete end point set, 105
Component, 40
Composition
    of a number, 306
    of sets, 23
Computer arithmetic, 249, 254
Conditional implication, 254
Conditional probability, 334
Congruence modulo n, 26
Conjugate partition, 307
Conjunction, 254
Connected
    strongly, 132
    unilaterally, 132
    weakly, 132
Connection table, 79
Connectivity
    edge, 41
    vertex, 41
Constant function, 231
Contradiction, 253
Coset, 313
Cotree, 47
Covering, 240
CPM, 155
CPS, 155
Critical path, 154

Crossing number, 63
Cut, 140
Cut capacity, 140
Cycle
    directed, 132
    undirected, 40
Cycle basis, 67
Cycle class, 303
Cycle indicator, 305
Cycle representation, 198
Cycle space, 68
Cycle vector, 62
Cyclomatic number, 47, 68

D

d-elimination, 120
d-equivalent, 120
d-introduction, 121
Definiendum, 119
Definiens, 119
Degree
    of an edge, 42
    of an undirected graph, 41
    of a vertex, 41, 132
Degree spectrum, 59
De Morgan's laws
    Boolean algebra, 227
    sets, 13
Denial, 253
Depth, 112
Depth function, 104
Determinant, 35
Diameter, 63
Digraph, 131
Disjunction, 254
Disjunctive normal form, 258
Distance, 63
Distribution
    binomial, 341
    normal, 347
    Poisson, 344
    uniform, 350
Distributive laws, 13
Domain, 22
Domino, 123
Don't care condition, 243
Duality
    Boolean, 227
    lattice, 211

E

Eccentricity, 63
Edge, 38
 multiple, 39
Element (of a set), 2
Elementary symmetric function, 292
Empty set, 3
End point, 111
Entry (of a matrix), 33
Enumerator, 291
Equality (of sets), 3
Equivalence, 254
Equivalence class, 27
Equivalence relation, 26
Explicit definition, 119
Exponential generating function, 298
Extension of a function, 25

F

Factorial expansion, 301
Ferrers graph, 307
Fibonacci numbers, 3
Finitary operator, 166
Finite automaton, 175
Finite state machine, 175
Flexibility, 161
Flow, 140
 maximal, 142
 minimal cost, 147
 value, 140
Flowchart, 136
Four color conjecture, 65, 97
Full adder, 250
Function, 20

G

Generating function, 291
Generator, 189, 190
Girth, 62
Grammar, 168
 ambiguous, 172
 context-free, 172
 context-sensitive, 172
 descriptive, 170, 171
 equivalent, 172
 generative, 170, 171
 inherently ambiguous, 172

 phrase structure, 170
 regular, 172
 Type 0, 170
 Type 1, 172
 Type 2, 172
 Type 3, 172
Graph
 bipartite, 45
 complete, 44
 complete bipartite, 45
 connected, 40
 cubic, 43, 83
 directed, 37, 131
 Euler, 48
 finite, 37
 Hamiltonian, 62
 infinite, 37
 labeled, 38
 linear, 37, 38
 planar, 63, 97
 regular, 42
 undirected, 37
 unlabeled, 38
Grapheme, 181
Group, 183
 Abelian, 183
 alternating, 202
 automorphism, 186
 commutative, 183
 dihedral, 188
 free, 191
 permutation, 202
 symmetric, 202

H

Half-adder, 250, 265
Height, 112
Hierarchy, 262

I

IBM 370/155, 181
Idempotency
 lattice, 213
 set, 13
Identity
 group, 183
 semigroup, 167

Identity function, 231
Identity map, 25
Incident, 38
  from, 131
  to, 131
Indegree, 132
Independence, 273
Independent set, 105
Inequivalence, 254
Infimum, 210
Information network, 158
Intersection, 12
Interval, 218
Invariant, 58
Inverse
  group element, 183
  matrix, 35
  relation, 21
Isograph, 134
Isomorphism
  graph, 58
  set, 22

J

Join
  lattice, 211
  set, 12

K

k-connected, 41
k-cycle, 40
k-regular, 42
Karnaugh map, 241
Kuratowski theorem, 64

L

Lagrange's theorem, 185
  corollary to, 191
Language
  Algol-like, 178
  natural, 181
  phrase structure, 169, 171
  Type 0, 171, 175
  Type 1, 175
  Type 2, 175
  Type 3, 174

Lattice 211
  atomic, 223, 229
  bounded, 217
  complemented, 218, 226
  complete, 220
  distributive, 219, 226
  modular, 220
  uniquely complemented, 218
Lattice of subsets, 31, 218
Length, 40
Lexicographic order, 105, 203
Linear bounded automaton, 175
Linear congruential generator, 353
Loop
  directed, 132
  undirected, 39
Lower bound, 210, 217

M

Mapping, 20
  into, 21
  onto, 22
Material implication, 254
Matrix, 33
  adjacency, directed, 135
  adjacency, undirected, 50
  bipartite, 51
  cycle, 51
  edge adjacency, 51
  incidence, directed, 135
  incidence, undirected, 49
  maximal complete subgraph, 51
  nonsingular, 35
  path, 51
  permutation, 202
  probabilistic adjacency, 162
  spanning tree, 51
  sparse, 78
  symmetric, 36
  zero-symmetric, 36
Matrix addition, 33
Matrix inverse, 35
Matrix multiplication, 33
Matrix transpose, 36
Max-flow min-cut theorem, 141
Maximal address, 106
Maximal complete subgraph, 72
Mean, 347

Meet
  lattice, 211
  set, 12
Member (of a set), 2
Minimal expression, 239
Minimization of Boolean functions, 239
Modus ponens, 274
Morphological application, 181
Multiprocessing system, 157

N

Nand, 263
Nand-gate, 264
Negation, 253
Network
  cyclic, 158
  decentralized, 159
  hierarchical, 159
  information, 158
  star, 159
  2-regular, 160
Nonterminal word, 170
Nor, 263
Nor-gate, 264
Normal density function, 347
Normal distribution, 347
Null set, 3
Numbering of a graph, 83

O

One-to-one relation, 21
Or, 254
Or-gate, 236
Order
  of an element, 184
  of a group, 185
Order (relation)
  lexicographic, 203
  linear, 26
  partial, 26, 105, 208, 213
  simple, 26
Order of precedence, 262
Outdegree, 132

P

Parsing
  bottom-up, 170
  top-down, 170

Parsing tree, 169
Partial order, 26, 105, 208
  in a lattice, 213
Partition hypothesis, 322
Partition
  conjugate, 307
  of a number, 306
  of a set, 27
Parts of discourse, 181
Pascal's triangle, 288
Path
  directed, 132
  undirected, 40
PDP-10, 181
Permutation, 197, 282
  cycle representation, 198
  even, 202
  odd, 202
Permutation generator, 203
Permutation group, 202
PERT, 155
Phoneme, 181
Phrase, 169
Planit, 29
Poisson distribution, 344
Polish notation, 108, 266
Precedence graph, 177
Predecessor, 104
  immediate, 104
Prefix representation, 114
Prime implicant, 240
Principle of inclusion and exclusion, 332
Probability, 328
Product rule, 282, 330
Product set, 17
Production, 170, 171
Projection function, 231
Projection mapping, 25
Proof, 273
Proposition, 252
Protosemata, 181
Push-down automaton, 175

Q

Quine–McCluskey technique, 244

R

Radius, 63
Ramification, 110

Random number, 351
Range, 22
Recurrence relation, 282
Relation, 18
    antisymmetric, 26
    binary, 18
    equivalence, 26
    n-ary, 19
    reflexive, 26
    symmetric, 26
    ternary, 18
    transitive, 26
Relation (group), 190
Relative complement, 7
Restriction of a function, 25
Rim representation, 308
Root address, 104

S

Scope, 111
Scope set, 111
SDR, 313
Self-loop, 39
Semantics, 169, 180
Semata, 181
Semigroup, 167
    free, 168
Sentence, 169
    syntactically correct, 182
Sentence meaning, 182
Separable, 41
Set, 2
    null, 3
    universal, 6
Set algebra, 12, 229
Simple form, 117
Simply stratified alphabet, 109
Sink, 133
Source, 133
Standard deviation, 347
Statistical independence, 335
Stirling numbers
    first kind, 301
    second kind, 302
Stirling's approximation, 338
Stratification function, 109
Subgraph, 39
    induced, 39
    spanning, 39

Subgroup, 185
Sublattice, 218
Subset, 3
    proper, 3
Subset lattice, 31, 218
Substitution, 274
Successor, 104
    immediate, 104
Sum of products form, 239, 258
Sum rule, 330
Supremum, 210
Symmetric difference, 12
Symmetric tabular expansion, 245
Syntax, 169, 180
System of distinct representatives, 313

T

Tautology, 253
Terminal word, 170
Theorem, 273
Thickness, 64
Trail
    directed, 132
    undirected, 40
Transformation rule, 170
Transposition, 201
Tree, 46
    Gorn, 109
    labeled, 109, 110
    parsing, 169
    rooted, 48
    spanning, 47
Tree domain, 106
Tree form, 117
    normal, 117
    simple, 117
Truth function, 253
Truth table, 255
Truth value, 252
Turing machine, 175

U

Uniform distribution, 350
Union, 12
Univac 1108, 181
Universal address set, 104
Universal set, 6
Upper bound, 210, 217

V

Vector, 32
Venn diagram, 7
Vertex, 38
  isolated, 41

W

Walk
  directed, 132
  undirected, 40
wff (well-formed formula), 261
Wiswesser notation, 82

Word, 169, 181, 189
  terminal, 170
  nonterminal, 70
Word class, 181
Word form, 181
Word meaning, 182
Word problem, 191
Wordset, 10, 280

Z

Zigzag graph, 307

Computer Science and Applied Mathematics
A SERIES OF MONOGRAPHS AND TEXTBOOKS

Editor
Werner Rheinboldt
University of Maryland

HANS P. KÜNZI, H. G. TZSCHACH, and C. A. ZEHNDER. Numerical Methods of Mathematical Optimization: With ALGOL and FORTRAN Programs, Corrected and Augmented Edition

AZRIEL ROSENFELD. Picture Processing by Computer

JAMES ORTEGA AND WERNER RHEINBOLDT. Iterative Solution of Nonlinear Equations in Several Variables

A. T. BERZTISS. Data Structures: Theory and Practice

AZARIA PAZ. Introduction to Probabilistic Automata

DAVID YOUNG. Iterative Solution of Large Linear Systems

ANN YASUHARA. Recursive Function Theory and Logic

JAMES M. ORTEGA. Numerical Analysis: A Second Course

G. W. STEWART. Introduction to Matrix Computations

CHIN-LIANG CHANG AND RICHARD CHAR-TUNG LEE. Symbolic Logic and Mechanical Theorem Proving

C. C. GOTLIEB AND A. BORODIN. Social Issues in Computing

ERWIN ENGELER. Introduction to the Theory of Computation

F. W. J. OLVER. Asymptotics and Special Functions

DIONYSIOS C. TSICHRITZIS AND PHILIP A. BERNSTEIN. Operating Systems

ROBERT R. KORFHAGE. Discrete Computational Structures

In preparation
PHILIP J. DAVIS AND PHILIP RABINOWITZ. Methods of Numerical Integration

A 4
B 5
C 6
D 7
E 8
F 9
G 0
H 1
I 2
J 3